METHODS IN MOLECULAR BIOLOGY

For further volumes:
http://www.springer.com/series/7651

Glucose Transport

Methods and Protocols

Edited by

Karin Lindkvist-Petersson

Department of Experimental Medical Science, Lund University, Lund, Sweden

Jesper S. Hansen

Department of Experimental Medical Science, Lund University, Lund, Sweden

Humana Press

Editors
Karin Lindkvist-Petersson
Department of Experimental Medical Science
Lund University
Lund, Sweden

Jesper S. Hansen
Department of Experimental Medical Science
Lund University
Lund, Sweden

ISSN 1064-3745 ISSN 1940-6029 (electronic)
Methods in Molecular Biology
ISBN 978-1-4939-7506-8 ISBN 978-1-4939-7507-5 (eBook)
https://doi.org/10.1007/978-1-4939-7507-5

Library of Congress Control Number: 2017959607

Cover caption: Front cover artwork by Nieng Yan

Printed on acid-free paper

This Humana Press imprint is published by Springer Nature
The registered company is Springer Science+Business Media, LLC
The registered company address is: 233 Spring Street, New York, NY 10013, U.S.A.

Preface

Glucose is one of the main energy sources for living organisms, and undoubtedly one of the most important compounds to life. In our body, glucose is the preferred energy resource for most cells. Mechanisms have consequently evolved to regulate its levels in the body. The first and limiting step in glucose metabolism is its transport across the plasma membrane. In eukaryotic cells, this transport is mediated by members of the GLUT protein family that are encoded by the *SLC2A* genes. They belong to the major facilitator superfamily that counts more than 5000 identified members to date. Humans have 14 different GLUT proteins, GLUT1–14, and they are expressed in virtually every cell type of the human body. Their central role in human physiology is reflected by their direct implications in disease, which also make them applicable drug targets. In particular, GLUTs have attracted attention as drug targets in cancer therapy, since the Warburg effect describes how most cancer cells exhibit high glucose consumption. The purpose of *Glucose Transport* is to combine protocols that encompass methods that can be used to scrutinize the structure-function relationship of GLUTs both in vitro and ex vivo. This collection of laboratory protocols will enable you as a researcher to determine the specific roles of the different GLUTs in various organisms.

Lund, Sweden

Jesper S. Hansen
Karin Lindkvist-Petersson

Contents

Contributors

CAROLINE B. ADIELS • *Department of Physics, University of Gothenburg, Gothenburg, Sweden*
PHILIP J. BILAN • *Cell Biology Program, The Hospital for Sick Children, Toronto, ON, Canada*
ECKHARD BOLES • *Institute of Molecular Biosciences, Goethe University, Frankfurt, Germany*
NIA J. BRYANT • *Department of Biology, University of York, York, UK*
JAMES G. BURCHFIELD • *School of Life and Environmental Sciences, The University of Sydney, Sydney, NSW, Australia; The Charles Perkins Centre, The University of Sydney, Sydney, NSW, Australia*
CHRIS I. CHEESEMAN • *Membrane Protein Research Group, University of Alberta, Edmonton, AB, Canada*
MERAV COHEN • *Alexander Grass Center for Bioengineering, The Rachel and Selim Benin School of Computer Science and Engineering, The Hebrew University of Jerusalem, Jerusalem, Israel; Department of Cell and Developmental Biology, Silberman Institute of Life Sciences, The Hebrew University of Jerusalem, Jerusalem, Israel*
DONG DENG • *Key Laboratory of Birth Defects and Related Diseases of Women and Children of MOE, State Key Laboratory of Biotherapy, West China Second Hospital, Sichuan University and National Collaborative Innovation Center, Chengdu, China*
DAVID DREW • *Centre for Biomembrane Research, Department of Biochemistry and Biophysics, Stockholm University, Stockholm, Sweden*
KARIN ELBING • *Department of Experimental Medical Science, Lund University, Lund, Sweden*
DANIEL J. FAZAKERLEY • *School of Life and Environmental Sciences, The University of Sydney, Sydney, NSW, Australia; The Charles Perkins Centre, The University of Sydney, Sydney, NSW, Australia*
TOM A. GEDDES • *School of Life and Environmental Sciences, The University of Sydney, Sydney, NSW, Australia; The Charles Perkins Centre, The University of Sydney, Sydney, NSW, Australia*
MATTIAS GOKSÖR • *Department of Physics, University of Gothenburg, Gothenburg, Sweden*
GWYN W. GOULD • *Institute of Molecular, Cell and Systems Biology, College of Medical, Veterinary and Life Sciences, University of Glasgow, Glasgow, UK*
CARLOTTA GRANCHI • *Department of Pharmacy, University of Pisa, Pisa, Italy*
JESPER S. HANSEN • *Department of Experimental Medical Science, Lund University, Lund, Sweden*
MONIQUE HEITMEIER • *Department of Pediatrics, Washington University School of Medicine, St. Louis, MO, USA*
STEFAN HOHMANN • *Department of Chemistry and Molecular Biology, University of Gothenburg, Gothenburg, Sweden; Department of Biology and Biological Engineering, Chalmers University of Technology, Gothenburg, Sweden*

GEOFFREY D. HOLMAN • *Department of Biology and Biochemistry, University of Bath, Bath, UK*
PAUL W. HRUZ • *Department of Pediatrics, Washington University School of Medicine, St. Louis, MO, USA*
ROBERT M. JACKSON • *Department of Biochemistry and Molecular Biology, University of Oklahoma Health Sciences Center, Oklahoma City, OK, USA*
JAVIER R. JALDIN-FINCATI • *Cell Biology Program, The Hospital for Sick Children, Toronto, ON, Canada*
DAVID E. JAMES • *School of Life and Environmental Sciences, The University of Sydney, Sydney, NSW, Australia; Sydney Medical School, The University of Sydney, Sydney, NSW, Australia; The Charles Perkins Centre, The University of Sydney, Sydney, NSW, Australia*
HAE JOO KANG • *Department of Life Sciences, Imperial College London, London, UK*
DIMITRIOS KIOUMOURTZOGLOU • *Department of Biology, University of York, York, UK*
AMIRA KLIP • *Cell Biology Program, The Hospital for Sick Children, Toronto, ON, Canada*
FRANCOISE KOUMANOV • *Department of Biology and Biochemistry, University of Bath, Bath, UK*
GWENAËL LABOUÈBE • *Center for Integrative Genomics, University of Lausanne, Lausanne, Switzerland*
VÉRONIQUE A. LACOMBE • *Department of Physiological Sciences, Oklahoma State University, Stillwater, OK, USA; Harold Hamm Diabetes Center, University of Oklahoma Health Science Center, Oklahoma City, OK, USA*
CHRISTOPHE LAMY • *Department of Medicine, University of Fribourg, Fribourg, Switzerland*
EUNICE E. LEE • *Department of Dermatology, UT Southwestern Medical Center, Dallas, TX, USA*
KARIN LINDKVIST-PETERSSON • *Department of Experimental Medical Science, Lund University, Lund, Sweden*
WENTONG LONG • *Membrane Protein Research Group, University of Alberta, Edmonton, AB, Canada*
ZAHRA MARIA • *Department of Physiological Sciences, Oklahoma State University, Stillwater, OK, USA; Harold Hamm Diabetes Center, University of Oklahoma Health Science Center, Oklahoma City, OK, USA*
FILIPPO MINUTOLO • *Department of Pharmacy, University of Pisa, Pisa, Italy*
BJÖRN MORÉN • *Department of Experimental Medical Science, Lund University, Lund, Sweden*
YAAKOV NAHMIAS • *Alexander Grass Center for Bioengineering, The Rachel and Selim Benin School of Computer Science and Engineering, The Hebrew University of Jerusalem, Jerusalem, Israel; Department of Cell and Developmental Biology, Silberman Institute of Life Sciences, The Hebrew University of Jerusalem, Jerusalem, Israel*
DOUGALL M. NORRIS • *School of Life and Environmental Sciences, The University of Sydney, Sydney, NSW, Australia; The Charles Perkins Centre, The University of Sydney, Sydney, NSW, Australia*
DEBBIE O'NEILL • *Membrane Protein Research Group, University of Alberta, Edmonton, AB, Canada*
ANN LOUISE OLSON • *Department of Biochemistry and Molecular Biology, University of Oklahoma Health Sciences Center, Oklahoma City, OK, USA*
MISLAV OREB • *Institute of Molecular Biosciences, Goethe University, Frankfurt, Germany*

MALA SHANMUGAM • *Department of Hematology and Medical Oncology, School of Medicine, Emory University, Atlanta, GA, USA*
KARIN G. STENKULA • *Department of Experimental Medical Science, Lund University, Lund, Sweden*
BERNARD THORENS • *Center for Integrative Genomics, University of Lausanne, Lausanne, Switzerland*
SABINA TSYTKIN-KIRSCHENZWEIG • *Alexander Grass Center for Bioengineering, The Rachel and Selim Benin School of Computer Science and Engineering, The Hebrew University of Jerusalem, Jerusalem, Israel*
TIZIANO TUCCINARDI • *Department of Pharmacy, University of Pisa, Pisa, Italy*
RAMINTA VENSKUTONYTĖ • *Department of Experimental Medical Science, Lund University, Lund, Sweden*
GRÉGORY VERDON • *Department of Life Sciences, Imperial College London, London, UK; Research Complex at Harwell, Rutherford Appleton Laboratory, Didcot, UK*
RICHARD C. WANG • *Department of Dermatology, UT Southwestern Medical Center, Dallas, TX, USA*
SEBASTIAN WASSERSTROM • *Department of Experimental Medical Science, Lund University, Lund, Sweden*
CHANGYONG WEI • *Department of Hematology and Medical Oncology, Winship Cancer Institute, School of Medicine, Emory University, Atlanta, GA, USA*
NIEK WELKENHUYSEN • *Department of Chemistry and Molecular Biology, University of Gothenburg, Gothenburg, Sweden*
NIENG YAN • *State Key Laboratory of Membrane Biology, School of Life Sciences, Tsinghua University, Beijing, China; Beijing Advanced Innovation Center for Structural Biology, Tsinghua-Peking Center for Life Sciences, School of Life Sciences and School of Medicine, Tsinghua University, Beijing, China*

Chapter 1

Expression and Purification of Rat Glucose Transporter 1 in *Pichia pastoris*

Raminta Venskutonytė, Karin Elbing, and Karin Lindkvist-Petersson

Abstract

Large amounts of pure and homogenous protein are a prerequisite for several biochemical and biophysical analyses, and in particular if aiming at resolving the three-dimensional protein structure. Here we describe the production of the rat glucose transporter 1 (GLUT1), a membrane protein facilitating the transport of glucose in cells. The protein is recombinantly expressed in the yeast *Pichia pastoris*. It is easily maintained and large-scale protein production in shaker flasks, as commonly performed in academic research laboratories, results in relatively high yields of membrane protein. The purification protocol describes all steps needed to obtain a pure and homogenous GLUT1 protein solution, including cell growth, membrane isolation, and chromatographic purification methods.

Key words Glucose transporter, GLUT1, *Pichia pastoris*, Protein expression, Protein purification, Membrane protein, Chromatography

1 Introduction

GLUT1 is one of the most studied glucose transporters and it is widely expressed in a variety of tissues, including erythrocytes, blood-brain barrier, and pancreas [1]. Besides the normal function in healthy tissues, GLUT1 is known to be overexpressed in tumors in several different types of cancer [2]. For this reason, GLUT1 is a possible drug target for cancer therapy, and various chemical compounds have been shown to be potential inhibitors of GLUT1 [3, 4]. Recently, the first X-ray crystal structure of the human GLUT1 was determined and it confirmed the 12-α-helical arrangement of the transporter [5]. In addition, three low-resolution X-ray structures of the GLUT1 in complex with inhibitor compounds have also been reported [6]. This is the first step towards an understanding of GLUT1 inhibition. However, much is to be done, including not only structural biology, but also various functional studies that would provide extended understanding of the

Karin Lindkvist-Petersson and Jesper S. Hansen (eds.), *Glucose Transport: Methods and Protocols*, Methods in Molecular Biology, vol. 1713, https://doi.org/10.1007/978-1-4939-7507-5_1,

GLUT1 protein structure and function. Therefore, an easy and efficient expression and purification protocol for this protein is important. The production of GLUT1 has been described in yeast, both in *Saccharomyces cerevisiae* and *Pichia pastoris*, and insect cells [5–7]. Here, we provide a protocol for the expression and purification of the rat GLUT1 in yeast *Pichia pastoris* [7]. *Pichia pastoris* is optimal for large-scale protein expression as it can be grown to a very high cell density and it carries a strong methanol promoter, which allows efficient induction of protein expression [8]. Hence, *P. pastoris* can be successfully used to express large amounts of membrane proteins while at the same time being a relatively cheap expression system, which makes it an attractive choice for membrane protein production [9].

2 Materials

Here we describe the expression of GLUT1 in *P. pastoris* cultivated in shaker flasks, which is commonly used in many academic laboratories. Materials and methods described here are adapted from a commercially available *Pichia* expression kit system (i.e., Pichia Expression Kit For expression of recombinant proteins in *Pichia pastoris*, Catalog Number K1710-01, Invitrogen). For the preparation of yeast expression media it is most convenient to first prepare stock solutions and use them for preparation of the final growth media.

2.1 Solutions for Yeast Expression Media

1. YNB solution, 10×: Dissolve 34 g of yeast nitrogen base and 100 g ammonium sulfate in 1 L of water and filter sterilize (*see* **Note 1**). Store at 4 °C. For 1 L of expression medium, 100 mL of 10× YNB solution is required.
2. Biotin solution, 500×: Dissolve 20 mg in 100 mL of water and filter sterilize. Store at 4 °C. For 1 L of expression medium, 2 mL of 500× biotin solution is required.
3. Methanol solution, 10×: Mix 5 mL methanol with 95 mL of water and filter sterilize (enough for 1 L BMMY medium).
4. Glycerol solution, 10×: Make a 10 % v/v glycerol solution by mixing 100 mL of glycerol with 900 mL of water, and autoclave. For 1 L of expression medium, 100 mL of 10× glycerol solution is required.
5. Potassium phosphate buffer: Prepare a 1 M potassium phosphate buffer by combining monobasic and dibasic potassium phosphate to give a pH of 6.0. Autoclave the buffer. For 1 L of expression medium, 100 mL of 1 M potassium phosphate buffer is required.
6. Glucose, 10×: Dissolve 200 g of glucose in 1 L of water and sterilize using an autoclave or filtration.

7. YPD medium agar plates: Make a desired volume of 1 % w/v yeast extract, 2 % w/v peptone, and 2 % w/v agar and autoclave the solution. After the autoclavation, add sterile glucose solution to give 2 %, divide the solution into petri dishes, and let it solidify (*see* **Note 2**).
8. YPDS medium agar plates: Prepare as YPD agar plates, but with the addition of 1 M sorbitol prior to autoclave sterilization. To prepare YPDS plates containing zeocin, add the desired amount of zeocin to the YPDS-agar solution after it has cooled down to 60 °C before dividing the medium into the plates (for the zeocin test it is convenient to use rectangular plates, e.g., OmniTray).

2.2 Expression Media

1. Growth medium: To prepare 1 L of growth medium dissolve 10 g of yeast extract and 20 g peptone in 700 mL water and autoclave. It is convenient to use 1 L blue-cap bottles for the growth medium.
2. BMMY medium: To prepare BMMY add 100 mL of 1 M potassium phosphate buffer, 100 mL 10× YNB solution, 2 mL of 500× biotin solution, and 100 mL of 10× methanol solution to the growth medium.
3. BMGY medium: For BMGY add 100 mL of 1 M potassium phosphate buffer, 100 mL 10× YNB solution, 2 mL of 500× biotin solution, and 100 mL of 10× glycerol to the growth medium.
4. The ingredients should be mixed using a sterile technique to avoid any contamination.

2.3 Equipment for Protein Expression

1. Consumables for cell work and incubation: Petri dishes, 96-well plates, 15 mL round-bottom tubes, 50 mL Falcon tubes, glass beads, inoculation loops. Glass beads for small test expression test.
2. Glassware for cell growth: 1 L conical flasks, 5 L conical baffled flasks.
3. Shaker incubator with temperature control, spectrophotometer for OD measurements, and centrifugation equipment for cell harvesting.

2.4 Buffers for Protein Purification

The pH of all buffers should be set at 4 °C, as protein purification is carried out in the cold and samples are held on ice at all times. Buffers used for chromatography with FPLC systems should be filtered through 0.2 μm filters. DTT should be added to buffers just prior to use. Also, detergents should be added immediately prior to use. Do not store buffers with detergents for longer periods of time.

1. Cell breaking buffer: 25 mM Sodium phosphate pH 7.4, 4 mM EDTA, 5 % v/v glycerol, 2 mM DTT (added at use), protease inhibitor cocktail (added at use).
2. Basic pH solution: 20 mM NaOH.
3. Solubilization buffer: 20 mM Tris HCl pH 8.3, 20 % v/v glycerol, 10 mM NaCl, 0.2 mM EDTA, 1 mM DTT, 1 % w/v DM (alternatively 2 % w/v DDM).
4. Phosphate buffer: 25 mM Sodium phosphate, pH 7.4.
5. Ion-exchange chromatography buffer A (buffer A): 20 mM Tris HCl pH 8.3, 10 % v/v glycerol, 0.2 mM EDTA, 1 mM DTT, 0.1 % w/v DM.
6. Ion-exchange chromatography buffer B (buffer B): 20 mM Tris HCl pH 8.3, 10 % v/v glycerol, 0.2 mM EDTA, 1 mM DTT, 0.1 % w/v DM, and 1 M NaCl.
7. Size-exclusion chromatography buffer: 20 mM Tris–HCl pH 8.3, 150 mM NaCl, 10 % v/v glycerol, 0.2 mM EDTA, 0.1 % w/v DM.

2.5 Equipment for Protein Purification

X-Press Disintegrator (AB Biox, Sweden): This apparatus is used for breaking yeast cells in our laboratory; however, other cell disruption methods, efficient enough to break *Pichia* cells, could also be used:

1. Dounce homogenizer.
2. Ultracentrifuge and rotors for small- and large-scale expression/purification.
3. Chromatography equipment, FPLC system.
4. Anion-exchange chromatography column and size-exclusion chromatography column.
5. Centrifugal concentrators with 100 kDa cutoff and 0.2 μm centrifugal filters.

3 Methods

The *P. pastoris* strain GS115 aqy1Δ [10] with a gene coding for *Rattus norvegicus* GLUT1 (rGLUT1) is used for the protein production. The DNA coding for the rGLUT1 should be subcloned into a pPICZ vector applying general molecular biology methods, which carries the recombination sites allowing the insertion of the gene of interest into the yeast genome. Then this vector is linearized and used for transformation in *Pichia* yeast cells, applying homologous recombination, according to standard protocols (for details see pPICZ A, B, and C *Pichia* expression vectors for selection on Zeocin™ and purification of recombinant proteins, Catalog no.

V190-20. Invitrogen). Once the yeast cells are transformed with the vector carrying the rGLUT1 gene, screening for the clone that gives the best protein expression should be performed. When such clones are identified, it is important to confirm the protein expression in small scale before proceeding to the large-scale protein production.

3.1 Screening for High Expressers

1. Aliquot 100 μL of sterile water into a 96-well plate for the number of transformants to be tested, and add small amounts of cells from one transformant by taking the cells from the top of one isolated colony.
2. Mix by pipetting up and down, withdraw 5 μL of each resuspended transformant, and spot it on an YPDS plate containing 0, 1000, and 1500 μg/mL zeocin, respectively. Incubate the plates at 30 °C for 2–3 days (*see* **Note 3**). Note that zeocin is light sensitive.
3. Inspect the plates to identify the transformants, which resulted in the best growth. High zeocin resistance indicates high protein expression (*see* Fig. 1).

3.2 Small-Scale Expression Test

1. Day 1: Using an inoculation needle or a pipette tip, take some cells from the selected clones from the zeocin screen and inoculate it into 5 mL of BMGY medium in 15 mL round-bottom culture growth tubes. Incubate at 30 °C with 250 rpm shaking for 24 h.
2. Day 2: Measure the OD_{600} of the 5 mL cultures and use these cultures to start a 10 mL BMGY culture in the afternoon. Calculate the dilution to reach an OD_{600} of around 6 the following morning (*see* **Note 4**).
3. Day 3: Measure the OD_{600}. Transfer the culture into sterile 50 mL centrifuge tubes and spin down the cells at 3,000 × *g* for 3 min at room temperature. Discard the supernatant and resuspend the cell pellet in 25 mL BMMY media. Then transfer the culture into a sterile 250 mL Erlenmeyer flask and incubate at 30 °C, 250 rpm, for 24 h. The starting OD_{600} of this culture should be approximately 1–1.5.
4. Day 4: In the morning feed the cultures with 2.5 mL of 10× methanol solution and incubate for additional 6–7 h. Then harvest the cells by centrifugation at 3,000 × *g* for 5 min at 4 °C. Resuspend the cells in 1 mL of cell breaking buffer and transfer to a 1.5 mL centrifuge tube. Spin the cells at 16,000 × *g* for 3 min at 4 °C and discard the supernatant. The cell pellet can be frozen at −80 °C at this point.
5. The cells can be lysed using a FastPrep® −24 Instrument (MP Biomedicals) for a small-scale experiment. Resuspend the cells in a small volume of breaking buffer and add the equivalent

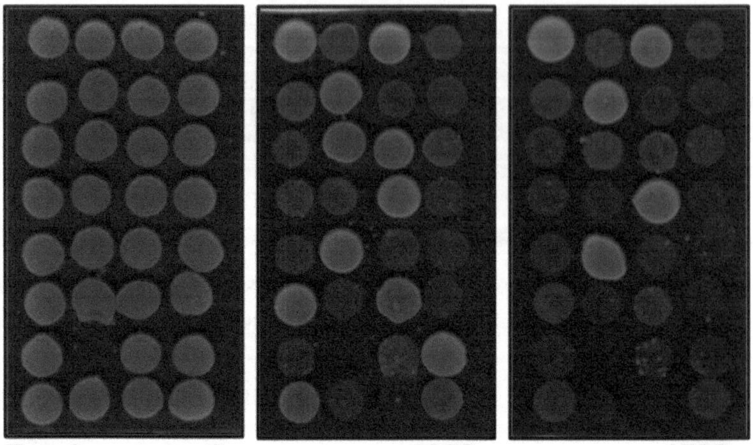

Fig. 1 Selecting the best clones for protein expression. Selection of the best clones for expressing GLUT1 is performed on YPDS agar plates with 0, 1000, and 1500 μg/mL zeocin

volume of glass beads (e.g., if the cell suspension is 0.5 mL, add glass beads up to 1 mL). Shake the cells in the FastPrep® –24 for 20 s, at the speed setting of 6. Repeat the shaking for 3 × times and cool the sample on ice in between the runs.

6. Transfer the suspension of broken cells to a new tube and centrifuge for 30 min at 16,000 × *g* at 4 °C. Collect the supernatant and centrifuge at 120,000 × *g* (*see* **Note 5**) for 90 min at 4 °C.
7. Resuspend the isolated membranes in 100 μL solubilization buffer without 1 % w/v DM.
8. Use equal amounts of the obtained membrane sample of each tested clone for Western blot analysis. Any standard Western blot protocol can be used for this purpose. Use a primary GLUT1 antibody for the detection of rGLUT1 and a suitable secondary antibody for your detection system.
9. Evaluate the Western blot results. Choose the clone showing the strongest signal for rGLUT1 protein for further large-scale expression.
10. Prepare a glycerol stock of the selected clones for long-term storage. This can be done by mixing equal amounts of culture of a specific clone in BMGY medium and 30 % sterile glycerol solution in a cryo tube. The glycerol stock should be stored at –80 °C and can be used for future expressions.

3.3 Large-Scale Expression of rGLUT1

Depending on how much protein is needed, large-scale production can be carried out in one to several liters. The expected yield of purified rGLUT1 is 1–3 mg/L culture. For each planned 1 L expression culture medium in BMMY, cells need to be pre-grown in 200 mL of BMGY medium.

1. Day 1: Spread some cells from a glycerol stock on a YPD petri dish and incubate at 30 °C for 3 days.
2. Day 4: Start several 5 mL BMGY pre-cultures in 15 mL tubes by inoculating with cells grown on a YPD plate. Incubate overnight at 30 °C with orbital shaking at 225–250 rpm.
3. Day 5: In the afternoon, measure the OD_{600} of pre-cultures and inoculate 200 mL BMGY culture medium in 1 L Erlenmeyer flask with a volume of the pre-culture that would result in OD_{600} of 6–7 after 16–18-h incubation with orbital shaking at 225–250 rpm (*see* **Note 6**).
4. Day 6: In the morning, measure the OD_{600} of 200 mL cultures. Centrifuge the cells at 3,000 × *g* for 5 min at room temperature in sterile centrifuge tubes, resuspend in BMMY medium, and transfer into a 5 L baffled Erlenmeyer flask bringing the volume of BMMY to 1 L. The starting OD_{600} of the 1 L culture should be around 1–1.5. Incubate the cultures for 24 h with orbital shaking at 225–250 rpm.
5. Day 7: Feed the cultures with methanol by adding 5 mL of 100 % methanol per L culture. Incubate for an additional 6–7 h with orbital shaking at 225–250 rpm. Harvest the cells by centrifugation at 4,000 × *g* for 30–45 min at 4 °C. Resuspend the cells from 1 L culture in approximately 30 mL phosphate buffer and transfer the suspension to a 50 mL centrifuge tube. Spin at 4,000 × *g* for 5 min at 4 °C. Remove the supernatant and store cells at −80 °C (*see* **Note 7**).

3.4 Cell Disruption

1. The cell pellet is mixed with breaking buffer to give a maximum of 30 mL of resuspended cell pellet. Let the cells melt completely on ice water. The suspension is poured into a precooled cell disintegrator and placed into −25 °C. Cell disintegrator with cell pellet should be equilibrated in the freezer for at least 20 h before proceeding with cell disruption (*see* **Note 8**).
2. The cells are pressed through the cell disintegrator three times. The cells are removed from the disintegrator, mixed with 30 mL of breaking buffer, and stirred on a magnetic stirrer. The total volume should be 50–60 mL of cell pellet in the breaking buffer. Cell pellets are left on a magnetic stirrer until the suspension is homogeneous and the pellet is fully resuspended.

3.5 Membrane Isolation

1. The cell lysate is subjected to a low-speed centrifugation to remove large cell debris at 6,000 × *g* for 10 min at 4 °C twice, and supernatant collected (*see* **Note 9**).

2. The supernatant from low-speed centrifugation is further centrifuged at 145,000 × *g* for 90 min at 4 °C.
3. The supernatant from the ultracentrifugation step is discarded and the pellet weighed. The pellet is transferred to a 50 mL Dounce tissue grinder (homogenizer) together with basic pH solution. The pellet is homogenized with the pestle manually or by connecting the pestle to an electrical homogenizer and diluted with more basic pH solution. The amount of basic pH solution for the wash in total should be approximately 1 mL per 50 mg membranes (*see* **Note 10**).
4. The washed membranes are subjected to ultracentrifugation again, at 145,000 × *g* for 90 min at 4 °C. The expected amount of the membranes at this step is approximately ten times less than the amount of the cell starting material.

3.6 Solubilization

1. The isolated membranes are transferred to a 50 mL homogenizer together with solubilization buffer without the detergent and homogenized with a pestle. Use enough of the buffer to achieve an easy homogenization, avoiding any foaming to occur.
2. The homogenized pellet solution is mixed with the solubilization buffer containing the detergent and protease inhibitors. The final concentration of the detergent in the membrane/buffer solution should be 1 % w/v for DM and 2 % w/v for DDM. Use a total of 25 mL of solubilization buffer per 10 g cell starting material. Solubilization should be carried out for 1 h at 4 °C with mild rotation (*see* **Notes 11** and **12**).
3. After solubilization the solution is centrifuged at 145,000 × *g* for 30 min at 4 °C to remove any insoluble material. The supernatant is collected and the pellet is discarded. Take a sample for the SDS-PAGE analysis before and after centrifugation.

3.7 Protein Purification

1. For ion-exchange chromatography, the solubilized sample is diluted six times with buffer A without DM detergent. The sample is then filtered through a 0.2 μm filter and loaded onto a pre-equilibrated anion-exchange column (*see* **Note 13**). The protein is eluted with a salt gradient, using a combination of buffer A and buffer B. The gradient of 20 column volumes up to 50 % buffer B is used and fractions of 1 mL collected.
2. Peak fractions are analyzed by SDS-PAGE and the fractions containing the rGLUT1 protein are pooled (*see* **Note 14**).
3. The pooled fractions are concentrated using a 15 mL centrifugal concentrator with a 100 kDa molecular cutoff to a final concentration of 10–15 mg/mL. Avoid high local protein

concentrations to build up, as well as increases in detergent concentration, by pipetting up and down every 10 min during the up-concentration process.

4. For size-exclusion chromatography, equilibrate the column with 1.5–2 column volumes of buffer and inject 0.5 mL of the concentrated sample from ion-exchange chromatography step. The sample should be filtered through a 0.2 μm syringe filter before injection. The rGLUT1 protein elution volume is approximately 13 mL when using the Superdex 200 30/100 GL column (GE healthcare). Collect 0.5 mL fractions and take out samples for SDS-PAGE (*see* Fig. 2). Pool the peak fractions of the rGLUT1 and concentrate until the desired concentration is reached (*see* **Note 15**).
5. The purified protein sample can be used directly for functional assays or it can be aliquoted into desired volumes and flash frozen in liquid nitrogen. The frozen samples should be stored at −80 °C.

4 Notes

1. Note that some solutions can be autoclaved, while others are sensitive to heat treatment and should be sterile filtered instead. A vacuum filtration with disposable sterile 0.2 μM filters and a bottle that has been autoclaved can be used.
2. The glucose solution can either be filtered or autoclaved. However, it is important to autoclave just the glucose solution; that is, glucose cannot be added to the media and then autoclaved, as it will caramelize. For example, if preparing 1 L of YPD medium, one should mix all the ingredients except the glucose and bring the volume to 900 mL. After the autoclavation, 100 mL of 20 % v/v glucose solution can be added to give a final concentration of 2 %.
3. It is recommended to monitor the plates regularly, twice a day, to observe the growth and to avoid false positives. The clones with high protein expression should show the best growth, but it is important to observe the colonies and to select the clones appearing earliest on the zeocin-containing medium. The plate without zeocin is used to see that approximately the same amount of cells were pipetted for each clone.
4. *P. pastoris* cells double approximately every 2 h; thus for example if 16-h growth is planned, one should estimate that the culture will double eight times and dilute the culture accordingly. Also, when measuring cell optical density one should dilute the sample if OD exceeds 1.0, as the measurement will not be accurate otherwise.

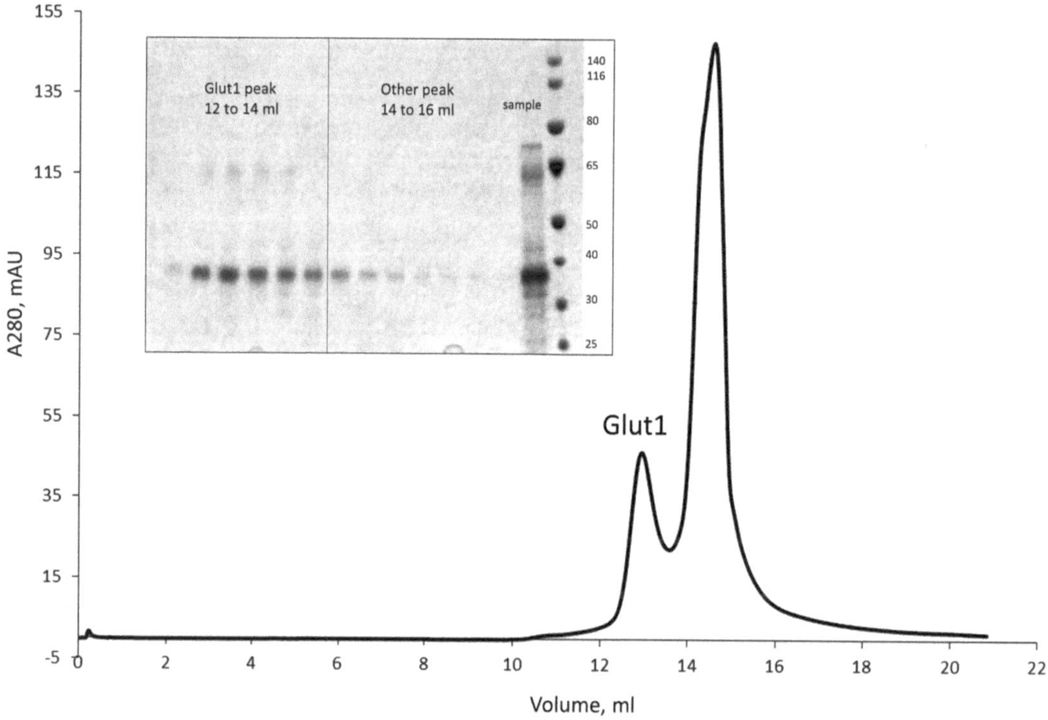

Fig. 2 Size-exclusion chromatography (SEC) profile of rGLUT1. The first peak is GLUT1 protein, while the second larger peak is most likely detergent micelles and/or impurities. The buffer used was 20 mM HEPES pH 7.0, 150 mM NaCl, and 10 % v/v glycerol and the sample was purified twice by ion-exchange chromatography and one SEC after which the purest fractions were collected and subjected to the SEC again resulting in the presented profile. The insert shows an SDS-PAGE of the peaks as well as a sample that was subjected to the SEC and a protein marker (PageRuler Prestained Protein Ladder, Thermo Fisher Scientific) with indicated MW in kDa

5. Depending on available centrifuge, rotors, and tubes, various solutions can be used to isolate the membranes at 120,000 × *g*. It is important to calculate the suitable rpm for a rotor used as well as to ensure that the tubes are compatible with high centrifugation speeds. In our laboratory, a TLA 120.1 rotor is used for this step, as it fits small centrifuge tubes.
6. Before starting the expression of the protein in a growth medium with methanol, the cells are grown in a medium with glycerol, to obtain enough biomass before the induction of protein expression with methanol.
7. If contamination is suspected, one could take a small sample of the culture and inspect it under the microscope to make sure that no bacterial contamination is present.
8. The cell disruptor should be assembled and placed into a freezer at least 1 day before the cells are poured into the disruptor. In this protocol, the X-Press disintegrator (AB Biox, Sweden) is used; however another method suitable for breaking yeast cells can also be used, for example, a

Bead Beater (Biospec) for cell lysis. The limitation of using the cell disintegrator is that the maximum volume of cell slurry is 30 mL, which is normally an amount from 2 L culture (20–25 g of cells).

9. This step is meant to clear the cell lysate before continuing with high-speed centrifugation for membrane isolation. The recommended centrifugation speed of 6,000 × *g* can be varied and it is possible to use lower speeds, compatible with tabletop centrifuges as well (e.g., 4,000 × *g*).
10. Many ultracentrifuge tubes are sensitive to high pH values and cannot be used for centrifugation of the NaOH solution. Therefore, after washing the membranes, the pH of the solution can be brought back to the pH value close to 8 before pouring the sample into the ultracentrifuge tubes. Alternatively, one-time tubes for ultracentrifugation, such as OptiSeal™ (Beckman Coulter), could be used at this step.
11. For homogenizing the membranes, use the solubilization buffer with protease inhibitor, but without the detergent, and after the membranes are homogenized mix them with the remaining solubilization buffer containing the detergent. The membranes in solubilization buffer can be poured into a 50 mL Falcon tube and placed on a rotator for such tube. Alternatively, the solubilization can be performed in a flask with magnetic stirring. Instead of DM, DDM can also be used as a detergent for solubilization.
12. We occasionally detect cleavage of rGLUT1 for some batches. If this occurs, including cholesterol esters to the solubilization buffer can resolve the problem. In our laboratory, solubilization with a buffer containing 2 % w/v DDM and 0.4 % w/v cholesteryl Hemisuccinate Tris Salt (Anatrace) works well in overcoming the cleavage issue.
13. In our laboratory, both 6 mL Resource Q and 5 mL HiTrap Q HP (GE Healthcare) anion-exchange columns have been used for the purification, and both are suitable. The flow rate should be adjusted according to the column's specifications.
14. The theoretical isoelectric point (pI) of the GLUT1 protein is 9 (calculated from the amino acid sequence with ProtParam tool [11]); however the pH for the binding to an anion-exchange column has been determined experimentally and the pH 8.3 is suitable for capturing the protein on a Resource Q or a HiTrap Q columns. It is recommended to analyze peak fractions on a SDS-PAGE, to check the purity of the protein after the ion-exchange chromatography, and to choose the purest peak fractions. Also, the samples of solubilized membranes can be included into SDS-PAGE analysis at this point to monitor the total solubilized protein. The level of impurities

can vary for different expressions. If large amounts of impurities, as judged from the SDS-PAGE, are seen after the ion-exchange step, the peak fractions can be diluted to a lower salt concentration (≤10 mM) with buffer A and subjected to an ion-exchange again (using a washed and pre-equilibrated column) before proceeding to the next step.

15. The protein sample used for size-exclusion chromatography should be almost pure. If no well-resolved peak is obtained, the sample can be collected, concentrated, and subjected to size-exclusion chromatography one more time. We often see a GLUT1 peak, eluting at around 13 mL followed by another peak, which does not show protein bands on the SDS-PAGE. We are not sure what this second peak is, but it might be that it corresponds to detergent micelles, which elute separately when concentrated samples are used. The peak fractions should be analyzed by SDS-PAGE to confirm the purity of the protein. It is common to see a monomeric as well as dimeric band on the gel. Moreover, the protein runs at approximately 40 kDa mark on a Bis-tris 4–12 % gel; thus it does not agree with its actual size (54 kDa). Also, a different buffer could be used for the SEC than suggested in this manual; for example, HEPES buffer at pH 7.0 is also suitable.

Acknowledgement

The Swedish Research Council (2011–2891), the Cancer Foundation (2010/1171 and 2014/575), and Novo Nordisk Foundation (9807) supported this work.

References

1. Mueckler M, Thorens B (2013) The SLC2 (GLUT) family of membrane transporters. Mol Asp Med 34(2–3):121–138. https://doi.org/10.1016/j.mam.2012.07.001
2. Szablewski L (2013) Expression of glucose transporters in cancers. Biochim Biophys Acta 1835(2):164–169. https://doi.org/10.1016/j.bbcan.2012.12.004
3. Barron CC, Bilan PJ, Tsakiridis T, Tsiani E (2016) Facilitative glucose transporters: implications for cancer detection, prognosis and treatment. Metab Clin Exp 65(2):124–139. https://doi.org/10.1016/j.metabol.2015.10.007
4. Granchi C, Fancelli D, Minutolo F (2014) An update on therapeutic opportunities offered by cancer glycolytic metabolism. Bioorg Med Chem Lett 24(21):4915–4925. https://doi.org/10.1016/j.bmcl.2014.09.041
5. Deng D, Xu C, Sun P, Wu J, Yan C, Hu M, Yan N (2014) Crystal structure of the human glucose transporter GLUT1. Nature 510(7503):121–125. https://doi.org/10.1038/nature13306
6. Kapoor K, Finer-Moore JS, Pedersen BP, Caboni L, Waight A, Hillig RC, Bringmann P, Heisler I, Muller T, Siebeneicher H, Stroud RM (2016) Mechanism of inhibition of human glucose transporter GLUT1 is conserved between cytochalasin B and phenylalanine amides. Proc Natl Acad Sci U S A 113(17):4711–4716. https://doi.org/10.1073/pnas.1603735113
7. Hansen JS, Elbing K, Thompson JR, Malmstadt N, Lindkvist-Petersson K (2015) Glucose transport machinery reconstituted in cell models. Chem Commun 51(12):2316–2319. https://doi.org/10.1039/c4cc08838g

8. Cereghino JL, Cregg JM (2000) Heterologous protein expression in the methylotrophic yeast Pichia pastoris. FEMS Microbiol Rev 24(1):45–66
9. Byrne B (2015) Pichia pastoris as an expression host for membrane protein structural biology. Curr Opin Struct Biol 32:9–17
10. Fischer G, Kosinska-Eriksson U, Aponte-Santamaria C, Palmgren M, Geijer C, Hedfalk K, Hohmann S, de Groot BL, Neutze R, Lindkvist-Petersson K (2009) Crystal structure of a yeast aquaporin at 1.15 angstrom reveals a novel gating mechanism. PLoS Biol 7(6):e1000130. https://doi.org/10.1371/journal.pbio.1000130
11. Wilkins MR, Gasteiger E, Bairoch A, Sanchez JC, Williams KL, Appel RD, Hochstrasser DF (1999) Protein identification and analysis tools in the ExPASy server. Methods Mol Biol 112:531–552

Chapter 2

Crystallization and Structural Determination of the Human Glucose Transporters GLUT1 and GLUT3

Dong Deng and Nieng Yan

Abstract

Overexpression, purification, and crystallization of eukaryotic membrane proteins represent a major challenge for structural biology. In recent years, we have solved the crystal structures of the human glucose transporters GLUT1 in the inward-open conformation at 3.17 Å resolution and GLUT3 in the outward-open and occluded conformations at 2.4 and 1.5 Å resolutions, respectively. Structural elucidation of these transporters in three distinct functional states reveal the molecular basis for the alternating access transport cycle of this prototypal solute carrier family. It established the molecular foundation for future dynamic and kinetic investigations of these GLUTs, and will likely facilitate structure-based ligand development. In this chapter, we present the detailed protocols of recombinant protein expression, purification, and crystallization of GLUT1 and GLUT3, which may help the pursuit of structural elucidation of other eukaryotic membrane proteins.

Key words Glucose transporters, Glut, GLUT1, GLUT3, Protein purification, Crystallization

1 Introduction

Structural biology aims to unveil the biological world at atomic scale. X-ray crystallography, electron microscopy (EM), and nuclear magnetic resonance (NMR) are the major experimental approaches for elucidating the three-dimensional structures of macromolecules. Among all the biological molecules and miniature machineries, integral membrane proteins, particularly eukaryotic membrane proteins, represent the most challenging targets for structural biology due to the technical difficulties associated with protein generation, purification, and crystallization. By October 4th 2016, in total 2052 structures of membrane proteins were reported, representing approximately 1.5 % of the total structure entries deposited to Protein Data Bank (PDB). Among these structures, only 647 are unique ones (http://blanco.biomol.uci.edu/mpstruc/).

Karin Lindkvist-Petersson and Jesper S. Hansen (eds.), *Glucose Transport: Methods and Protocols*, Methods in Molecular Biology, vol. 1713, https://doi.org/10.1007/978-1-4939-7507-5_2,

The recent technological breakthrough of cryo-EM has drastically promoted structural elucidation of membrane proteins with a large molecular mass. However, for those proteins with a molecular weight below 150 kDa, the primary method for high-resolution structure resolution remains to be X-ray crystallography.

The SLC2A family glucose transporters, exemplified by GLUT1, 2, 3, and 4, have been the prototype in the investigation of solute transport. GLUT1–4 catalyze facilitative diffusion of glucose across biomembranes, responsible for the supply of glucose to brain and other organs [1]. Owing to their fundamental physiological significance, structures of GLUTs have been pursued for decades. In the past several years, we have solved four X-ray structures of human GLUT1 and GLUT3 at three distinct transport states, including GLUT1 in the inward-open conformation at 3.17 Å, and GLUT3 in the maltose-bound outward-open state at 2.6 Å and occluded conformations in the presence of glucose or maltose at 1.5 and 2.4 Å resolutions, respectively. With these structures, a morph of a nearly complete alternating access transport cycle can be generated.

To capture the structures of GLUT1 and GLUT3 at different functional states, we aimed to set up efficient strategies for (1) the overexpression of GLUTs in insect cells, (2) the purification with appropriate detergents, (3) crystallization, and (4) structure determination. We hereby also describe a proteoliposome-based counter-flow assay following a modified protocol [2] to qualitatively examine the transport activity of GLUTs. In retrospect, the successful crystallization and structural determination of GLUT1 and GLUT3 in different conformations may be attributed to the following elements:

1. Introduction of point mutations that remove glycosylation. We have introduced glycosylation-eliminated variants of GLUT1 (N45T) and GLUT3 (N43T).
2. Introduction of point mutation to GLUT1 that may lock the protein in the inward-open conformation. By literature search, we identified one single-point mutation GLUT1 (E329Q) that may lock the protein in an inward-open state. The mutation was originally identified as a disease-related variation in GLUT4.
3. The recombinant GLUT1 protein was purified and crystallized in the presence of the detergent β-nonyl-D-glucopyranoside, which may further stabilize an inward-open conformation of the transporter.
4. Crystallization of GLUT1 was carried out at 4 °C, which may lower the mobility of the transporter, facilitating crystallization.

2 Materials

1. Modified vectors from pFastBac1 (Thermo).
2. *E. coli* strains: DH5α and DH10Bac.
3. Insect cell lines: Sf-9 (*Spodoptera frugiperda*), High Five (*Trichoplusia ni*).
4. Luria-Bertani (LB) medium: LB Broth, Miller.
5. LB agar plates (for plasmid generation): LB agar plates containing 100 μg/mL ampicillin (Amresco).
6. LB agar plate (for bacmid generation): LB agar plates containing 100 μg/mL x-gal, 40 μg/mL IPTG, 50 μg/mL kanamycin, 7 μg/mL gentamicin, 10 μg/mL tetracycline.
7. LB medium (for bacmid generation): LB medium with 50 μg/mL kanamycin, 7 μg/mL gentamicin, 10 μg/mL tetracycline.
8. Ethanol solution: 70 % v/v Ethanol.
9. Antibiotic solution: 10 × Penicillin-streptomycin solution.
10. HyClone™ SFX-Insect cell culture medium (GE Healthcare).
11. Cellfectin II (Thermo).
12. SIM HF cell culture medium (Sino Biological Inc.).
13. FBS solution: 5 % v/v FBS (Thermo).
14. Lysis buffer 1: 25 mM Tris–HCl pH 8.0 and 150 mM NaCl.
15. Lysis buffer 2: Lysis buffer 1 containing protease inhibitors (0.8 μM aprotinin, 2 μM pepstatin, 5 μg/mL leupeptin) and 2 % w/v n-dodecyl-b-D-maltoside (DDM).
16. Ni-NTA agarose.
17. Wash buffer: 25 mM MES pH 6.0, 150 mM NaCl, 30 mM imidazole, 5 % v/v glycerol, and 0.05 % w/v n-dodecyl-β-D-maltopyranoside (DDM) (Anatrace).
18. Elution buffer: 25 mM MES pH 6.0, 150 mM NaCl, 300 mM imidazole, 5 % v/v glycerol, and 0.05 % w/v DDM.
19. Centrifugal filters: 50 kDa Molecular cutoff.
20. Size-exclusion buffer: 25 mM MES pH 6.0, 150 mM NaCl, 5 % v/v glycerol, and 0.4 % w/v n-Nonyl-β-D-glucopyranoside (β-NG) (Anatrace).
21. Anti-His tag mouse polyclonal antibody, goat anti-mouse IgG, HRP-conjugated IgG (CWBio), and TMB substrate.
22. Phusion DNA Polymerase (Thermo).
23. T4 DNA ligase (NEB).
24. dNTP solution: 10 mM dNTP.

25. TIANgel Mini Purification Kit (TIANGEN).
26. DNA product Purification Kit.
27. Plasmid Purification Kit.
28. Reservoir solution (for hanging-drop vapor diffusion crystallization of GLUT1): 30 % w/v PEG400, 0.1 M MES pH 6.0, and 0.1 M $MgCl_2$.
29. Mother liquor (for lipid cubic phase (LCP) crystallization of GLUT3): 28 % v/v PEG400, 0.1 M HEPES pH 6.8, and 50 mM ammonium citrate.
30. Precipitant solution (outward-occluded conformation): 38–40 % v/v PEG 400, 100 mM $(HCOO)_2Mg$, 50 mM maltose, and 100 mM ADA pH 6.5.
31. Precipitant solution (outward-open conformation): 34 % v/v PEG 400, 400 mM $(NH_4)_2HPO_4$, 50 mM maltose, and 100 mM ADA pH 6.9.
32. MicroMesh (M3-L18SP-50; MiTeGen).
33. Radioactive glucose: 1 μCi D-[2-^{3}H] glucose (specific radioactivity 21.5 Ci/mmol, 0.46 mM external D-[2-3H] glucose) (Perkin Elmer).
34. *E. coli* polar lipids extract (Avanti Polar Lipids Inc.).
35. Mixed liposome solution: 3:1 v/v chloroform and methanol.
36. KPM buffer: 50 mM Potassium phosphate pH 6.5, 2 mM MgSO4.
37. PC Membranes 0.4 μm (Satorius), GSTF membrane filter 0.22 μm (Merck-Millipore).
38. Bio-Beads SM2 (Bio-Rad).
39. Optiphase HISAFE 3 (PerkinElmer).
40. MicroBeta JET (PerkinElmer).
41. Glass sandwich plates (Shanghai FAstal BioTech).
42. Robot arm Gryphon (ARI).
43. SPEXTM 6770PLUS (SPEX SamplePrep).
44. SDS-polyacrylamide gel equipment.
45. Superdex-200 10/300 GL.
46. HiTrap Desalting column, 5 mL.
47. ÄKTA pure chromatography system (GE Healthcare).

3 Methods

The general steps required for the structure determination of GLUTs. The methods comprise (1) the establishment of Bac-to-Bac® Baculovirus Expression System for expression of GLUTs in

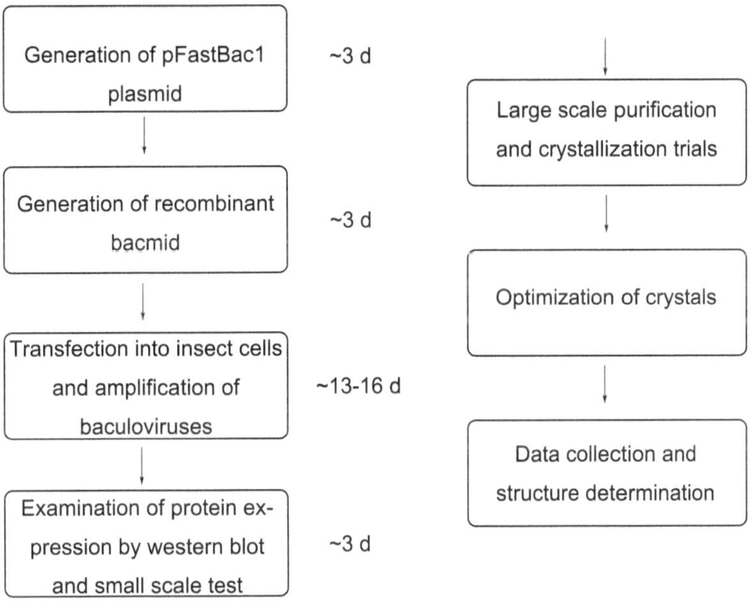

Fig. 1 General steps for structure determination of GLUTs. This flow chart presents the procedure for determining the structure of GLUT1/GLUT3. Each step will be described in the following parts. Also shown is the approximate time cost for major steps involved

insect cells, (2) protein expression and purification, (3) crystallization, and 4) data collection and structure determination (*see* Fig. 1). In this chapter, we only mention the case using insect cells; however GLUT1 was also purified from the overexpression system of yeast (*Saccharomyces cerevisiae* [3] and *Pichia pastoris* [4]) and human erythrocyte membrane [5].

3.1 Generating the Recombinant pFastBac Vector

For GLUT1, the full-length human GLUT1 cDNA is subcloned into the NdeI and XhoI sites of a modified pFastBac vector with a standard PCR-based strategy. A C-terminal 10× His-tag immediately follows the sequence. For GLUT3, the synthesized and codon-optimized cDNA of human GLUT3 (N43T) is subcloned into a modified pFastBac vector with an N-terminal 10× His-tag. The pFastbac vectors used in our study were modified to introduce the His-tag to GLUTs (*see* **Note 1** and Fig. 2).

1. The cDNAs of GLUTs are amplified using a standard PCR protocol. The PCR reaction mixture, 50 μL in total, is composed of 1 μL (200 ng) of DNA template, 50 pmol of each primer, 0.5 μL of Phusion enzyme, 2 μL dNTP, and 1 μL DMSO in the phusion buffer.
2. The PCR products are purified with the TIANgel Mini Purification Kit.

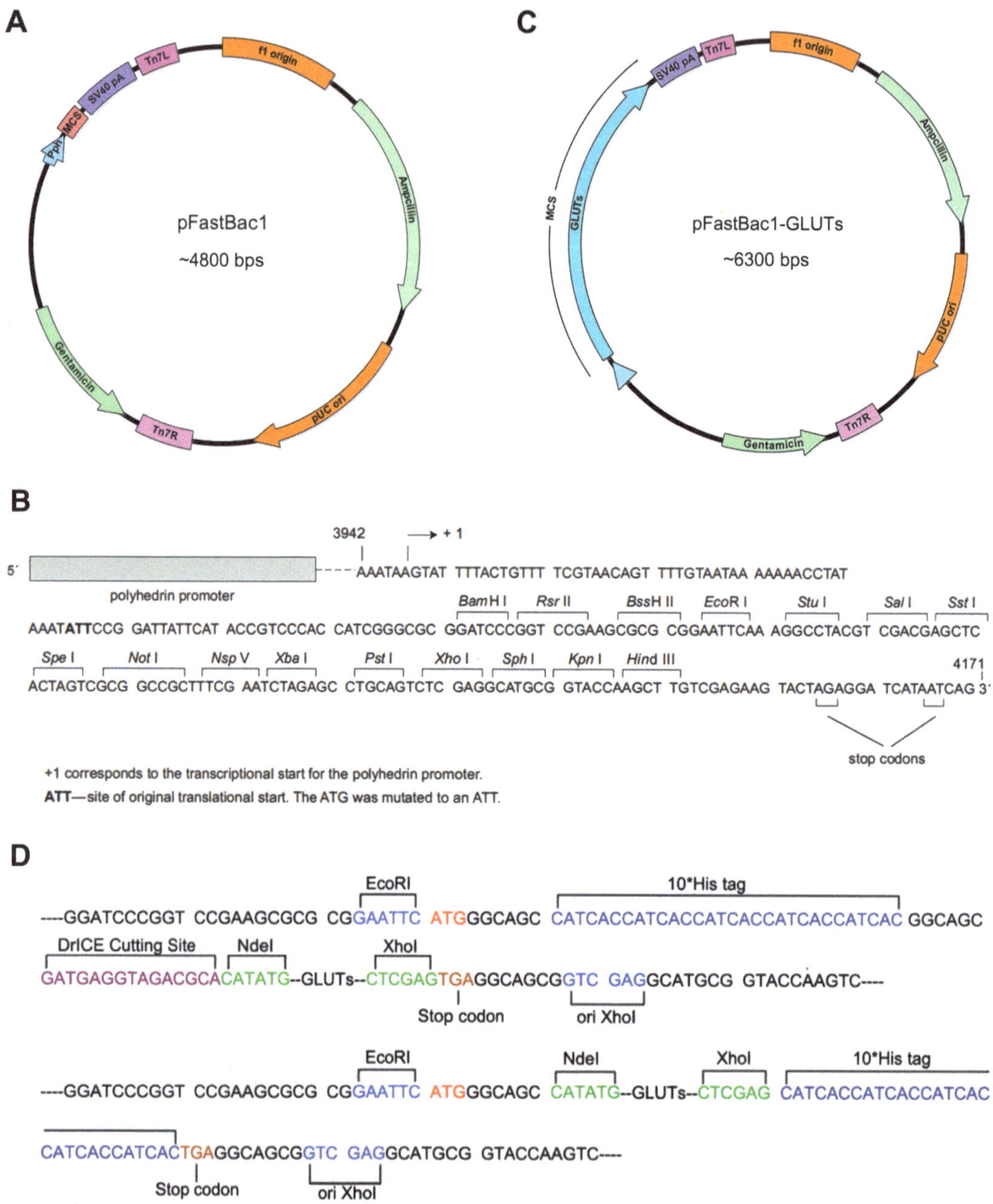

Fig. 2 Maps of plasmids used in the experiments. (**a**) Map of commercial pFastBac1 plasmid. (**b**) Sequence of original MCS in pFastBac1. (**c**) Map of recombinant plasmid (pFastBac1-GLUTs) with vector reconstructed. (**d**) Detailed reconstruction of the pFastBac1-GLUTs at MCS. The original RE cutting site, the constructed RE cutting site, and the 10× His tag and protease cutting site are shown in *light blue*, *green*, *dark blue*, and *purple*, respectively

3. The purified PCR products and vectors are digested by *Nde*I and *Xho*I for 1.5 h at 37 °C. The restricted digestion reaction, 60 μL in total, is composed of 52 μL PCR products or 2 μg vectors, 1 μL NdeI, and 1 μL *Xho*I in Cutsmart buffer.
4. The digested PCR products are purified with the DNA product purification kit. The vectors are purified with the TIANgel Mini Purification Kit.
5. The purified PCR products and vectors are ligated using T4 ligase. The ligation reaction, 10 μL in total, is composed of 1 μL T4 ligase, 1 μL digested vectors (20 ng), and 7 μL digested DNA (~100–200 ng) insert in ligation buffer.
6. Once the insert is integrated into the vector, the ligation reaction is transformed into DH5α *E. coli* and selected for ampicillin-resistant transformants on LB agar plates (amp+).
7. To identify the correct colony, positive transformants are analyzed by PCR and further checked with sequencing.
8. The colony is purified with the Plasmid Purification Kit to generate the pFastBac-GLUTs plasmids.
9. For mutagenesis of GLUT1 N45T/E329Q, and GLUT3 N43T, the quick-change methods were carried out with complementary primers comprising mutations following a modified protocol (*see* **Note 2**).

3.2 Generating the Recombinant Bacmids

1. For transformation, 50 μL DH10Bac competent cells are thawed on ice in advance.
2. 1 μL pFastBac-GLUT plasmids (about 400 ng) are added into the competent cells and mixed gently.
3. Incubate the transformed cells on ice for 30 min. Heat shock the cells at 42 °C for 90 s and then immediately chill on ice for 2–5 min.
4. Add 600 μL of LB medium without antibiotics and recover the cells at 37 °C at 220 rpm for 3 h.
5. Plate 25 μL, 50 μL, and 100 μL, respectively, of the cells on a LB agar plate (for bacmid generation). Incubate the plate for at least 48 h at 37 °C. To identify the correct colonies, single white bacterial colonies are analyzed by PCR with pUC/M13 forward and reverse primers (*see* **Note 3**).
6. Incubate the correct colonies in 3 mL LB medium (bacmid) at 37 °C overnight in a shaking incubator. Collect the bacterial cells by centrifugation and purify the recombinant bacmids with plasmid purification kit following the manufacturer's manual, but instead of using the column to purify the bacmid, precipitate the DNA with isopropanol. Do this by adding 640 μL isopropanol into 800 μL supernatant after centrifugation, and incubate the mixture in −20 °C for 60 min.

7. To collect the precipitated DNA, the solution is centrifuged at 13,000 × *g* at 4 °C for 15 min.
8. Wash the pellet with 1 mL ethanol solution followed by 1 mL 100 % ethanol.
9. Dissolve the recombinant bacmids in 50 μL sterile ddH_2O. Do not pipette or vortex the solution.
10. Run a 0.5 % w/v agarose gel to check the purified bacmid DNA (large amounts of plasmids can also be found) as well as the additional PCR analysis with the pUC/M13 primers (pUC/M13 forward: 5′-CCCAGTCACGACGTTGTAAAACG-3′; pUC/M13 reverse: 5′-AGCGGATAACAATTTCACACAGG-3′) to confirm that the bacmids contain the target genes.

3.3 Producing Recombinant Baculovirus

Transfecting the recombinant bacmids into the Sf-9 cells produces recombinant baculovirus.

1. Sf-9 cells are cultured in Hyclone SFX insect cell culture medium. Add 1 mL Sf-9 cells (~1.5 × 10^6 cells/mL) and 1.5 mL medium in 60 mm tissue culture plates, allowing cells to attach for 10 min at 27 °C. Wash the attached cells once with 2 mL Hyclone SFX insect cell culture medium and add another 2 mL Hyclone SFX insect cell culture medium (*see* **Note 4**).
2. Dilute 10 μL of recombinant bacmids into 100 μL of Hyclone SFX insect cell culture medium, and 6 μL of cellfectin II in 100 μL of Hyclone SFX insect cell culture medium, respectively. Combine the diluted DNA and cellfectin II and mix gently. Incubate the mixture for 15–30 min at room temperature.
3. Add the mixture dropwise onto the cells and incubate for 4 h at 27 °C. Remove the mixture and add 2.5–3 mL of the Hyclone SFX insect cell culture medium with antibiotic solution and FBS solution. Incubate the cells at 27 °C for 72–96 h.
4. Collect the medium containing the virus. Centrifuge the medium at 800 × *g* for 5 min to collect the supernatant and remove the precipitates. Store the clarified supernatant (P1 viral stock) in a fresh conical tube at 4 °C away from light.
5. To obtain higher titer, the baculoviral stock needs to be amplified from P1. Add 5–7 mL Sf-9 cells (~1.5 × 10^6 cells/mL) in 150 mm tissue culture plates with 35–40 mL fresh Hyclone SFX insect cell culture medium with antibiotic solution and add P1 baculoviral stock at the ratio of 1:20. Incubate the cells at 27 °C for 48–72 h in humidified incubator. Collect the medium containing the virus.

6. Centrifuge the medium at 800 × *g* for 5 min in a sterile 50 mL conical tube to remove the precipitates.
7. Store the clarified supernatant (P2 viral stock) with additional 2–3 % serum in a fresh 50 mL conical tube at 4 °C, away from light (*see* **Note 5**).
8. The P3 or P4 viral stocks are generated from P2 or P3, respectively, and accomplished by repeating this protocol.

3.4 Expressing Recombinant GLUTs Proteins

Before large-scale expression of GLUTs, it is necessary to detect the expression by Western blotting for a preliminary analysis. Other screening systems, such as the fluorescence-detection size-exclusion chromatography for GFP-tagged proteins, are also available [6]. Here, we only mention the protocol for Western blotting analysis and the following large-scale expression.

1. After transfection with bacmid, the Sf-9 cells generate virus and meanwhile start to express the GLUT proteins.
2. Collect the Sf-9 cells in the step of generating P3 or P4 viral stock and lyse the cells with 2× SDS loading buffer.
3. Analyze the cell lysates by SDS-PAGE and Western blotting. Use the anti-His antibody to detect the GLUT protein which carries an N- or a C-terminal 10× His-tag.
4. For large-scale expression of GLUT1 (N45T/E329Q) and GLUT3 (N43T) we recommend to grow cells in suspension using the High Five and Sf-9, respectively.
5. Maintain the cells in a 27 °C shaker. Determine the viable cell count from a 3-day-old suspension culture and dilute the cell suspension to 0.8×10^6 cells/mL in the SIM HF medium for High Five cells or SFX medium for Sf-9 cells in a 2 L flask.
6. Shake the cells until their density reaches 1.5×10^6 cells/mL.
7. Add 15–25 mL baculoviral stock (P3 or P4) to 800 mL culture to infect the cells and incubate for 48 or 72 h in a 27 °C shaker (*see* **Note 6**).
8. Harvest the cells by centrifugation at 3,000 × *g* for 10 min at 4 °C and remove the media. The cell pellets are ready for protein purification.

3.5 Protein Purification

Purification of GLUT1 and GLUT3 (*see* Fig. 3) follows the same approach, except with minor modifications (see **Note 7** and **Note 8**).

1. Forty-eight hours after viral infection, the High Five cells of 4 L cultures are collected.
2. Resuspend the cells in 100 mL lysis buffer. The collected cells can either be immediately used or frozen in liquid nitrogen and then stored at −80 °C.

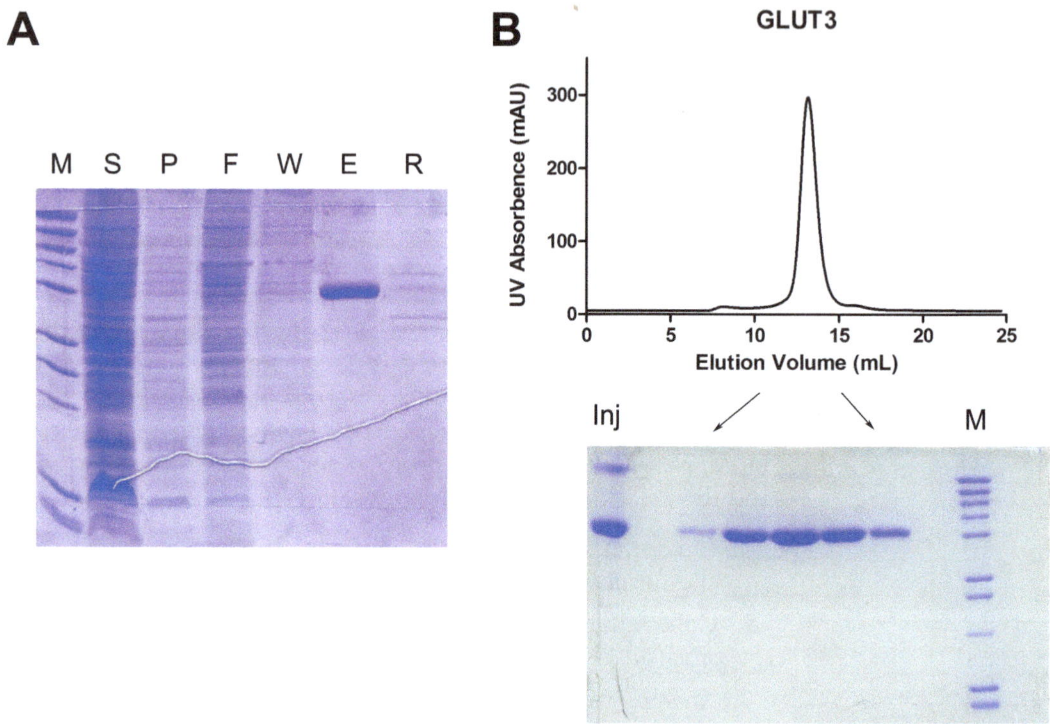

Fig. 3 A typical example for purification of GLUT3 by Ni-NTA and size-exclusion chromatography. (**a**) Samples from the Ni-NTA purification were applied for SDS-PAGE followed by Coomassie blue staining. *M* protein marker, *S* supernatant after ultracentrifugation, *P* pellet after ultracentrifugation, *F* flow through, *W* wash, *E* elution, *R* resin of Ni-NTA. (**b**) Result of size-exclusion chromatography, the sample before injection (INJ) and peak fractions were applied for SDS-PAGE analysis

3. Disrupt the cells using Dounce homogenizer for 80 cycles on ice. After centrifuging the total extract at 5,000 × *g* for 10 min followed with ultracentrifugation of the supernatant at 150,000 × *g* for 1 h at 4 °C, the membrane fraction is solubilized in lysis buffer II at 4 °C for 2 h.
4. Ultracentrifuge the membrane fraction at 150,000 × *g* for 30 min, and then incubate the detergent-soluble fraction with nickel affinity resin (Ni-NTA) at 4 °C for 30 min.
5. Rinse the resin with the wash buffer three times.
6. Elute the protein from the affinity resin with the elution buffer (*see* **Note 7**).
7. Concentrate the protein to about 10 mg/mL using a 50 kDa centrifugal filter.
8. Further purification of GLUT1 protein is carried out in the size-exclusion buffer by gel filtration. Collect the peak fractions and flash freeze the protein in liquid nitrogen and then store at −80 °C for following crystallization trials and biochemical assay (*see* **Note 8**).

3.6 Protein Crystallization

GLUT1 and GLUT3 were crystallized in different manners. We provide here the individual protocols that led to the successful crystallization of GLUT1 and GLUT3, respectively.

3.6.1 Crystallization of GLUT1 (N45T/E329Q)

For crystallization of GLUT1, the hanging-drop vapor diffusion method is carried out.

1. Pipette 0.3 mL crystallization buffer into the reservoir of the 24-well plate (such as the VDX plate).
2. Pipette 0.8 μL of the GLUT1 solution into the center of a siliconized 22 mm square cover slide.
3. Pipette 0.8 μL of buffer from the reservoir into the drop on the cover slide.
4. Invert the cover slide so that the drop will be hanged from cover slide and position the cover slide onto the bead of grease on the reservoir. Press the slide down gently and ensure a complete seal.
5. The initial crystallization screening of GLUT1 is set up using commercial screening kits, including MemGold I/II, MemStart, MemSys, and MemPlus (Molecular Dimensions).
6. Most of the extensive crystallization trials failed to yield crystals for GLUT1 purified in various detergents. The crystals did however appear in the reservoir solution (GLUT1) at 4 °C after 2 days and reached full size in 5–7 days. Collect the crystals using Mounted Cryoloop with 20 μm diameter nylon (Hampton Research) and flash freeze the crystals in liquid nitrogen immediately.

3.6.2 Crystallization of GLUT3 (N43T)

GLUT3 (N43T) in complex with D-glucose or maltose was crystallized with the lipidic cubic phase approach. The following protocol explains this approach.

1. The protein is concentrated to 30–40 mg/mL, and then mixed with monoolein in ratio 1:1.5 w/w proteins:lipid ratio using a syringe lipid mixer.
2. For crystallization of GLUT3 (N43T) in complex with D-glucose, the robot arm Gryphon (ARI) was used to mix the 40 nL meso phase with 900 nL crystallization buffer for each condition on glass sandwich plates.
3. The crystals appeared within 1 week with a typical size of 70 μm × 50 μm × 10 μm.
4. These crystals diffracted X-rays to approximately 2.5 Å at SSRF beamline BL17U. Mother liquor is used to optimize the crystals, which in our hands gives rise to crystals with an approximate size of 140 μm × 100 μm × 20 μm.

5. For crystallization of GLUT3 (N43T) in complex with maltose, 30–45 nL meso phase was overlaid with 800 nL of precipitant solution. In our hands, crystals appeared overnight and grew to a maximum size of about 20 μm × 20 μm × 5 μm at 20 °C within 1 week.
6. Collect the crystals using MicroMesh and flash freeze the crystals in liquid nitrogen immediately.

3.7 Data Collection and Structure Determination

All data sets of GLUT1 and GLUT3 were collected at BL17U at SSRF and the microfocus beamline BL32XU at SPring-8, respectively (*see* **Note 9**).

3.8 Preparation of Liposomes and Proteoliposomes

Liposomes can be prepared according to general protocols.

1. Dissolve 20 mg *E. coli* polar lipid extract into chloroform to a final concentration of 50 mg/mL in a glass vial. Evaporate the solvent and make sure that the lipids are distributed over the wall and bottom evenly using a stream of ultrapure nitrogen gas to prevent the lipids from oxidation.
2. Resuspend the lipids with the KPM buffer plus 50 mM glucose and vortex to make an emulsion at 20 mg/mL.
3. Freeze the liposomes in liquid nitrogen and thaw again. Repeat this for ten cycles. Then extrude the liposomes through a membrane filter.
4. For the counter-flow assay of GLUTs, proteoliposomes can be prepared using the 20 mg/mL pre-extruded liposome suspension.
5. Incubate the liposomes with 1 % w/v n-octyl-β-D-glucopyranoside (β-OG) for 60 min at 4 °C.
6. Add the purified GLUT1 (N45T) or GLUT3 (N43T) (10 μg protein per mg lipid) and incubate for an additional 60 min at 4 °C.
7. To remove β-OG, the sample is incubated with 240 mg/mL Bio-Beads SM2 overnight and then with 120 mg/mL Bio-Beads for an additional 2 h. Removal of detergent reconstitutes the protein into the liposomes to produce proteoliposomes.
8. Freeze the proteoliposomes in liquid nitrogen and thaw. Repeat for five cycles, and extrude the sample through a membrane filter.
9. Collect the proteoliposomes by ultracentrifugation at 100,000 × *g* for 1 h, and wash with ice-cold KPM buffer to remove the excessive glucose. Resuspend the proteoliposomes with KPM buffer to a final concentration of 100 mg/mL (phospholipids).

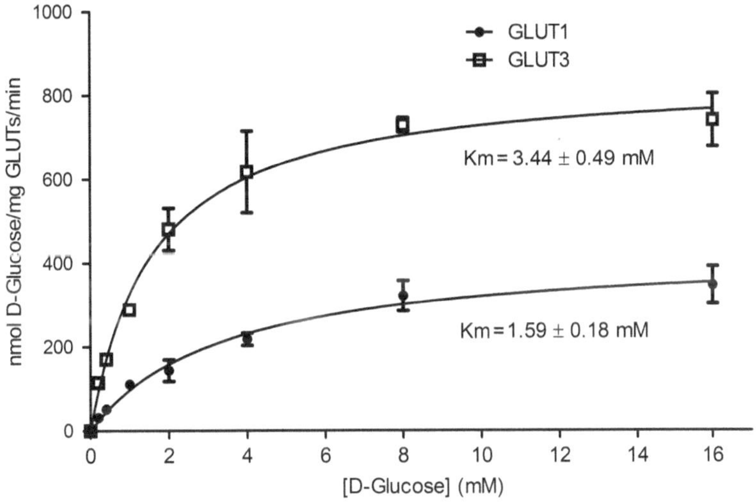

Fig. 4 Detected kinetic parameters of GLUT1 and GLUT3 by in vitro proteoliposome-based assay. K_m and V_{max} values of these GLUTs were detected with the described protocol. Different concentrations of unlabeled glucose ("cold") were added in the reaction buffer. Samples were taken each 10 s to monitor the V_0 of GLUTs. Data points were done in triplicates. Error bars represent standard deviations

3.9 Counter-Flow Assay

1. Add 2 μL prepared proteoliposomes of GLUT1 or GLUT3 into 100 μL KPM buffer plus radioactive glucose, 1 μCi D-[2-^{3}H] glucose.
2. After incubation, stop the uptake of radiolabeled substrates by filtering the sample through 0.22 μm Millipore filters.
3. Wash the filter membranes with 2 mL ice-cold KPM buffer immediately and solubilize the membranes with 0.5 mL Optiphase HISAFE 3. The solution is then used for liquid scintillation counting with MicroBeta JET.
4. The counter-flow assays are performed at 25 °C and repeated at least three times.
5. For determination of V_{max} and K_m, the initial velocities are measured at 10-s time intervals with a concentration gradient of D-glucose, which consists of a mixture of isotope-labeled and non-labeled ("cold") D-glucose, in the outside buffer (*see* Fig. 4).
6. The data are then fitted to the Michaelis–Menten equation using GraphPad Prism 5.0.

4 Notes

1. It is more convenient to adjust the restriction enzyme (RE) cutting site and ideal tags in commercial vectors to unify the cloning system. Thus, we introduced the *Nde*I and *Xho*I site

to MCS of the pFastBac vector (digested with *EcoR*I and *Sal*I), mediated by *EcoR*I and *Sal*I (block original *Xho*I). For N-tag vector, 10× His-tag was added just before the *Nde*I cutting site with linker and a protease-cutting site (GS-10× His-SGDEVDA). For C-tag, 10× His-tag was added after *Xho*I cutting site followed with a stop codon. The recombinant plasmid is shown in Figure for a direct overview.

2. Complementary oligonucleotides containing the desired mutation are synthesized, flanked by 15 nt unmodified nucleotide sequence. 200–300 ng modified pFastBac-GLUT1 or GLUT3 is used as a template. Run the 50 μL PCR reaction for 30 cycles and use a 4-min extension. Generally, a higher annealing temperature (like 58 °C) gives a better result. Purify the PCR product using a kit. Add 1 μL of the DpnI restriction enzyme to 50 μL purified product with Cutsmart buffer, mix gently, and then incubate the reaction at 37 °C for 3–4 h to digest the methylated parental plasmids. The Dpn1-treated DNA (10 μL/reaction) is then directly used for transformation into separate aliquots of the DH5α *E. coli* competent cells.
3. It is necessary to double check the recombinant bacmid generated in the DH10Bac *E. coli* using PCR analysis. It happens that white clones are false positive. The correct size of the PCR product should be ~2300 bp plus the size of the GLUT1 or GLUT3 for the bacmid transposed with our modified pFastBac. Moreover, the pair of primers for the target insert is not recommended for PCR analysis for the DH10Bac cell containing the recombinant plasmid pFastBac-GLUT1 or -GLUT3.
4. We recommend using Sf-9 cells for generating virus for the expression of GLUTs. Other cells may have less efficient transfection based on our experience. However, once the baculovirus stock is generated, Sf-9, High Five™, or Mimic™ Sf-9 cells can be used for expression trials.
5. The viral stocks are stored in aliquots at −80 °C for later re-amplification. Repeated freeze and thaw of the viral stock will result in dramatically decreased virus titer. Routinely used viral stocks should be stored at 4 °C, away from light.
6. The optimal time of protein expression is between 48 and 72 h. The cell morphology and density should be monitored to confirm the progress of infection.
7. The detergent used for GLUT3 is 0.05 % w/v DDM in the wash and the eluting buffer is replaced with 0.06 % w/v 6-cyclohexyl-1-hexyl-b-D-maltoside (CYMAL-6).
8. For crystallizing the GLUT3 in complex with D-glucose or maltose, 50 mM D-glucose or maltose is added throughout

the purification procedure. Concentrated GLUT3 protein is applied to HiTrap Desalting 5 mL in the 25 mM MES pH 6.0, 150 mM NaCl, 0.06 % w/v CYMAL-6 plus 50 mM D-glucose, or maltose.

9. Extended information about the data collection and structure determination can be found in refs [7, 8].

References

1. Thorens B, Mueckler M (2010) Glucose transporters in the 21st century. Am J Phys Endocrinol Metab 298(2):E141–E145. https://doi.org/10.1152/ajpendo.00712.2009
2. Sun L, Zeng X, Yan C, Sun X, Gong X, Rao Y, Yan N (2012) Crystal structure of a bacterial homologue of glucose transporters GLUT1-4. Nature 490(7420):361–366. https://doi.org/10.1038/nature11524
3. Kapoor K, Finer-Moore JS, Pedersen BP, Caboni L, Waight A, Hillig RC, Bringmann P, Heisler I, Muller T, Siebeneicher H, Stroud RM (2016) Mechanism of inhibition of human glucose transporter GLUT1 is conserved between cytochalasin B and phenylalanine amides. Proc Natl Acad Sci U S A 113(17):4711–4716. https://doi.org/10.1073/pnas.1603735113
4. Alisio A, Mueckler M (2010) Purification and characterization of mammalian glucose transporters expressed in Pichia pastoris. Protein Expr Purif 70(1):81–87. https://doi.org/10.1016/j.pep.2009.10.011
5. Boulter JM, Wang DN (2001) Purification and characterization of human erythrocyte glucose transporter in decylmaltoside detergent solution. Protein Expr Purif 22(2):337–348. https://doi.org/10.1006/prep.2001.1440
6. Kawate T, Gouaux E (2006) Fluorescence-detection size-exclusion chromatography for precrystallization screening of integral membrane proteins. Structure 14(4):673–681. https://doi.org/10.1016/j.str.2006.01.013
7. Deng D, Xu C, Sun P, Wu J, Yan C, Hu M, Yan N (2014) Crystal structure of the human glucose transporter GLUT1. Nature 510(7503):121–125. https://doi.org/10.1038/nature13306
8. Deng D, Sun P, Yan C, Ke M, Jiang X, Xiong L, Ren W, Hirata K, Yamamoto M, Fan S, Yan N (2015) Molecular basis of ligand recognition and transport by glucose transporters. Nature 526(7573):391–396. https://doi.org/10.1038/nature14655

Chapter 3

Screening and Scale-Up of GLUT Transporter Constructs Suitable for Biochemical and Structural Studies

Grégory Verdon*, Hae Joo Kang*, and David Drew

Abstract

Identifying membrane proteins that can be produced and isolated in homogenous form in detergent is a lengthy trial-and-error process that can be facilitated by fluorescence-based screening approaches. We describe (1) the strategy and protocol of cloning by homologous recombination, (2) whole-cell and in-gel fluorescence measurements to estimate GLUT-GFP fusion protein yields, (3) use of size-exclusion chromatography monitored by fluorescence (FSEC) for assessing the homogeneity of the GLUT-GFP fusion proteins, and (4) the protocol for large-scale production and purification of the *Bos taurus* GLUT5 construct that enabled its crystal structure determination.

Key words *Saccharomyces cerevisiae*, GLUT, GLUT5, Glucose transporters, GFP, Purification

1 Introduction

We have previously reported a protocol to screen eukaryotic membrane proteins for expression in yeast *Saccharomyces cerevisiae* and isolation in various detergents [1, 2]. With time and consumable costs similar to those associated with the use of *E. coli*, and the possibility to clone into standard vectors by homologous recombination, *S. cerevisiae* represents an efficient and cheap production host when compared to insect and mammalian cell lines. Although it was thought that hyper-mannose glycosylation would be a significant issue for the heterologous proteins, we have not encountered any problem for the large number of polytopic membrane proteins that we have screened so far in this host [2]; we suspect that this is more likely to be a concern for secreted soluble proteins.

GLUTs are major facilitator superfamily (MFS) transporters that are made up of two 6-transmembrane bundles that are connected by a large cytosolic loop [3]. Here, we describe in details

*These authors contributed equally.

Karin Lindkvist-Petersson and Jesper S. Hansen (eds.), *Glucose Transport: Methods and Protocols*, Methods in Molecular Biology, vol. 1713, https://doi.org/10.1007/978-1-4939-7507-5_3,

the practical steps that constitute our *S. cerevisiae* GFP-based pipeline [1, 4, 5], with results from our screen for expression and purification of glucose (GLUT) transporter constructs, with an emphasis on the *bovine* fructose transporter GLUT5, for which we have solved a crystal structure in apo inward-facing conformation [6]. Although bovine GLUT5 is stable in detergent it is a very poorly expressing membrane protein, producing ~0.3 mg per L cell culture. Thus, the overexpression and purification protocol outlined herein is ideal for maximizing yields and purity of membrane proteins where the expression levels are particularly low.

2 Materials

1. Expression vectors such as 2 μ GFP-fusion vector pDDGFP-2.
2. Materials for yeast transformation (*see* **Note 1**).
3. PCR reagents.
4. SmaI restriction enzyme.
5. Growth medium without uracil (−URA medium), for 1 L: 6.7 g Yeast nitrogen base without amino acids, 2 g yeast synthetic dropout medium supplement without uracil, and either 2 % w/v glucose (for pre-culture) or 0.1 % w/v glucose (for expression culture). For plates add 20 g of bacto agar.
6. 20 % w/v Galactose.
7. Yeast suspension buffer (YSB): 50 mM Tris–HCl pH 7.6, 5 mM EDTA, 10 % v/v glycerol, 1× complete protease inhibitor cocktail tablets.
8. 96-Well black optical-bottom plates.
9. Microplate spectrofluorometer such as SpectraMax M2e microplate reader (Molecular Devices).
10. Glass beads: 500 μm diameter. Acid-washed.
11. Heavy-duty mixer-mill disruptor such as TissueLyser mixer.
12. Benchtop ultracentrifuge.
13. 1.5 mL Polyallomer microcentrifuge tubes.
14. Sample buffer (SB) for in-gel fluorescence: 50 mM Tris–HCl pH 7.6, 5 % v/v glycerol, 5 mM EDTA (pH 8.0), 4 % w/v SDS, 50 mM DTT, 0.02 % v/v bromophenol blue.
15. Tris-glycine SDS-PAGE gels.
16. Fluorescent protein standard.
17. Pre-stained protein standard.
18. LAS-1000-3000 charge-coupled device (CCD) imaging system (Fujifilm).
19. Coomassie brilliant blue R-250.

20. Cell resuspension buffer (CRB): 50 mM Tris–HCl pH 7.6, 1 mM EDTA, 0.6 M sorbitol.
21. Constant Systems TS series cell disruptor (Constant Systems).
22. Membrane resuspension buffer (MRB): 20 mM Tris–HCl pH 7.6, 0.3 M sucrose, 0.1 mM CaCl2
23. Bicinchoninic acid (BCA) protein assay kit.
24. Phosphate-buffered saline (PBS), for 1 L: 1.44 g $Na_2HPO_4{\cdot}2H_2O$ (8.1 mM phosphate), 0.25 g KH_2HPO_4 (1.9 mM phosphate), 8 g NaCl, 0.2 g KCl, adjust pH to 7.4 using 1 M NaOH or HCl.
25. Detergents: Lauryl maltose neopentyl glycol (LMNG); N,N-dimethyldodecylamine-N-oxide (LDAO); n-Dodecyl- β-D-maltopyranoside (DDM), n-undecyl-β-D-maltopyranoside (UDM).
26. Cholesteryl hemisuccinate Tris salt (CHS).
27. Superose 6 10/300 or Superdex 200 10/300 gel filtration column (GE Healthcare).
28. ÄKTA FPLC system with fraction collector (GE Healthcare), or HPLC system with fluorescence in-line detector such as Prominence HPLC (Shimadzu).
29. 50 mL Aerated capped tubes.
30. Centrifuges Beckman (for harvesting cultures, ultracentrifugation) BioRad columns.
31. Purified GFP standard.

3 Methods

3.1 Target Selection and Cloning by Homologous Recombination

1. We used disorder prediction servers such as RONN (https://www.strubi.ox.ac.uk/RONN) to identify human GLUTs and orthologous transporters from mammals that show the least predicted disorder, in particular in their cytoplasmic loop region.
2. We used synthetic genes encoding for the GLUT transporter orthologous that are predicted to show the least fraction of disorder. Alternatively, or in addition, consider making short N- and C-terminal truncation mutants, especially if a particular GLUT transporter of interest is from a specific organism that contains a significant disorder in the terminal region(s). To reduce potential heterogeneity, predicted and known glycosylation sites can also be removed by introducing a point mutation(s) (e.g., N to A in rat GLUT5, N to Y in bovine GLUT5).
3. Amplify the GLUT gene of interest with primers that contain approximately 35 bp complementary 5′ and 3′ overhangs to a

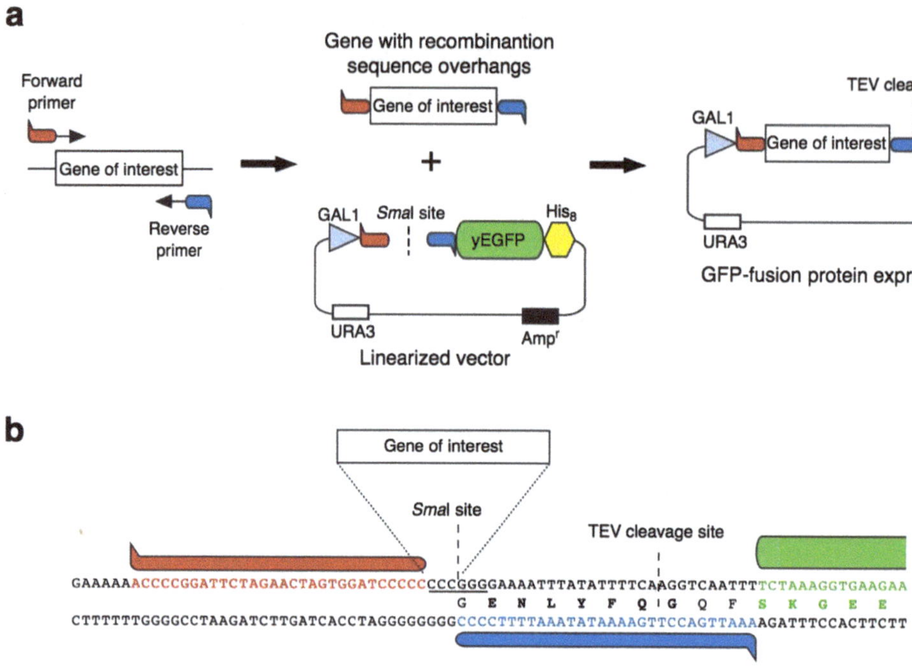

Fig. 1 Cloning of GLUTs. (**a**) GLUT genes were cloned into a vector pDDGFP-2 by homologous recombination. Genes were amplified using primers containing recombination site overhangs (shown in *red* and *blue* bars in the diagram) and transformed with *Sma*I-linearized vector into *S. cerevisiae* competent cells. The transformants were grown in -URA medium and induced with galactose. (**b**) Cloning site used in 2 μ GFP-fusion vector pDDGFP-2. TEV, tobacco etch virus; yEGFP, yeast-enhanced green fluorescent protein. The figure is reproduced with permission from ref. [1]

*Sma*I-linearized 2 μ GFP-fusion vector (*see* Fig. 1); alternatively, the overhangs can be added during the gene synthesis to bypass this step.

4. Transform 100 μL of *S. cerevisiae* competent cells with 750 ng of the gene of interest and 75 ng of linearized vector (*see* **Note 1**).
5. Spread transformed cells onto an agar plate of the appropriate selection media (i.e., without uracil; –URA medium). Place parafilm around each plate, and incubate at 30 °C for 2 days.

3.2 Small-Scale Isolation of Crude Membranes and in-Gel Fluorescence

See Fig. 2 for an overview of the screening process starting from cloned genes into *S. cerevisiae*.

1. Pick a colony for each construct and inoculate in a 5 mL –URA medium supplemented with 2 % w/v glucose. Because colonies can be of different sizes and the cells can grow slowly, to ensure saturation of cultures, inoculation in the morning is recommended. Incubate the cultures in an orbital shaker at 280 rpm and 30 °C overnight (*see* **Note 2**).

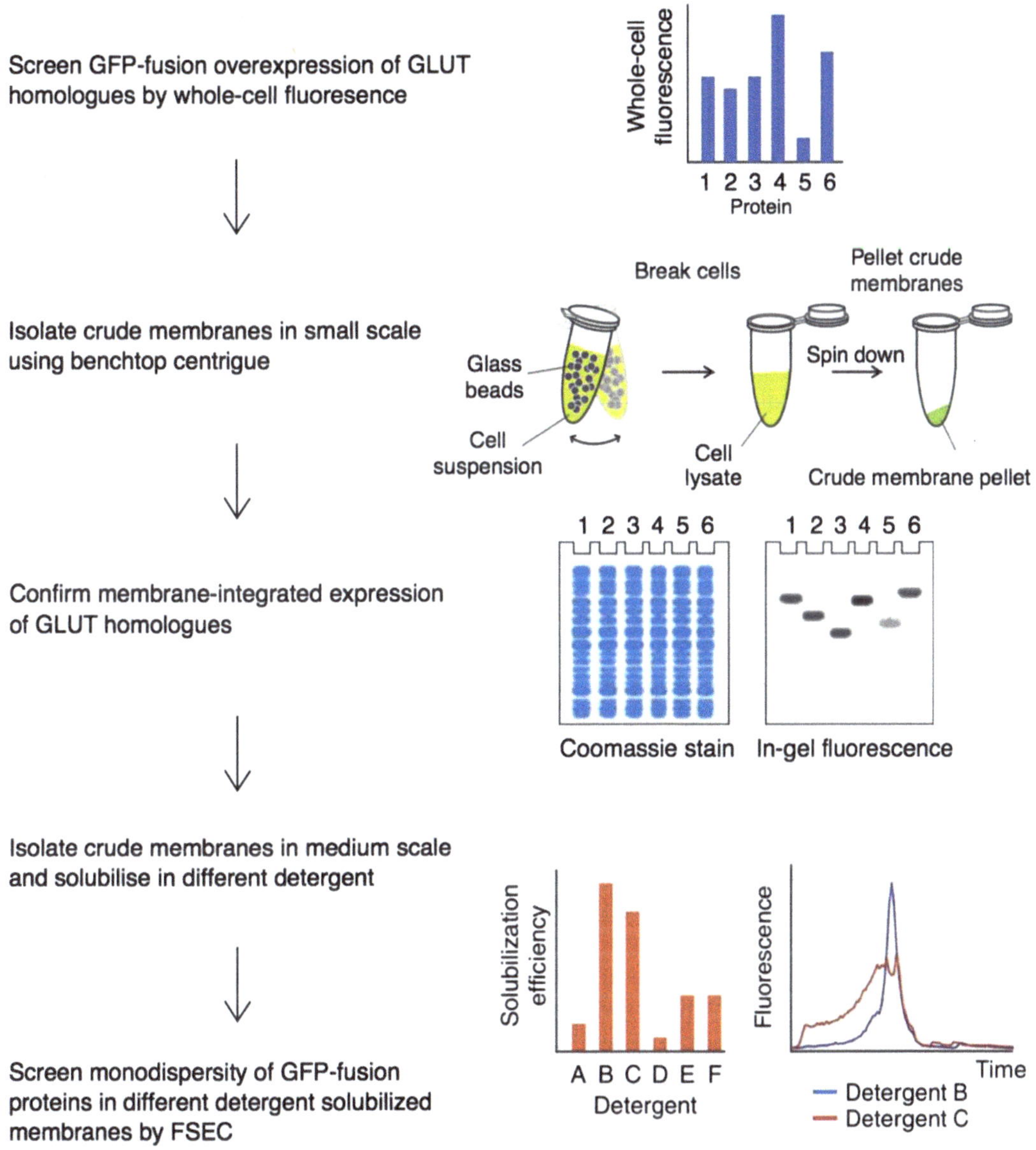

Fig. 2 Flowchart of screening process for the overexpression and monodispersity of GLUT-GFP fusion membrane proteins in *S. cerevisiae*. The figure is reproduced with permission from ref. [1]

2. Spot 10 μL or more from each of the overnight cultures onto a fresh –URA plate. After drying the spots at room temperature, transfer the plates to a 30 °C incubator and grow for 1–2 days. The plates can be stored at 4 °C and can be used to inoculate cultures (*see* **Note 3**).
3. Dilute the overnight culture from **step 1** to OD_{600} 0.12 in a 50 mL aerated tube containing 10 mL –URA medium plus 0.1 % w/v glucose. This can be done in duplicates to take into account of any irregularities during growth.

4. Incubate the cultures in an orbital shaker at 30 °C and 280 rpm. At OD_{600} 0.6 and after approximately 7 h, induce production of membrane protein–GFP fusion by adding 20 % w/v galactose to the final concentration of 2 %.
5. After 22 h of induction, centrifuge the cells at 3,000 × *g* for 5 min, remove the supernatant, and resuspend the cell pellets in 200 μL of YSB.
6. Transfer 200 μL of the cell suspension to a black 96-well optical bottom plate. Measure the GFP fluorescence emission at 512 nm by excitation at 488 nm in a microplate spectrofluorometer with a bottom-read option.
7. Transfer the cell suspension from the 96-well plate into a 1.5 mL capped tube. To the cell suspension, add glass beads to a final volume of 500 μL and additional 500 μL of YSB.
8. Using a mixer-mill disruptor set at 30 Hz for 7 min at 4 °C break the yeast cells. Vortex can also be used instead, but using a heavy-duty disruptor is highly recommended in order to achieve more efficient and reproducible cell breakage.
9. Remove unbroken cells and debris by centrifuging at 22,000 × *g* in a benchtop centrifuge for 5 s at 4 °C. Transfer the supernatant into a new tube.
10. Repeat **steps 8** and **9** by adding 500 μL YSB to the mixture of unbroken cell pellet and glass beads. Combine the supernatant with that from **step 9**.
11. Crude membrane can be pelleted by centrifuging the 1 mL supernatant from **step 10** at 20,000 × *g* in a benchtop centrifuge at 4 °C for 1 h. Ultracentrifugation is not necessary, as the crude membrane pellet from a benchtop centrifuge is sufficient for this analysis.
12. Remove the supernatant and resuspend crude membranes in 50 μL of YSB. Transfer 15 μL into a tube containing 15 μL SB and load 10 μL per well for SDS-PAGE in-gel fluorescence. Including both nonfluorescent and fluorescent protein standards is recommended. The SB composition works well with standard SDS denaturing gels for in-gel fluorescence (*see* **Note 4**).
13. After running the gel, rinse the gel with deionized H_2O and detect the fluorescent bands with a CCD camera-based system. Expose the gel to blue light source set at 460 nm with a cutoff filter of 515 nm and capture the images.
14. Analyze the gel images and compare the size of the bands to the protein standards. If two closely spaced bands are present, this could be due to the GLUT protein being glycosylated (*see* **Note 5**).
15. Stain the gel with Coomassie Brilliant Blue and transfer to destain.

3.3 Medium-Scale Isolation of Membranes and Size-Exclusion Chromatography Monitored by Fluorescence (FSEC)

1. Select GLUT homologues based on the in-gel fluorescence results from **step 14** in Subheading 3.2 (*see* **Note 6**). Inoculate 10 mL –URA medium containing 2 % w/v glucose with the spotted yeast culture from **step 2** in Subheading 3.2 and incubate overnight in an orbital shaker at 280 rpm and 30 °C.
2. Use the overnight culture to inoculate a 500 mL shake flask containing 150 mL –URA medium supplemented with 2 % w/v glucose. Incubate the culture overnight in an orbital shaker at 280 rpm, 30 °C.
3. In the following morning, dilute the 150 mL overnight culture to OD_{600} 0.12 in 1 L –URA medium containing 0.1 % w/v glucose in a 2.5 L baffled shake flask. Incubate the culture in an orbital shaker at 280 rpm, 30 °C. At OD_{600} 0.6 induce production of membrane protein–GFP fusion by adding 20 % w/v galactose to the final concentration of 2 % (*see* **Note 7**).
4. Harvest the cells after 22 h by centrifugation at 4,000 × *g* in 4 °C for 10 min. Decant the supernatant and resuspend the cell pellet in 25 mL CRB per liter of cell culture. Transfer 100 μL of the cell suspension into a 96-well plate and measure GFP fluorescence as outlined in **step** 7 in Subheading 3.2.
5. Disrupt the cells with a cell disruptor at 4–15 °C. Typically, four passes at incremental pressures of 25, 30, 32, and 35 kpsi (approximately 1.7–2.4 × 10^3 atm) are required to ensure efficient breakage.
6. Remove the unbroken cells and debris by centrifugation at 10,000 × *g* at 4 °C for 10 min and collect the supernatant containing the crude membranes. Transfer 100 μL of supernatant to a 96-well plate and measure GFP fluorescence as outlined in **step** 7 in Subheading 3.2. Calculate the yeast cell breakage efficiency by comparing the GFP fluorescence to that measured in **step 4** in Subheading 3.3 (*see* **Note 8**).
7. To collect the membranes, centrifuge the cleared supernatant at 150,000 × *g* at 4 °C for 120 min. Discard the supernatant and resuspend the pellet to a final volume of 6 mL MRB per L of cell culture using a disposable 10 mL syringe with a 21-gauge needle. Transfer 100 μL membrane suspension to a 96-well plate and measure GFP fluorescence. Calculate the amount of total protein using the BCA protein assay kit following the manufacturer's instructions.
8. Adjust the volume of the membrane suspension with PBS to achieve a protein concentration of 3.5 mg/mL. Transfer 900 μL aliquots of this membrane suspension into 1.5 mL microcentrifuge tubes.
9. Add 100 μL freshly prepared 10 % w/v detergent stocks of LMNG, DDM, DM, NM, or LDAO to the tubes containing

900 μL membrane suspension, resulting in final concentrations of 1 % w/v detergent and 3.2 mg/mL protein. Incubate the mixtures at 4 °C for 1 h with gentle agitation (*see* **Note 9**).

10. Transfer 100 μL of the detergent-solubilized membrane protein solution into 96-well plate and measure GFP fluorescence as outlined in **step 7** in Subheading 3.2. To remove insoluble material, centrifuge the remaining 900 μL in a benchtop ultracentrifuge at 100,000 × *g* in 4 °C for 45 min.
11. Transfer the clarified supernatant to a new 1.5 mL tube. Transfer 100 μL to a 96-well plate and measure the GFP fluorescence as outlined in **step 7** in Subheading 3.2. Calculate the detergent solubilization efficiency by comparing the GFP fluorescence measurement with that in **step 7** in Subheading 3.3.
12. For FSEC, inject 0.5 mL of the detergent-solubilized sample onto a gel filtration column equilibrated in 20 mM Tris–HCl pH 7.5, 0.15 M NaCl, and 0.03 % DDM. If the system is equipped with an in-line fluorescence detector, the fluorescence can be measured directly. Alternatively, after elution of the first 6 mL, collect 0.2 mL fractions row by row into a 96-well plate and measure GFP fluorescence using the plate reader. Plot the GFP fluorescence in each well against the fraction number. Use the FSEC traces to compare monodispersities of GLUT homologues in different detergents. As shown in Fig. 3, GLUT #1 (bovine GLUT5) was selected as it has the best FSEC profile despite the poorest expression levels.

3.4 Large-Scale Production and Purification of Bovine GLUT5 for Structure Determination

Large-scale production, typically, 36 flasks of 1 L each, total of ~40 L:

1. Inoculate 50 mL –URA medium containing 2 % w/v glucose with the spotted yeast culture from **step 2** in Subheading 3.3 and incubate all day in an orbital shaker at 280 rpm, 30 °C. If fresh spot is not available, scrap some of the frozen glycerol stock into 10 mL –URA medium containing 2 % w/v glucose and grow for ~7–8 h at 30 °C with shaking at 280 rpm before spotting 50 μL aliquots on a 2 % w/v agar plate and incubate for ~2 days at 30 °C.
2. Add 50 mL pre-culture to 1 L of –URA medium containing 2 % w/v glucose and grow overnight at 30 °C with shaking at 250 rpm (prepare 2 × 1 L for 40 L culture).
3. Next morning, back dilute to OD_{600} = 0.12 in –URA media in 36 flasks containing 2 % w/v glucose and grow at 30 °C with shaking at 250 rpm (*see* **Note 10**).
4. When the OD_{600} reaches ~0.7–0.75 (check several flasks), add the inducer galactose at a final concentration of ~2.0 % (i.e., add 100 mL/L of culture of a 22 % w/v galactose solution in

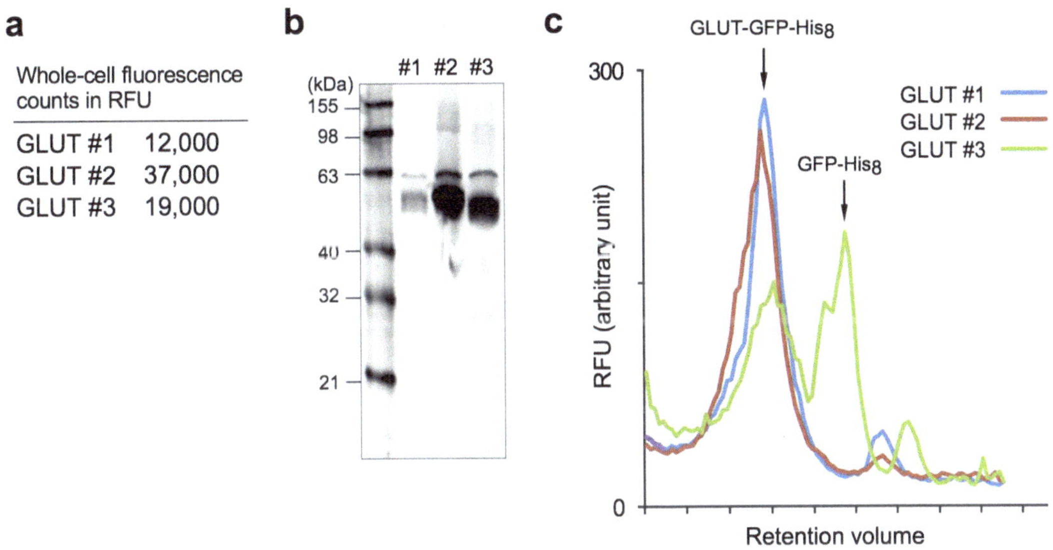

Fig. 3 Example of screening results from three mammalian GLUT homologues. (**a**) Whole-cell fluorescence was measured in relative fluorescence unit (RFU). (**b**) Crude membranes were prepared and analyzed in-gel fluorescence. (**c**) Monodispersity of GLUT-GFP fusions in DDM was examined by fluorescence size-exclusion chromatography (FSEC). GLUT #1 showed a relatively poor expression level, but its monodispersity was slightly better than GLUT #2 whose expression level was much higher. Only GLUT #1, bovine GLUT5, could be used for successful large-scale purifications and crystallizations

-URA medium with 0.1 % w/v glucose). Grow at 30 °C with shaking at 250 rpm for 22–24 h.

5. Next day, harvest the cell culture by centrifugation for 10 min at 1,000 × *g*. The typical yield is ~6–8 g of wet cells per L of culture in 2 L baffled flasks.

3.5 Cell Lysis and Membrane Isolation for Large Scale

1. Add the frozen cell pellet to CRB buffer at room temperature using ~25 mL per L of culture in a large beaker of adequate size and place the beaker in wet ice.
2. Load cell solution into disruptor through sieve and process solution at 35,000 psi. Let outflow pour over the frozen cell pellet in a beaker to help thawing. Perform 4–8 disruption cycles and avoid foaming by letting outflow drip onto the beaker wall.
3. Centrifuge the cell lysate for 20–30 min at low speed of ~1,000 × *g* at 4 °C to pellet unbroken cells and large cellular debris.
4. Centrifuge the supernatant from low-speed centrifugation run for 2 h at 41,000 × *g* using an ultracentrifuge at 4 °C. Discard supernatant, pool membrane pellets, resuspend in 9 mL MRB buffer/L of culture, and freeze in liquid nitrogen for storage at −80 °C (*see* **Note 11**).

3.6 Purification for Large Scale

See Fig. 4 *for an overview of the large-scale purification process of bovine GLUT5*

1. Place membranes in beaker; add ice-cold solubilization buffer containing 1× PBS, 150 mM NaCl, and 10 % v/v glycerol to a final volume of ~400 mL; and add 2 % w/v DDM (final) (*see* **Note 12**).
2. Stir solubilization solution gently for 2 h and then centrifuge at 41,000 × *g* for 1 h at 4 °C.
3. Retrieve the supernatant in an ice-cold glass beaker and add Ni-NTA agarose, equilibrated in water, as slurry using 30 mL of resin and imidazole to a final concentration of 55 mM.
4. Incubate for 2 h while stirring at 4 °C, pour the slurry into a gravitational drip column, and collect flow through.
5. Wash the Ni-NTA agarose beads with 20 column volumes (CVs) of 1× PBS, 150 mM NaCl, 10 % v/v glycerol, 0.05 % w/v DDM, and 55 mM imidazole, pH 7.3.

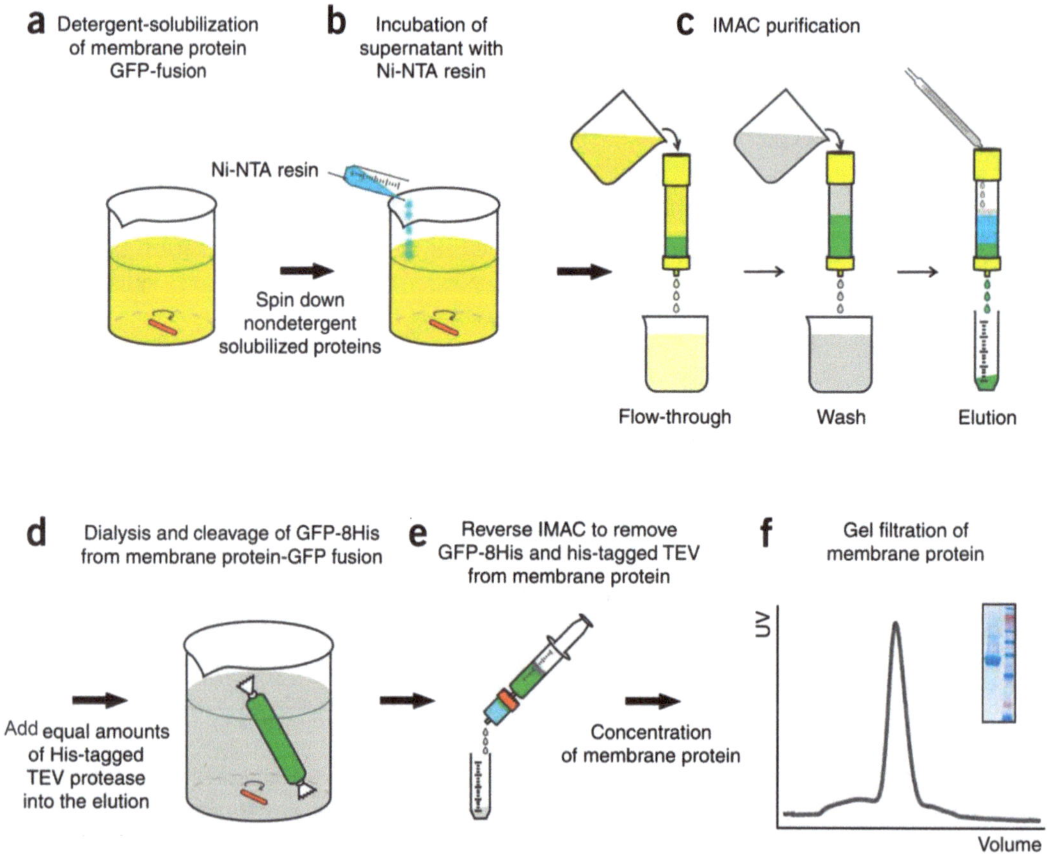

Fig. 4 Flowchart of the purification process for GLUT5-GFP fusions. The figure is reproduced with permission from ref. [1]

6. Elute with ~4 CVs of 1× PBS, 150 mM NaCl, 10 % v/v glycerol, 0.05 % w/v DDM, 250 mM, and imidazole, pH 7.3, and load only half of the CV at a time onto the column.
7. Calculate the amount of GFP using the fluorescence yield from the purified GFP standard as reference, add an equal amount of TEV–His_6 protease to the eluate, transfer the solution into dialysis tubing (molecular cutoff ~3000 kDa), and incubate overnight with very slow stirring in 2 L of 20 mM Tris–HCl, pH 7.5, 150 mM NaCl, 10 % v/v glycerol, and 0.05 % w/v DDM.
8. Use 4× CVs of dialysis buffer to equilibrate the previously used Ni-NTA column, add TEV-digested material to column, and collect flow through. Wash column with 3× CVs of dialysis buffer and collect flow through.
9. Concentrate flow through-containing protein using 100 K Millipore Amicon Ultra 15 mL concentrators to 2 mL.
10. Load the sample onto a Ni-NTA agarose medium gravitational drip column using 1–2 mL of Ni-NTA resin equilibrated in dialysis buffer and collect flow through. Wash column with 3× CVs of dialysis buffer and collect flow through.
11. Concentrate flow through using 100 K Millipore Amicon Ultra 4 mL concentrators.
12. Split sample into 2–3 injections on Superdex 200 10/300 column in buffer containing 10 mM Tris–HCl pH 7.5, 150 mM NaCl, and 0.09 % UDM.
13. Store peak fractions at 4 °C overnight for next day concentration and crystallization experiment setup.

4 Notes

1. Apply general protocol for yeast transformation and preparation of competent cells [1].
2. One can pick 2–3 colonies from each construct and proceed to check if there are any variable levels of expression among different colonies.
3. The plates can be stored for up to several weeks. However, this can result in reduced expression at times, as cells on the plates get older. Please rejuvenate the yeast spot for large-scale culturing as outlined in Subheading 3.4.
4. We have also tested pre-cast Criterion (Bio-Rad) and Tris–glycine gels with equal success. We have found that the NuPAGE gels (Thermo Fisher Scientific) are not compatible with in-gel fluorescence.

5. For bovine GLUT5, we substituted the N-linked glycosylation site Asn51 with an alanine. However, it is also known that GLUT1 produced in *S. cerevisiae* is not glycosylated [7].
6. *S. cerevisiae* has endogenous fluorescent protein of ~100 kDa in size, which can be used to confirm that equal amounts of material were loaded between samples [1].
7. Selection by in-gel fluorescence is only to confirm that the full-length protein is expressed, which is required. Degradation of the protein can still lead to "free" GFP, but the fusion partner is very stable and quite resistant to proteolysis. Further, do not select clones based on the intensity of the in-gel fluorescent band alone. Detergent-solubilized stability based on FSEC is a more reliable parameter than expression levels and the two do not correlate [8]. Indeed, as shown in the screening of bovine GLUT5 in Fig. 3, the in-gel fluorescence band was the poorest, but gave the best FSEC profile, and this is the most important parameter for obtaining crystals. It is recommended to dissolve 20 % w/v galactose solution in pre-sterilized –URA medium just before it is used for induction.
8. Breakage efficiency should be greater than 80 %. If lower than this, consider diluting the sample in a larger CRB before breakage to improve efficiency or increase the number of passes up to 8 as outlined in Subheading 3.4.
9. The addition of cholesterol hemisuccinate (CHS) to the detergent mixture (at a final concentration of 0.2 %) can also be tested, as this can be essential for isolating monodisperse mammalian proteins such as the GLUTs.
10. In our hands, the expression levels between growth of cultures in shaker flasks or 15 or 50 L fermenters are often similar [1] and the decision between the two is mainly based on which is the most convenient for the setup currently used in your laboratory.
11. Alternatively, membranes can be frozen as a pellet and resuspended as outlined in Subheading 3.6.
12. The detergent solubilization efficiency decreases with increasing amounts of total membranes. The recommended dilution is to reach a final total protein concentration of 3.5 mg/mL in 1 % w/v DDM. For 12 L of yeast culture, the final amount of total membranes in 400 mL in ~3.5 mg/mL. For 24 L, we typically keep the final volume the same to fit one Ti45 rotor, but increase the final detergent concentration to 2 % w/v DDM. Using GFP fluorescence the amount of dilution required can be adjusted on a case-by-case basis to maximize solubilization efficiency with the least amount of DDM required; that is, it is possible to still have high solubilization efficiency of your target in total final protein concentration of ~7 mg/mL.

References

1. Drew D, Newstead S, Sonoda Y, Kim H, von Heijne G, Iwata S (2008) GFP-based optimization scheme for the overexpression and purification of eukaryotic membrane proteins in Saccharomyces cerevisiae. Nat Protoc 3(5):784–798. https://doi.org/10.1038/nprot.2008.44
2. Newstead S, Kim H, von Heijne G, Iwata S, Drew D (2007) High-throughput fluorescent-based optimization of eukaryotic membrane protein overexpression and purification in Saccharomyces cerevisiae. Proc Natl Acad Sci U S A 104(35):13936–13941. https://doi.org/10.1073/pnas.0704546104
3. Drew D, Boudker O (2016) Shared molecular mechanisms of membrane transporters. Annu Rev Biochem 85:543–572. https://doi.org/10.1146/annurev-biochem-060815-014520
4. Drew D, Kim H (2012) Screening for high-yielding Saccharomyces cerevisiae clones: using a green fluorescent protein fusion strategy in the production of membrane proteins. Methods Mol Biol 866:75–86. https://doi.org/10.1007/978-1-61779-770-5_8
5. Drew D, Kim H (2012) Large-scale production of membrane proteins in Saccharomyces cerevisiae: using a green fluorescent protein fusion strategy in the production of membrane proteins. Methods Mol Biol 866:209–216. https://doi.org/10.1007/978-1-61779-770-5_18
6. Nomura N, Verdon G, Kang HJ, Shimamura T, Nomura Y, Sonoda Y, Hussien SA, Qureshi AA, Coincon M, Sato Y, Abe H, Nakada-Nakura Y, Hino T, Arakawa T, Kusano-Arai O, Iwanari H, Murata T, Kobayashi T, Hamakubo T, Kasahara M, Iwata S, Drew D (2015) Structure and mechanism of the mammalian fructose transporter GLUT5. Nature 526(7573):397–401. https://doi.org/10.1038/nature14909
7. Kapoor K, Finer-Moore JS, Pedersen BP, Caboni L, Waight A, Hillig RC, Bringmann P, Heisler I, Muller T, Siebeneicher H, Stroud RM (2016) Mechanism of inhibition of human glucose transporter GLUT1 is conserved between cytochalasin B and phenylalanine amides. Proc Natl Acad Sci U S A 113(17):4711–4716. https://doi.org/10.1073/pnas.1603735113
8. Sonoda Y, Newstead S, Hu NJ, Alguel Y, Nji E, Beis K, Yashiro S, Lee C, Leung J, Cameron AD, Byrne B, Iwata S, Drew D (2011) Benchmarking membrane protein detergent stability for improving throughput of high-resolution X-ray structures. Structure 19(1):17–25. https://doi.org/10.1016/j.str.2010.12.001

Chapter 4

GLUT Characterization Using Frog *Xenopus laevis* Oocytes

Wentong Long, Debbie O'Neill, and Chris I. Cheeseman

Abstract

Xenopus laevis oocytes are a useful heterologous expression system for expressing glucose transporters (GLUTs) and examining their functions. In this chapter, we provide a detailed protocol on oocyte extraction and preparation for GLUT9 protein expression. Furthermore, we describe the determination of GLUT9 overexpression level by biotinylation and Western blotting analysis. Finally, we also describe how GLUT9-expressing oocytes can be used to measure urate kinetics by radioisotopes as well as two-microelectrode voltage clamping techniques.

Key words *Xenopus laevis* oocytes, Glucose transporter (GLUT), GLUT9, Isotope flux, Electrophysiology, Biotinylation

1 Introduction

Xenopus laevis oocytes are a heterologous expression system that can be used for expressing glucose transporters (GLUTs) as well as investigating their functions. With very low endogenous GLUT expression, oocytes are suitable for radiolabeled hexoses kinetics studies when they are overexpressing GLUTs [1]. Therefore, GLUT-expressing oocytes have been commonly used for isotope sugar uptake or efflux studies to examine the structure-function relationships of GLUTs [2–4]. For instance, investigators have extensively used oocytes to try to reveal the glucose-binding site in GLUTs [5–9]. In addition, the large size of these oocytes makes them appropriate for electrophysiology experiments. One example is glucose transporter 9 (GLUT9) [10, 11], an electrogenic transporter that can transport both hexoses and the organic anion urate, allowing for the examination of the structure-function relationship of GLUTs through not only isotope flux measurements but also two microelectrode voltage clamp methods. Hence, in this chapter, we provide details on how to extract *X. laevis* oocytes to overexpress GLUT protein. We also document how to determine the membrane-overexpressing GLUT9 protein expression levels

Karin Lindkvist-Petersson and Jesper S. Hansen (eds.), *Glucose Transport: Methods and Protocols*,
Methods in Molecular Biology, vol. 1713, https://doi.org/10.1007/978-1-4939-7507-5_4,

by biotinylation and Western blotting. Lastly, we demonstrate the methods of applying GLUT9-expressing oocytes to measure urate kinetics by both radioisotopes and two-microelectrode voltage clamping techniques [10, 12, 13, 14].

2 Materials

2.1 Preparation for Xenopus laevis Oocyte Harvest and mRNA Injection

1. Euthanasia solution: 0.3 % w/v Tricaine methane sulfonate, pH 7.4. Prepare 4 L in a bucket.
2. A dissecting tray.
3. A 100 mm petri dish.
4. A plastic Pasteur pipette.
5. 8 mL Glass vials.
6. An orbital shaker.
7. Two forceps.
8. A dissecting microscope.
9. A micro-injector.
10. 70 % v/v Ethanol.
11. Modified Barth's medium (MBM): 88 mM NaCl, 1 mM KCl, 0.33 mM $Ca(NO_3)_2$, 0.41 mM $CaCl_2$, 0.82 mM $MgSO_4$, 2.4 mM $NaHCO_3$, 10 mM Hepes, 2.5 mM sodium pyruvate, 0.1 mg/mL penicillin, 0.05 mg/mL gentamycin sulfate, pH 7.5.
12. Collagenase solution: 2 mg/mL Type I Collagenase in MBM.
13. Phosphate buffer: 100 mM K_2HPO_4 in ddH_2O.

2.2 GLUT Protein mRNA Preparation

1. *NheI* enzyme kit.
2. T7 polymerase mMESSAGE mMACHINE® kit (Ambion).

2.3 Preparation for Determination of GLUT Protein Surface Expression Levels

1. Transfer pipette
2. Microcentrifuge tubes
3. Phosphate-buffered saline (PBS): 137 mM NaCl, 2.78 mM KCl, 4.3 mM Na_2HPO_4 (dibasic), 1.5 mM KH_2PO_4 (monobasic), pH 7.4.
4. Biotinylation solution: 2 mM Sulfo-NHS-LC-Biotin (Pierce) in PBS.
5. Quenching buffer: 192 mM Glycine and 25 mM Tris–HCl in PBS, pH 7.4.
6. RIPA buffer: 150 mM NaCl, 1 % v/v Triton X-100, 1 % w/v deoxycholic acid, 0.1 % w/v SDS, 1 mM EDTA, 10 mM Tris–HCl, pH 7.5.
7. Streptavidin agarose beads.

8. Benchtop centrifuge.
9. 2× SDS-page loading sample buffer.
10. PBS-T: 0.05 % v/v Tween-20 in PBS, pH 7.4.
11. Blocking buffer: 3 % w/v Skim milk powder in PBS-T.
12. Electrophoresis chamber.
13. 10 % SDS-page gel.
14. Nitrocellulose membranes.
15. Antibodies: Primary hSLC2A9 monoclonal antibody and secondary anti-rabbit secondary antibody.

2.4 Preparation for Functional Studies by Radiotracer Flux Measurements

1. Pasteur pipettes.
2. 12 × 75 mm Glass test tubes 5 mL.
3. Scintillation vials with caps.
4. An orbital shaker.
5. A Beckman LS6500 liquid scintillation counter.
6. Modified Barth's medium (MBM): 88 mM NaCl, 1 mM KCl, 0.33 mM $Ca(NO_3)_2$, 0.41 mM $CaCl_2$, 0.82 mM $MgSO_4$, 2.4 mM $NaHCO_3$, 10 mM Hepes, 2.5 mM sodium pyruvate, 0.1 mg/mL penicillin, 0.05 mg/mL gentamycin sulfate, pH 7.5.
7. Urate solutions: Prepare six different concentrations of urate ranging from 0.1 to 7.5 mM in MBM supplemented with 250 μCi/mL ^{14}C (a specific activity of 54 mCi/mmol) labelled urate.
8. SDS solution: 5 % w/v SDS in ddH_2O.

2.5 Preparation for Two-Microelectrode Voltage Clamping

1. Sodium-containing transport medium (STM): 100 mM NaCl, 2 mM KCl, 1 mM $CaCl_2$, 1 mM $MgCl_2$, 10 mM Hepes, pH 7.5. Adjust pH with Tris base.
2. Urate solutions: Prepare eight urate concentrations in the range from 0.05 to 5 mM in STM.
3. GeneClamp 500B (Molecular Devices Inc., Sunnyvale, CA, USA).
4. Glass electrodes.
5. Digidata 1320A converter and pClamp8 (Axon Instruments, Union City, CA).

3 Methods

3.1 *Xenopus laevis* Oocyte Harvest and mRNA Injection

1. Two-year-old female *Xenopus laevis* frogs are removed from their vivarium holding tank and placed into a bucket containing 4 L of euthanasia solution.

2. After the frog is completely euthanasized, place it onto a dissecting tray and wash the abdomen with 70 % v/v ethanol prior to removal of the egg sacs.
3. Remove the egg sacs via a midline incision along the abdomen and place them into a 100 mm petri dish containing 19 °C MBM.
4. Open the egg sacs using two forceps, and then break down the sacs into smaller bunches of oocytes (*see* **Note 1**). Place these bunches into 8 mL glass vials containing 5–6 mL of 19 °C MBM (*see* **Note 2**).
5. Wash the oocytes three times with fresh MBM to discard broken and underdeveloped ones, and then remove most of the MBM via a plastic Pasteur pipette leaving just enough MBM to cover the oocytes.
6. Treat the oocytes with 5 mL collagenase solution per vial (*see* **Note 3**). Cap the vials and tape them down on their sides and place on to an orbital shaker using standard laboratory labeling tape.
7. Gently shake the oocytes for 90 min at medium speed on the orbital shaker. The oocytes should be moving freely when the vials are swirled after digestion.
8. Decant the collagenase solution and rinse the oocytes six times with fresh MBM. This is best achieved by filling the vial to the brim with fresh MBM, then letting the oocytes settle to the bottom before decanting, and refilling with fresh MBM.
9. After the final rinse, sort the oocytes in a 100 mm petri dish containing 19 °C MBM under a dissecting microscope. Only intact stage V and VI oocytes are retained and then placed in clean 8 mL glass vials (approximately 100 oocytes per vial) containing fresh MBM. Place the oocytes in the 19 °C refrigerator overnight.
10. The following day, decant off the MBM and add 5–6 mL of 100 mM phosphate buffer to each vial. Place the vials onto the orbital shaker with gentle shaking for 30–35 min to remove most of the follicular membranes from the oocytes.
11. Decant off the phosphate buffer, and rinse the oocytes 7 × with fresh MBM. Place the vials back into the 19 °C fridge for 15–20 min to allow the oocytes to firm up.
12. Re-sort the oocytes under the dissecting microscope, and remove the remaining follicular membrane from each egg manually using forceps. Place the defolliculated oocytes into clean vials containing fresh MBM and put them back into the 19 °C fridge overnight.
13. The following day, re-sort oocytes into clean vials containing fresh MBM and keep only unmarked, intact, round, stage V and VI oocytes for subsequent mRNA injection.

14. Inject 10–20 nL (50 ng) mRNA or water (control) into the isolated oocytes. Then incubate the injected oocytes in fresh MBM for 3–5 days in 19 °C fridge.
15. Oocytes are sorted each day with fresh changes of MBM.
16. After 3–5 days of incubation, oocytes are ready for experimental assays.

3.2 GLUT Protein Transcription mRNA

1. GLUT plasmids are constructed into an optimized pGEM-HE vector for protein overexpression in oocytes (*see* **Note 4**).
2. Linearize 1 μg pGEMHE-GLUT plasmids by 1 unit *NheI* enzyme (*see* **Note 5**). For 100 μL reaction volume, add 10 μL of 10× buffer 2 from the restriction enzyme kit, 1 μL of *NheI* (1000 units), 20 μL of 50 ng/μL pGEMHE-GLUT plasmid, and 69 μL of nuclease-free water to a final volume of 100 μL.
3. Incubate the reaction mixture at 37 °C for more than 1 h to allow complete linearization.
4. Clean up linearized plasmids by phenol-chloroform purification.
5. Dissolve extracted plasmids in 6.5 μL of nuclease-free water. The plasmids are now ready for transcription.
6. Set up a 20 μL reaction by adding 6 μL plasmids, 10 μL NTP/CAP from the mMessage mMACHINE kit, 2 μL 10× reaction buffer, and 2 μL enzyme mix. Incubate the reaction at 37 °C for 2 h.
7. Extract GLUT-RNA by phenol-chloroform extraction.
8. After extraction, dilute mRNA to desired concentration, and the mRNA is ready for injection.

3.3 Biotinylation and Western Blot Determination of GLUT Protein Surface Levels

1. Transfer 20 GLUT protein-expressing oocytes from the incubation vial to a well in a 12-well plate.
2. Wash oocytes three times with PBS (pH 8.0) with gentle shaking for 30 s.
3. Gently remove the PBS from the last wash and incubate the oocytes with 1 mL 2 mM biotinylation solution in PBS (pH 9.0) at room temperature for 30 min with gentle shaking in the dark.
4. Gently remove the biotin solution and stop the reaction by adding 1 mL of quenching buffer.
5. With gentle shaking, wash the oocytes in quenching buffer for 30 s.
6. Remove quenching buffer and wash the oocytes three times with PBS (pH 7.4) with gentle shaking. From now on, all steps are performed on ice.
7. In an icebox, transfer the oocytes to a prechilled 1.5 mL centrifuge tube and gently remove all of the PBS solution.

8. Lyse oocytes with RIPA buffer containing 100× dilution of protease inhibitor cocktail.
9. Using a syringe fitted with a 25-gauge needle repeatedly draw up and then push out the oocytes to completely break them up and forming a cell lysate. Let the lysate sit on ice for 20 min in the dark.
10. Spin the lysate at 1,000 × *g* at 4 °C.
11. Transfer the lysate supernatant into a new 1.5 mL Eppendorf tube without disturbing the pellet and the top yolk layer.
12. Repeat **steps 9** and **10** three more times (or until the yolk layer is completely removed).
13. After the last spin, transfer the cleared lysate to a new 1.5 mL Eppendorf tube and keep it on ice.
14. Mix the lysate well with 100 μL of streptavidin agarose beads. Incubate the mixture on a tube rotator at 4 °C overnight.
15. On the next day, centrifuge the beads with bound protein at 1,000 × *g* and wash two times with protease-containing RIPA buffer.
16. After the last wash, resuspend the biotinylated proteins using 2× SDS loading sample buffer and incubate at room temperature for 20 min (*see* **Note 6**). Then, the proteins are subjected to SDS-PAGE and transferred to a nitrocellulose membrane for Western blotting.
17. Block the membrane with blocking buffer for 1 h at room temperature.
18. Probe the membrane with primary GLUT antibody overnight at 4 °C, gently on a shaker (*see* **Note** 7).
19. Wash the membrane three times with blocking buffer. Then incubate the membrane with anti-rabbit secondary antibody.
20. Wash the membrane three times with blocking buffer, then develop the membrane, and capture protein signals on a film or by using a densitometer (*see* **Note 8**).
21. Relative protein levels on the film are determined by measuring the band intensities using the software ImageJ or with the densitometer. For example, Fig. 1 shows published data with representative GLUT9 and its mutants' total protein, unbound protein, and biotinylated protein [12].

3.4 Functional Studies by Radiotracer Flux Measurements

All urate radiotracer kinetics flux studies are conducted at room temperature, 20–22 °C.

1. Place 10–12 pre-sorted oocytes for each urate concentration (in a 12 × 75 mm glass test tube).
2. Remove excess MBM from the oocytes with a plastic Pasteur pipette.

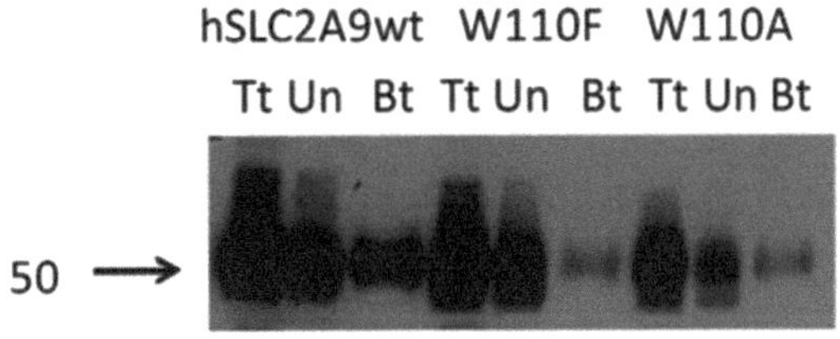

Fig. 1 A representative picture of Western blot analysis of protein expression of water-injected, hSLC2A9 WT, I335V, W110F, and W110A expressing oocytes. Adopted from ref. [12]

3. Add 200 μL of each concentration urate solution containing 250 μCi/mL ^{14}C-labeled urate to the oocytes.
4. Incubate oocytes on an orbital shaker with moderate shaking.
5. Wash oocytes five times with ice-cold MBM to stop the uptake.
6. Place individual oocytes in scintillation vials and add 200 μL of SDS solution to each vial.
7. Place the vials on an orbital shaker with vigorous shaking for 30 min for digestion (*see* **Note 9**).
8. Add 4 mL of scintillation fluid to each vial.
9. Cap the vials and mix by vortexing before placing the vials in the liquid scintillation counter for counting.
10. Measure radioactivity using a liquid scintillation counter (*see* **Note 10**). Please *see* Fig. 2 as an example from our previously published data [12].

3.5 Two-Microelectrode Voltage Clamping

1. Turn on GeneClamp 500B and Digidata 1320A converter. Then, turn on computer and open pClamp8 for data recording.
2. Set up the chamber by connecting tubes from solutions to the solution switcher. Tubes that come out from the switcher are merged into a single tube, which is connected to one end of a perfusion chamber. Another tube coming out from the other end of the perfusion chamber is connected to a vacuum pump.
3. Use a glass electrode puller to pull two electrodes, and then place one glass electrode in left headstage holder and another glass electrode on the right headstage.
4. Place a GLUT-expressing oocyte in a STM buffer-filled chamber.
5. Slowly move two electrodes toward the oocyte until both electrodes are just above the oocyte membrane.
6. Set membrane potential value to zero on GeneClamp 500B (or from your Clamplex program). With a gentle and fast motion, place two electrodes on the oocyte membrane and seal

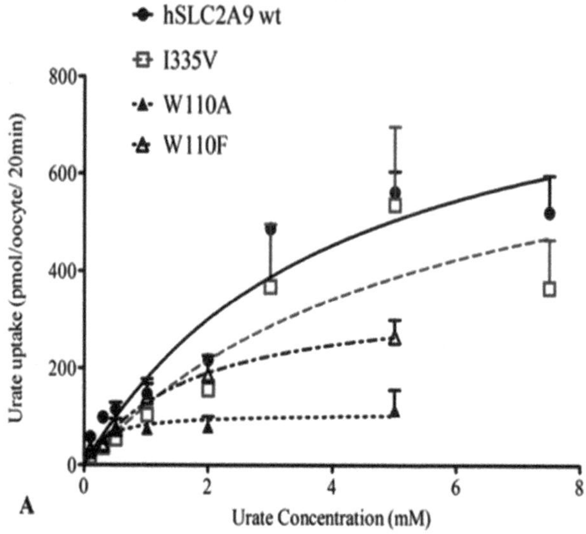

A

	hSLC2A9 wt	I335V	W110A	W110F
$K_{M\,(mM)}$	4.5	6.6	0.3	1.7
SEM	11.6	3.1	0.1	0.7
V_{MAX} (pmol/oocyte/20min)	981.9	937.7	108.5	355.6
SEM	136.5	278.1	11.1	61.9

B

Fig. 2 ^{14}C Urate kinetic measurements in oocytes expressing hSLC2A9. (**a**) Michaelis-Menten curves of ^{14}C urate kinetics of hSLC2A9 WT (*filled circle*) and its mutants I335V (*open square*), W110A (*filled triangle*), and W110F (*open triangle*). Urate uptake was measured by incubating protein-expressing oocytes in 200 μL urate solution ranging from 100 μM to 5 mM for 20 min. Uptake activity was corrected for nonspecific transport measured in control water-injected oocytes from the same batch of oocytes. (**b**) ^{14}C urate kinetic constants of the three isoforms (n = 4) [12]

the openings of both electrodes (*see* **Note 11**). Run seal test, and it should appear as giga-Ω seal with resistance over 1 G omega for both electrodes.

7. A membrane potential value (normally −50 mV to −60 mV) should appear on the GeneClamp 500B screen (*see* **Note 12**).
8. Start recoding current and voltage using pClamp8 program with a GAP-free protocol.
9. Open the perfusion switcher and vacuum for a constant perfusion of STM (at a flow rate of 1 mL STM/2.5 min) over the oocyte (*see* **Note 13**).
10. Once membrane potential reaches a stable state, turn clamping mode on to clamp the oocyte membrane at constant membrane

potential of −30 mV. Allow current reading to reach a stable state (*see* **Note 14**).

11. Switch perfusion to 0.5 mM urate STM. Allow current reading to reach a stable state, and then wash the oocyte with fresh STM buffer. Allow current reading to reach a stable state (*see* **Note 15**).
12. Repeat the same process (**steps 10** and **11**) for other concentrations of urate-containing STM.
13. Urate kinetics curve is expressed as urate-induced mean peak current (*see* **Note 16**). Please see Fig. 3 as an example from our previous published data [12].

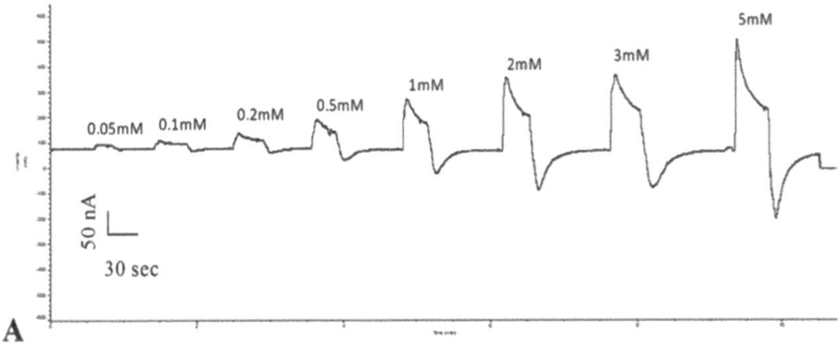

A

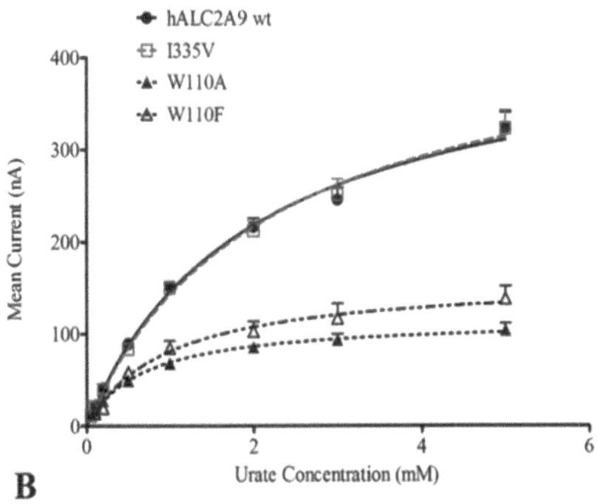

B

	hSLC2A9 wt	I335V	W110A	W110F
$K_{M\ (mM)}$	1.97	2.2	0.7	1
SEM	0.2	0.3	0.1	0.3
V_{MAX} (Current)	433.4	451	115	158.9
SEM	21.5	26.5	5.8	14.5

C

Fig. 3 Urate-induced currents in oocytes measured with two-microelectrode voltage clamp (TEVC). (**a**) Provides a representative trace from the GAP-free protocol of WT hSLC2A9-expressed oocytes. Single oocytes were clamped at −30 mV and super-perfused with different concentrations of urate for 30 s (range from 0.1 to 5 mM) followed by a 1-min wash with urate-free buffer in between. (**b**) Michaelis-Menten curves of urate kinetics of hSLC2A9 WT (*filled square*) and its mutants I335V (*open square*), W110A (*filled triangle*), and W110F (*open triangle*). Data were collected at the peak of each urate-induced current. Points represent the mean of urate-induced peak outward current for each concentration. (**c**) ^{14}C urate kinetic constants of the WT and mutant isoforms ($n > 15$ oocytes from 3 frogs) [12]

4 Notes

1. Care must be taken to work quickly and not be overly aggressive when breaking down the egg sacs into smaller bunches.
2. The ratio of oocytes to media is also important; approximately 2 mL of oocytes per vial with 5–6 mL of media is ideal.
3. Overloading the oocytes in the vials will make it difficult to achieve good digestion with the collagenase.
4. The pGEM-HE vector appears to work best for GLUT expression in oocytes.
5. 1 μg of plasmid DNA is optimal for this current mRNA protocol.
6. Do not heat/boil the GLUT protein after the addition of SDS-page loading buffer, just let them sit at room temperature for 15–20 min (do not exceed 30 min).
7. GLUT proteins bind better to their primary antibodies at 4 °C.
8. It is a good idea to wash the membrane two additional times with PBS-T before developing the Western blot, as this will further reduce the background and increase the signal to noise ratio.
9. Make sure that oocytes were lysed completely before radioactivity measurements.
10. All experiments are performed four to six times and the results are corrected for the flux values obtained with water-injected oocytes obtained from the same frog.
11. Make sure that glass electrodes are just touching the oocyte membrane, but not penetrating into the cytoplasm of the oocyte.
12. An outward current will appear on the computer screen and the current value will increase on the GeneClamp 500B screen with the application of urate-containing buffer.
13. An inward current will appear on the computer screen and the current value will decrease on the GeneClamp 500B screen when switching from a urate-containing buffer to fresh STM buffer.
14. Oocytes with low membrane potential (10–20 mV) might have non-stable base lines before urate treatment.
15. Oocytes with high membrane potential (80–90 mV) might have smaller urate induced currents.
16. Digidata 1320A converter and pClamp8 (Axon Instruments) were used for experiments and data analysis.

References

1. Gould GW, Lienhard GE (1989) Expression of a functional glucose transporter in Xenopus oocytes. Biochemistry 28(24):9447–9452
2. Long W, Cheeseman CI (2015) Structure of, and functional insight into the GLUT family of membrane transporters. Cell Health Cytoskelet 7:167–183. https://doi.org/10.2147/CHC.S60484
3. Mueckler M (1994) Facilitative glucose transporters. Eur J Biochem 219(3):713–725
4. Thorens B, Mueckler M (2010) Glucose transporters in the 21st century. Am J Physiol Endocrinol Metab 298(2):E141–E145. https://doi.org/10.1152/ajpendo.00712.2009
5. Hruz PW, Mueckler MM (2001) Structural analysis of the GLUT1 facilitative glucose transporter (review). Mol Membr Biol 18(3):183–193
6. Mueckler M, Makepeace C (2006) Transmembrane segment 12 of the Glut1 glucose transporter is an outer helix and is not directly involved in the transport mechanism. J Biol Chem 281(48):36993–36998. https://doi.org/10.1074/jbc.M608158200
7. Mueckler M, Makepeace C (2009) Model of the exofacial substrate-binding site and helical folding of the human Glut1 glucose transporter based on scanning mutagenesis. Biochemistry 48(25):5934–5942. https://doi.org/10.1021/bi900521n
8. Mueckler M, Thorens B (2013) The SLC2 (GLUT) family of membrane transporters. Mol Asp Med 34(2-3):121–138. https://doi.org/10.1016/j.mam.2012.07.001
9. Uldry M, Ibberson M, Horisberger JD, Chatton JY, Riederer BM, Thorens B (2001) Identification of a mammalian H(+)-myo-inositol symporter expressed predominantly in the brain. EMBO J 20(16):4467–4477. https://doi.org/10.1093/emboj/20.16.4467
10. Caulfield MJ, Munroe PB, O'Neill D, Witkowska K, Charchar FJ, Doblado M, Evans S, Eyheramendy S, Onipinla A, Howard P, Shaw-Hawkins S, Dobson RJ, Wallace C, Newhouse SJ, Brown M, Connell JM, Dominiczak A, Farrall M, Lathrop GM, Samani NJ, Kumari M, Marmot M, Brunner E, Chambers J, Elliott P, Kooner J, Laan M, Org E, Veldre G, Viigimaa M, Cappuccio FP, Ji C, Iacone R, Strazzullo P, Moley KH, Cheeseman C (2008) SLC2A9 is a high-capacity urate transporter in humans. PLoS Med 5(10):e197. https://doi.org/10.1371/journal.pmed.0050197
11. Phay JE, Hussain HB, Moley JF (2000) Cloning and expression analysis of a novel member of the facilitative glucose transporter family, SLC2A9 (GLUT9). Genomics 66(2):217–220. https://doi.org/10.1006/geno.2000.6195
12. Long W, Panwar P, Witkowska K, Wong K, O'Neill D, Chen XZ, Lemieux MJ, Cheeseman CI (2015) Critical roles of two hydrophobic residues within human glucose transporter 9 (hSLC2A9) in substrate selectivity and urate transport. J Biol Chem 290(24):15292–15303. https://doi.org/10.1074/jbc.M114.611178
13. Witkowska K, Smith KM, Yao SY, Ng AM, O'Neill D, Karpinski E, Young JD, Cheeseman CI (2012) Human SLC2A9a and SLC2A9b isoforms mediate electrogenic transport of urate with different characteristics in the presence of hexoses. Am J Physiol Renal Physiol 303(4):F527–F539. https://doi.org/10.1152/ajprenal.00134.2012
14 Long W, Panigrahi R, Panwar P, Wong K, O'Neill D, Chen XZ, Lemieux ML, Cheeseman CI (2017) Identification of Key Residues for Urate Specific Transport in Human Glucose Transporter 9 (hSLC2A9). Scientific Reports 7:41167. http://doi.org/10.1038/srep41167

Chapter 5

Glucose Uptake in Heterologous Expression Systems

Eunice E. Lee and Richard C. Wang

Abstract

Understanding the physiological regulation of glucose transport requires the analysis of transporters, like GLUT1, in diverse tissue types. We document the utility of viral vectors for the stable expression of wild-type and modified GLUT1 transporter in different types of mammalian cells. Once expression of the alleles has been confirmed by Western blotting, the effect of specific mutations on the regulation of glucose transport can be measured through a previously described radiolabeled glucose transport assay. Although this discussion is focused on the GLUT1 transporter, these techniques are easily transferrable to other glucose transporters.

Key words Retroviral transduction, Glucose uptake, GLUT1, 2-Deoxyglucose, 2DG uptake, Radiolabeled glucose, Mammalian cells

1 Introduction

In most tissues, a family of integral membrane glucose transport proteins (GLUTs) catalyzes the facilitated diffusion of glucose across the cell membrane. The ubiquitously expressed GLUT1 was the first of 14 GLUT proteins to be identified [1] and has been studied extensively [2]. Despite this extensive characterization, the mechanisms that regulate GLUT1-mediated glucose transport are still not completely understood [3]. Because the regulation of facilitative glucose transporters, including GLUT1, may be tissue specific, our understanding of transporter regulation and function has been advanced through the stable expression of wild-type and modified transporters in a variety of cell types cultured in vitro [4].

In this chapter, we describe methods that allow for the stable heterologous expression of wild-type and mutant GLUT1 in a variety of mammalian cell types to study the regulation of glucose transport. Included in this chapter are methods for generating and expressing wild-type and mutant *Glut1* constructs that are expressed in target cells through retroviral transduction. Confirmation of *Glut1* allele expression in transduced cells is

Karin Lindkvist-Petersson and Jesper S. Hansen (eds.), *Glucose Transport: Methods and Protocols*, Methods in Molecular Biology, vol. 1713, https://doi.org/10.1007/978-1-4939-7507-5_5,

confirmed through Western blotting. Finally, we describe a protocol for the measurement of glucose uptake in this system using the radiolabeled glucose analog 2-deoxyglucose (2DG). Although the protocols and techniques in this chapter are optimized for GLUT1 in mammalian cell lines, they can be adapted for use in a diverse range of heterologous expression systems.

2 Materials

2.1 Retroviral Construct Preparation

1. DNA vectors: prGT3 vector with rat GLUT1 gene insert and pWZL Hygro empty vector backbone (Addgene).
2. Restriction enzyme: *Eco*R1 restriction enzyme (New England Biolabs).
3. QuikChange II XL Site-Directed Mutagenesis Kit (Agilent Technologies).
4. Oligonucleotide primers: Sense and antisense for mutagenesis.
5. PCR tubes: Thin-walled 0.2 mL.
6. Selection plates; LB agar plates with ampicillin.
7. Agarose gel electrophoresis system.
8. NucleoSpin Gel and PCR Clean-up kit (Macherey-Nagel).

2.2 Retroviral Transduction

1. Tissue culture dishes: 6 and 10 cm dishes suitable for cell culture.
2. T-75 tissue culture flasks.
3. 15 mL Polypropylene conical tubes.
4. Mammalian cells: Lenti-X 293 T cells (Clontech) and Rat2 cells (ATCC).
5. Cell culture medium: Dulbecco's modified Eagle's medium (DMEM).
6. Trypsin-EDTA solution (1X): 0.05 % w/v Trypsin, 0.02 % w/v EDTA.
7. Phosphate-buffered saline (PBS) solution: 137 mM NaCl, 2.7 mM KCl, 10 mM Na_2HPO_4, 1.8 mM KH_2PO_4, pH 7.4.
8. Serum: Fetal bovine serum (FBS).
9. Antibiotic-antimycotic solution: 100×, containing 10,000 units/mL of penicillin, 10,000 μg/mL of streptomycin, 25 μg/mL of Amphotericin B (Gibco).
10. Transfection agent: Lipofectamine 3000 (ThermoFisher Scientific).
11. Gag/pol packaging plasmid (Addgene).
12. VSV-G envelope plasmid (Addgene).

13. Hexadimethrine bromide stock solution: 8 mg/mL in PBS.
14. Hygromycin B solution: 50 mg/mL solution.
15. 10 mL Syringes.
16. 0.45 μm Syringe filters.

2.3 Western Blot Confirmation of GLUT1 Expression

1. Blotting membrane: Polyvinyl difluoride membranes (PVDF).
2. Gels: Pre-cast 4–15 % 15-well protein gels (Bio-Rad).
3. Extra thick blot filter paper.
4. TBS-T buffer solution: Tris-buffered saline containing 0.05 % v/v Tween-20 (TBS-T).
5. Blocking buffer: 5 % w/v nonfat dry milk diluted in TBS-T.
6. Sample loading buffer: 2× Laemmli Sample Buffer.
7. Primary antibodies: GLUT1 (Millipore 07-1401; 1:1000) and phosphoGLUT1 (Ser226) (Millipore ABN991; 1:200).
8. Secondary antibody: Donkey anti-mouse IgG horseradish peroxidase conjugate.
9. Phorbol 12-myristate 13-acetate (TPA, Cayman Chemical).
10. SDS-PAGE running buffer: 10× Tris–glycine–SDS, diluted to 1× in water (Bio-Rad).
11. Tris–glycine transfer buffer: 10× Diluted to 1× in water (Bio-Rad).

2.4 Measurement of Glucose Uptake Using Radiolabeled 2-Deoxyglucose

1. 12-Well cell culture plates.
2. 2-Deoxyglucose (2DG) solution: 0.5 M 2DG dissolved in water.
3. Radiolabeled glucose: 20–50 Ci/mmol [^{3}H] 2-Deoxy-D-glucose (Perkin Elmer).
4. Glucose solution: 0.5 M D-glucose diluted in water.
5. Cytochalasin B solution: 10 mg/mL Solution in dimethyl sulfoxide (DMSO).
6. Krebs Ringer's HEPES buffer with pyruvate (KRHP): 136 mM NaCl, 4.7 mM KCl, 1.25 mM $CaCl_2$, 1.25 mM $MgSO_4$, 10 mM HEPES, 1 mM sodium pyruvate, 0.1 % w/v BSA, pH 7.4.
7. Albumin solution: 10 % w/v Bovine serum albumin (BSA).
8. 500 mM Sodium hydroxide diluted in water.
9. 500 mM Hydrochloric acid diluted in water.
10. 5 mL Plastic scintillation vials.
11. Commercial scintillation counting fluid.
12. Liquid scintillation counter (Beckman Coulter).
13. BCA Protein Assay Kit (ThermoFisher).

3 Methods

3.1 Retroviral Generation of Stable Cell Lines Expressing GLUT1

3.1.1 Retroviral Construct Preparation: Mutagenesis and Subcloning

1. Design and synthesize two mutagenic primers for the desired mutation using the Web-based QuikChange Primer Design Program. Primers should be purified by polyacrylamide gel electrophoresis (PAGE) or fast polynucleotide liquid chromatography (FPLC) (*see* **Note 1**). Reconstitute primers in sterile water to a final concentration of 100 ng/μL.
2. In a thin-walled 0.2 mL tube, assemble the PCR reaction as follows:
   5 μL 10× reaction buffer.
   X μL (15 ng) DNA template (*see* **Note 2**).
   1.25 μL (125 ng) primer #1.
   1.25 μL (125 ng) primer #2.
   1 μL of dNTP mix.
   3 μL of QuikSolution.
   Adjust with ddH_2O to a final volume of 50 μL.
   1 μL *Pfu Ultra* HF DNA polymerase (2.5 U/μL). Add this last.
3. Perform PCR amplification with the following cycling conditions: denaturation at 95 °C for 1 min, 18 cycles of 95 °C for 50 s, 60 °C for 50 s, 68 °C for 1 min/kb of plasmid size, and extension at 68 °C for 7 min (*see* **Note 3**).
4. Verify amplification by agarose gel electrophoresis using 8 μL of PCR product (*see* **Note 4**).
5. Purify the remaining PCR product using a DNA purification column (e.g., the Macherey-Nagel NucleoSpin Gel and PCR Clean-up kit according to the manufacturer's protocol.
6. Add 1 μL of *Dpn*I restriction enzyme to the purified PCR product. Incubate at 37 °C for 90 min.
7. Transform the final reaction into XL10–Gold ultracompetent cells according to the manufacturer's instructions.
8. Subclone the mutated *Glut* alleles (e.g., *R. norvegicus* Glut on prGT3 vector) into the pWZL Hygro vector by standard restriction digestion cloning (*see* **Note 5**).
9. Select clones from each transformation and sequence the entire insert to verify the presence of the desired mutation and confirm that other unwanted mutations have not been introduced (e.g., Genewiz).

3.1.2 Retroviral Transduction

Exercise due caution when handling and producing retroviruses. Refer to local and institutional biosafety guidelines of your institution for handling and waste decontamination. Perform all incubation steps at 37 °C in a humidified incubator with 5 % CO_2.

1. **Day 0:** Seed the Lenti-X 293 T packaging cells in 6 cm cell culture dishes at two different densities: 1.0×10^6 and 1.25×10^6 cells/dish using 4 mL of antibiotic-free DMEM supplemented with 10 % FBS per dish (*see* **Note 6**). Optimal cell density for transfection is 40–60 %. Gently swirl the dish to mix. Place cells in the incubator for 12–16 h. Thaw target cells in complete media.
2. **Day 1:** Check the density of the packaging cells. If cells have reached the desired density, transfect the cells with a mixture of transfer vector, gag/pol packaging plasmid, and VSV-G envelope plasmid in a 4:3:1 ratio using Lipofectamine 3000 according to the manufacturer's instructions (*see* **Note 7**). Add the mixture dropwise onto the cells. Gently swirl and rock the dish to evenly distribute the transfection mixture and incubate overnight.
3. **Day 2:** Carefully aspirate the medium on the packaging cells and add 4.2 mL of medium of the target cells (containing antibiotics and serum) 24 h after transfection. Seed the target cells on 100 mm cell culture dishes to target a density of 30–40 % at the time of infection. Seeding density may vary with the cell line. Include an extra dish of target cells to be used as a control for antibiotic selection (*see* **Note 8**).
4. **Day 3:** Prepare the polybrene infection/transfection reagent stock at 8 mg/mL in PBS. Aliquot and store at 4 °C or −20 °C for long-term storage. Check the density of the target cells. If target cells have reached the desired density (40–50 % confluent), proceed with the first virus collection. Forty-eight hours after transfection, initiate the first infection by collecting 4 mL of viral supernatant from the packaging cells using a serological pipet. Filter through a 0.45 μm syringe filter to remove any detached cells and collect the supernatant in a 15 mL conical tube (*see* **Note 9**). Add 4 mL of fresh medium to the packaging cells. Add 4–8 μg/mL polybrene infection/transfection reagent and 2 mL of fresh target media to the filtered supernatant. Invert the tube to mix. Remove existing media from the target cells and infect with 6 mL of the virus containing supernatant prepared earlier. Return all cells to the incubator.
5. **Day 4:** In the morning, perform the second of three infections as outlined (*see* **Day 3**). In the late afternoon or evening, perform the last infection (*see* **Day 3**). Bleach and discard packaging cells (*see* **Note 10**).
6. **Day 5:** Remove the viral supernatant from the target cells and replace with virus-free medium without selection antibiotic 24 h after the last infection.
7. **Day 6:** Split infected target cells using high-dose selection medium containing 250 μg/mL hygromycin B or other

appropriate selection antibiotic (*see* **Note 11**). Change medium on the uninfected control cells using the same high-dose selection media.

8. **Day 7:** Perform media changes on both uninfected and infected target cells using high-dose selection medium.
9. **Day 8–13:** Replace medium with 150 μg/mL hygromycin B or appropriate selection antibiotic. Continue routine media changes and monitor control cells daily. Most uninfected control cells should be nonviable within 3 days after addition of hygromycin B. Complete selection of cells infected with vectors conferring antibiotic resistance should be complete in approximately 7 days. Different selection antibiotics require different time courses for selection. Expand cells and freeze early-passage cells for future culture (*see* **Note 12**).

3.2 Western Blot Detection of GLUT1

After selection is complete, confirm transduction efficiency and GLUT1 expression levels before proceeding with uptake assays or other downstream experiments (*see* **Note 13**).

1. Seed 2 wells each of parental and transduced cells in a 6-well plate (0.3×10^6 cells/well).
2. When cells have reached confluence, wash cells once with PBS and replace the standard growth medium with serum-free medium supplemented with 0.1 % w/v BSA for 1 h.
3. Treat one well of each cell type with 500 nM TPA diluted in DMSO. Add an equal volume of DMSO to the untreated cells as a vehicle control.
4. Incubate cells for 20–30 min at 37 °C.
5. Place cells on ice and rinse once with ice-cold PBS.
6. Scrape cells in 2× sample buffer (*see* **Note 14**).
7. Heat samples for 10 min at 50 °C. Do not boil (*see* **Note 15**).
8. Separate proteins on a SDS-PAGE gel.
9. Transfer proteins to a PVDF membrane.
10. Block the membrane in blocking buffer for 30 min at room temperature.
11. Incubate the membrane overnight at 4 °C with 1:1000 dilution of anti-total GLUT1 or other desired antibodies diluted in blocking buffer (*see* **Note 16**) (*see* Fig. 1).

3.3 Measurement of Glucose Uptake Using Radiolabeled 2-Deoxyglucose

All incubation steps should be performed at 37 °C in a humidified incubator with 5 % CO_2.

1. Seed cells in triplicate in a 12-well plate 18–24 h before the assay using DMEM supplemented with 10 % FBS or other standard cell culture medium.

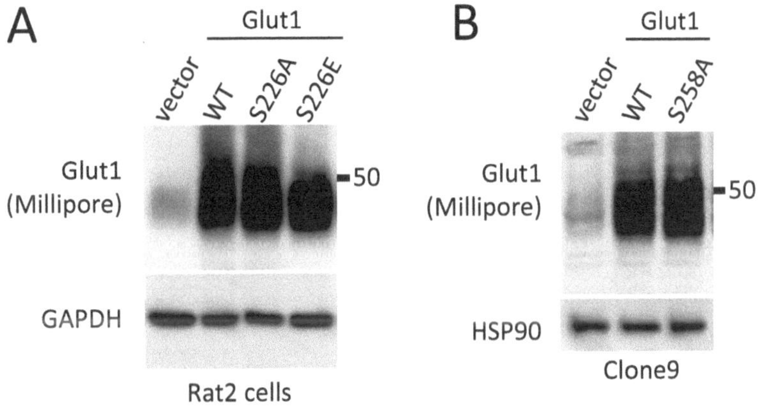

Fig. 1 Western blots confirm Glut1 allele expression after retroviral transduction. (**a**) Stable overexpression of wild-type (WT), S226A, and S226E Glut1 alleles in Rat2 cells. (**b**) Expression of WT and GLUT1-T258A in Clone9 cells. Position T258 of GLUT1 is a putative 5′ AMP-activated protein kinase (AMPK) site

2. Optional: Replace media with reduced 5 % FBS 6–8 h after seeding. Incubate cells overnight (*see* **Note 17**).
3. Wash the cells twice with 2 mL PBS.
4. Incubate cells in serum-free media supplemented with 0.1 % w/v BSA for 2 h before treatment is added.
5. Prepare the treatment solution by dissolving TPA or desired treatment stock in KRHP buffer (*see* **Note 18**).
6. Dissolve an equal concentration of DMSO in KRHP buffer for untreated control wells.
7. Wash the cells twice with 2 mL PBS.
8. Add exactly 450 μL of the treatment solution or KRHP buffer containing DMSO to each well.
9. Incubate at 37 °C for 30 min (*see* **Note 19**).
10. During the incubation step, prepare uptake solution containing 1 μCi ^{3}H 2DG/well and 0.1 mM unlabeled 2DG in KRHP buffer (*see* **Note 20**).
11. Initiate transport by adding 50 μL of uptake solution to each well.
12. Incubate for 5 min at 37 °C (*see* **Note 21**).
13. Terminate transport by rapidly removing the uptake solution and washing 3× with 900 μL ice-cold PBS containing 25 mM D-glucose (*see* **Note 22**).
14. Add 500 μL of 0.5 M NaOH to each well to lyse the cells (*see* **Note 23**).
15. Add 500 μL of 0.5 M HCl to each well to neutralize.
16. Pipet up and down several times to disperse any cell clumps.

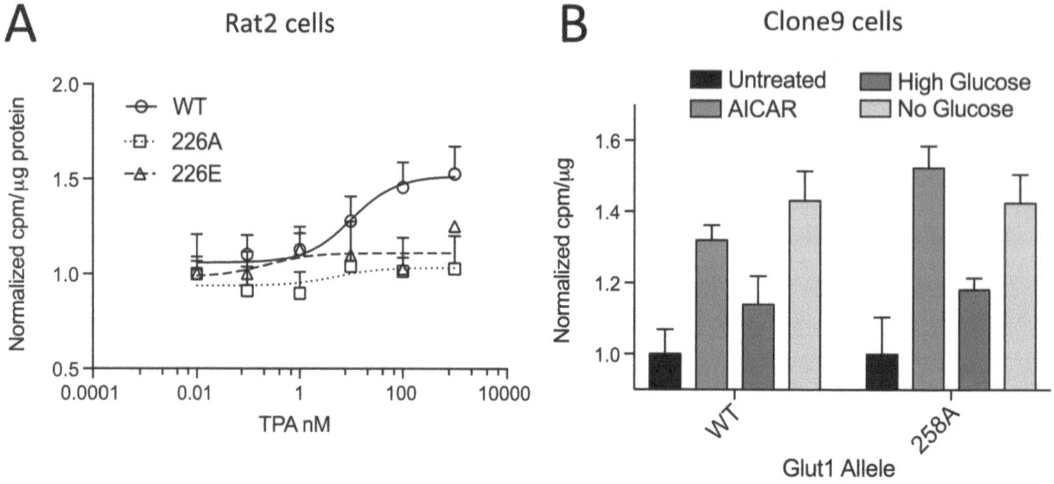

Fig. 2 Radiolabeled 2DG uptake assays can be used to assess the effects of specific mutations on glucose uptake. (**a**) WT GLUT1, but not the non-phosphorylatable, mutants, GLUT1–S226A or GLUT1–S226E, expressing Rat2 cells show a dose-dependent increase in glucose uptake. Uptake values are normalized against untreated cells of the same genotype to allow for the analysis of uptake over multiple independent experiments. (**b**) Both WT and GLUT1-T258A-expressing Clone9 cells respond to 5-amino-imidazole-4-carboxamide ribonucleotide (AICAR) or glucose starvation by increasing the glucose uptake

17. Pipet 250 μL of the cell lysate into a scintillation vial containing 5 mL of scintillation fluid.
18. Cap the tube and shake well.
19. Quantify the internalized radiolabeled glucose by liquid scintillation counting (*see* **Note 24**).
20. Determine the protein concentration by BCA and normalize uptake to the amount of protein present in each well (*see* **Note 25** and *see* Fig. 2).

4 Notes

1. The QuikChange Primer design software can be found on the manufacturer's website. To maximize mutation efficiency, Stratagene strongly suggests using purified mutagenic primers. This kit is optimal for generating point mutations or short insertions/deletions.
2. If the PCR is unsuccessful using 15 ng of template DNA, performing a series of reactions using a range of dsDNA template (5–50 ng) is recommended.
3. Exceeding 18 cycles may increase the possibility of introducing undesired second-site mutations.
4. If a PCR band is not detectable, the manufacturer's protocol suggests continuing with *Dpn*I digestion and transformation.

In our experience, mutagenesis is often unsuccessful when a band cannot be visualized.

5. Restriction enzyme-based strategies for cloning reduce the rate of mutation. However, when restriction enzyme or recombination-based cloning strategies are not possible, it is possible to use a PCR-based strategy to amplify and clone your allele of interest. While retroviral expression vectors are useful for a broad range of mammalian cell lines, the *Glut1* allele or insert of interest may also be subcloned into other expression vectors (e.g., pGEMHE for expression in *Xenopus oocytes* [5], pWPXL for lentiviral packaging).
6. We recommend seeding the packaging cells at two different densities for optimal cell density at the time of transfection. It is important to thoroughly mix the cells in order to produce a single-cell suspension and ensure an even distribution of cells. Since 293T cells detach easily from the cell culture dish surface, change media by gently dispensing against the side of the dish to prevent dislodging cells. It is helpful to include a small amount (< 0.1 μg) of an eGFP expression vector to ensure high transfection efficiency.
7. Ensure that your viral production plasmids include an envelope plasmid, Gag-pol plasmid, and transgene transfer vector. Both two and three vector viral packaging systems exist. For example, the retroviral packaging plasmid pCL-Eco (Addgene) expresses both envelope and Gag-pol proteins and may be used to transduce mouse and rat cell lines. It is helpful to process a vector control in parallel to measure transfection efficiency and to exclude against allele toxicity.
8. For retroviruses, target cells must be actively dividing to ensure infection. Cells that are too dense or sparse may be poorly transduced by retroviruses. For quiescent or slowly dividing cells, lentiviruses may allow for more efficient cell transduction.
9. Filtering or centrifugation is essential to prevent contamination of target cells with the 293T Lenti-X cells. If target cells have not yet reached the optimal density for infection, viral supernatant can be stored at 4 °C for up to 24 h or snap frozen in liquid nitrogen and stored at −80 °C.
10. All tissue culture reagents (pipettes, syringe filters, syringes, and tubes) coming in contact with viral supernatant should be treated with bleach prior to disposal.
11. Perform a dose titration or "kill" curve using the appropriate selection antibiotic if the optimal concentration is unknown. The optimal concentration depends on the cell type.
12. Although efficient viral infections reduce the risk of clonal artifacts in downstream assays, it is recommended that several

independent infections be used for biological replicate experiments.

13. For some mutant alleles (small deletions, point mutations without specific antibodies), it will not be possible to distinguish between the WT allele and mutant alleles by Western blotting. For confirmation of these cell lines, PCR amplification and sequencing of the region of interest will confirm successful transduction. Designing primers that span introns can prevent the amplification of endogenous alleles.
14. Whole-cell lysates may be prepared at a concentration of ~10,000 cells/μL sample buffer. Precise quantitation of the amount of protein loaded is not essential at this step, only verification that the allele of interest is expressed. Lysates should be sheared by passage through a 31-gauge syringe needle or sonication prior to loading.
15. Glycosylation of the many glucose transporters, including GLUT1, causes them to run as a smear. Boiling the protein lysate results in aggregation of GLUT1, which appears as a higher molecular weight smear above the correct expected size (55 kDa).
16. While many antibodies can be used repeatedly, freshly diluted phospho-specific antibodies should be used for each new Western blot. Signal intensity may significantly decrease with repeated use and freeze-thaw of a diluted stock. Phospho-specific antibodies should be stored in small aliquots at −20 °C to avoid repeated freeze-thaw cycles.
17. Although growth factors in FBS can affect glucose uptake, a "step-down" in FBS concentration does not appear to strongly impact glucose uptake.
18. Prepare fresh dilutions of treatment stock in KRHP for each experiment. The final concentration of DMSO in solution should be kept below 1 % if possible. It is *critical* to perform TPA treatment and glucose uptake measurements in KRHP media. Performing these experiments in culture media (e.g., DMEM without glucose), which contains a complex mixture of nutrients, strongly blunts the effects of treatments on glucose uptake.
19. Incubation times may vary depending on the treatment, cell type, and aim of the experiment. For example, AICAR treatment (1 mM) occurs for 2 h in serum-free media.
20. If the measurement of nonspecific glucose transport is needed, treat cells with 40 μM cytochalasin B before adding the pulse.
21. 2DG uptake in Rat2 and Clone9 cells is linear for <10 min under the specified conditions. Time course control experiments may need to be performed for each new cell line.
22. Washes should be rapid, yet gentle. PBS should be applied to the side of the well rather than the base of the plate. A repeater

pipette facilitates the washing of multiple wells. Confirm that the majority of cells remain adherent after completing washes as the loss of significant numbers of cells impacts uptake values despite protein normalization.

23. 1 mL of 0.2 % SDS may be substituted for NaOH followed by HCl to achieve cell lysis. Cells should be viewed under the microscope to confirm complete lysis.
24. If Lumex values are high (above 5–10 %), leave the samples at room temperature overnight and repeat the readings the following day.
25. For the microplate option of the BCA assay, 50 μL of cell lysate was sufficient. Before protein normalization, subtract background values obtained from cytochalasin B-treated cells if necessary. Because we have noted daily variability in the absolute levels of 2DG uptake, it is necessary to normalize uptake to an untreated control sample to combine independent biological replicates done over multiple experiments.

Acknowledgement

This work was supported by NIH K08 CA164047 and a Burroughs Wellcome Fund CAMS (1010978) Award to R.C.W.

References

1. Kasahara M, Hinkle PC (1977) Reconstitution and purification of the D-glucose transporter from human erythrocytes. J Biol Chem 252(20):7384–7390
2. Thorens B, Mueckler M (2010) Glucose transporters in the 21st century. Am J Physiol Endocrinol Metab 298(2):E141–E145. https://doi.org/10.1152/ajpendo.00712.2009
3. Cura AJ, Carruthers A (2012) Role of monosaccharide transport proteins in carbohydrate assimilation, distribution, metabolism, and homeostasis. Compr Physiol 2(2):863–914. https://doi.org/10.1002/cphy.c110024
4. Kanai F, Nishioka Y, Hayashi H, Kamohara S, Todaka M, Ebina Y (1993) Direct demonstration of insulin-induced GLUT4 translocation to the surface of intact cells by insertion of a c-myc epitope into an exofacial GLUT4 domain. J Biol Chem 268(19):14523–14526
5. Lee EE, Ma J, Sacharidou A, Mi W, Salato VK, Nguyen N, Jiang Y, Pascual JM, North PE, Shaul PW, Mettlen M, Wang RC (2015) A protein kinase C phosphorylation motif in GLUT1 affects glucose transport and is mutated in GLUT1 deficiency syndrome. Mol Cell 58(5):845–853. https://doi.org/10.1016/j.molcel.2015.04.015

Chapter 6

Evaluating the Efficacy of GLUT Inhibitors Using a Seahorse Extracellular Flux Analyzer

Changyong Wei, Monique Heitmeier, Paul W. Hruz, and Mala Shanmugam

Abstract

Glucose is metabolized through anaerobic glycolysis and aerobic oxidative phosphorylation (OXPHOS). Perturbing glucose uptake and its subsequent metabolism can alter both glycolytic and OXPHOS pathways and consequently lactate and/or oxygen consumption. Production and secretion of lactate, as a consequence of glycolysis, leads to acidification of the extracellular medium. Molecular oxygen is the final electron acceptor in the electron transport chain, facilitating oxidative phosphorylation of ADP to ATP. The alterations in extracellular acidification and/or oxygen consumption can thus be used as indirect readouts of glucose metabolism and assessing the impact of inhibiting glucose transport through specific glucose transporters (GLUTs). The Seahorse bioenergetics analyzer can measure both the oxygen consumption rate (OCR) and extracellular acidification rate (ECAR). The proposed methodology affords a robust, high-throughput method to screen for GLUT inhibition in cells engineered to express specific GLUTs, providing live cell read-outs upon GLUT inhibition.

Key words Glycolysis, GLUT inhibitor, Seahorse XF analyzer, Oxidative phosphorylation, ECAR

1 Introduction

Glucose is a major carbon source that plays a fundamental role in cellular bioenergetics, biosynthesis, redox homeostasis, signaling, and epigenetics [1]. Glucose enters a cell through specific glucose transporters (GLUTs) by facilitative diffusion and is retained within the cell upon phosphorylation by hexokinase. Once phosphorylated the major routes for glucose catabolism are via the glycolytic and pentose phosphate pathways. The product of glycolysis, pyruvate can then be used via anaerobic respiration to generate lactate. Alternatively, pyruvate is decarboxylated to acetyl-CoA that fuels the TCA cycle and aerobic respiration [2]. Thus, lactate production and oxygen consumption are indirect readouts of how imported glucose is metabolized within the cell.

GLUT1, 3, and 4 are high-affinity glucose transporters [3]. GLUT1 is ubiquitously expressed and mediates basal glucose

Karin Lindkvist-Petersson and Jesper S. Hansen (eds.), *Glucose Transport: Methods and Protocols*,
Methods in Molecular Biology, vol. 1713, https://doi.org/10.1007/978-1-4939-7507-5_6,

uptake in a number of cell types [4, 5]. GLUT4, expressed primarily in adipocytes and muscle cells, is the principal glucose transporter mediating insulin/exercise-responsive glucose uptake and is important in maintaining whole-body glucose homeostasis [6]. GLUT1, GLUT4, and other GLUTs have been shown to be required for the growth and survival of specific cancer cells [7, 8]. In order to identify inhibitors of specific GLUTs, we have optimized the use of the Seahorse XF Analyzer with cell lines engineered to express specific GLUTs. Similar strategies can be applied to the interrogation of other targets if ECAR and/or oxygen consumption are impacted.

Protons generated during pyruvate production and lactate generated from pyruvate, both contribute to the acidification of the extracellular medium [9]. The Seahorse extracellular flux (XF) analyzer (Seahorse Bioscience) directly measures the extracellular acidification rate (ECAR) in real time, providing quantifiable data that can be used as an indirect readout of cellular glycolytic flux. The change in ECAR upon addition of glucose can be used as a surrogate measure of the rate of glycolysis. To assess the activity of a given GLUT, we have over-expressed specific GLUTs in a cellular background devoid of the other major endogenous GLUTs. Additionally, in cells cultured in the presence of glucose, inhibiting glucose uptake with specific inhibitors should decrease both the protons produced by glycolysis and lactate production, leading to a reduction in ECAR. One caveat to keep in mind is that OXPHOS produces CO_2 that can also change the pH of the media and impact ECAR [10]. To further tie the ECAR to glycolysis one can therefore include treatments with the ATP synthase inhibitor, oligomycin. Oligomycin inhibits mitochondrial ATP production and shifts the reliance on cellular energy production to glycolysis, resulting in further increase in the ECAR. The impact of testing a GLUT inhibitor on maximum glycolytic capacity (which eliminates the impact of OXPHOS-derived CO_2 on pH) can better reflect the ability of the inhibitor to impact glycolysis. We thus propose that measuring ECAR and/or glycolytic capacity using the Seahorse XF Analyzer can be utilized to evaluate GLUT inhibition.

We verified this method with a GLUT1 inhibitor phloretin [11] and a GLUT4 inhibitor (compound **20**) developed in our laboratory [12, 13]. We have combined the use of GLUT knockdown and GLUT over-expressing cell lines to increase sensitivity of this platform for assessment of selective GLUT inhibition. This assay can be performed in 2 days and requires low cell numbers. Four injection ports also provide the option to further interrogate the metabolic readouts by introduction of additional inhibitors or supplements.

2 Materials

1. Seahorse XFe96 Analyzer.
2. Incubators: CO_2 and non-CO_2; 37 °C.
3. Inverted-phase contrast microscope.
4. pH meter.
5. XFe96 FluxPAK.
6. Seahorse XF Calibrant.
7. Seahorse XF Base Medium.
8. Oligomycin solution: 20 mM Oligomycin.
9. 2-Deoxy-D-glucose (2DG) solution: 1 M 2DG.
10. RPMI1640 medium: RPMI1640 supplemented with 10 % v/v FBS and 1 % v/v Pen/Strep.
11. Minimum essential medium (MEM): MEM supplemented with 10 % v/v FBS and 1 % v/v Pen/Strep.
12. Trypsin.
13. Glucose: 1 M Glucose solution in water.
14. Phloretin: 20 mM Phloretin solution in ethanol.
15. Compound **20**: 50 mM Compound **20** solution in DMSO.
16. P10 and P200 pipettes and tips.
17. A549 cells: A non-small cell lung cancer cell line.
18. HEK G1KD cells: HEK293 cells with GLUT1 knockdown.
19. HEK G1OE cells: HEK293 cells overexpressing human GLUT1.
20. HEK G1KDG4OE cells: HEK293 cells overexpressing human GLUT4 and endogenous GLUT1 knockdown.

3 Methods

3.1 The Day Prior to Assay

3.1.1 Seeding Cells in XF96 Cell Culture Microplates

1. Harvest cells from the culture plate/flask using trypsin and count cells in suspension. We used four cell lines HEK G1KD, HEK G1OE, HEK G1KDG4OE, and A549 (*see* **Note 1**).
2. Take the desired number of cells (based upon the number of wells to be plated) and resuspend in growth medium. For HEK293 cells, 15,000 cells are suspended in 80 μL MEM medium/well; for A549, 12,000 cells are suspended in 80 μL RPMI medium/well.
3. Seed 80 μL of cell suspension per well. Apply medium only (without cells) in the four corner wells of the plate to use for background measurements.

4. Keep plate at room temperature in the cell culture hood for 1 h to promote even cell distribution and reduce edge effects.
5. Transfer plate to a 37 °C incubator with 5 % CO_2 and allow cells to incubate overnight. Monitor growth and health of the cells using a microscope.

3.1.2 Hydrating the XF Sensor Cartridge

1. Place the sensor cartridge upside down next to the utility plate.
2. Add 200 μL of Seahorse Calibrant to each well of the utility plate.
3. Place the XF Sensor Cartridge on top of the utility plate, wrap with parafilm and place in a non-CO_2 incubator at 37 °C overnight.

3.2 Day of Assay

For one plate prepare 200 mL of assay medium:

3.2.1 Preparing Assay Medium

1. Warm 200 mL Seahorse XF Base Medium to 37 °C.
2. Add 2 mL glucose solution to a final concentration of 10 mM.
3. Adjust pH to 7.35 +/− 0.05 using 1 M NaOH.
4. Filter sterilize with a 0.2 μM filter.
5. Keep the assay medium at 37 °C.

3.2.2 Preparing Compounds for Injection

1. Add 80 μL of phloretin solution to 2 mL assay medium (*see* **Note 2**).
2. Add 3.2 μL of compound **20** to 1 mL assay medium (*see* **Note 2**).
3. Prepare medium with vehicle only as control. 80 μL Ethanol in 2 mL assay medium (for phloretin) and 3.2 μL DMSO in 1 mL assay medium (for compound **20**).
4. Add 5.4 μL oligomycin solution to 4 mL assay medium.
5. Prepare 2DG solution in sterile cell culture water.

3.2.3 Performing the Medium Exchange

1. Remove all but 20 μL of the culture medium from each well.
2. Rinse cells two times with 200 μL of assay medium (*see* **Note 3**).
3. Add 155 μL assay medium to each well for a final volume of 175 μL/well.
4. Place the plate in an incubator at 37 °C without CO_2 for 1 h prior to the assay.

3.2.4 Loading Injection Ports

1. The injection volume is 25 μL. For port A, assay medium with either vehicle or phloretin was loaded for HEK G1OE and A549 cells, and assay medium with either vehicle or compound **20** was loaded for HEK G1KDG4OE cells. Oligomycin was loaded for all port B and 2DG was loaded for all port C wells. For each group, there were at least five replicates (*see* **Note 4**).
2. Once all compounds have been loaded, transfer the cartridge to the Seahorse XF Analyzer to start calibration.

3.2.5 Running the Assay

1. Open the XF software.
2. Create a new experiment.
3. Enter the information for each group and injection port.
4. Edit protocol as follows: three measurement cycles (3 mins Mix, 0 min Wait, 3 mins Measure) – port A injection – three measurement cycles (3 mins Mix, 3 mins Wait, 3 mins Measure) – port B injection – (3 mins Mix, 0 min Wait, 3 mins Measure) – port C injection – (3 mins Mix, 0 min Wait, 3 mins Measure).
5. Place the cartridge and calibration plate with loaded injection ports on the sliding tray.
6. Click Continue to start calibration.
7. When prompted, replace the calibration plate with the cell plate.
8. Continue to run.
9. When the run is over, follow the prompts in the software and remove the cartridge and cell plate and discard.

3.2.6 Analyzing the XF Glycolysis Stress Test Data

Data can be analyzed in Wave Desktop Software provided by Seahorse Bioscience (http://www.agilent.com/en-us/support/cell-analysis-(seahorse)/seahorse-xf-software).

1. The glycolytic capacity is determined by subtracting non-glycolytic acidification (the extracellular acidification rate (ECAR) measured after injecting 2DG) from the ECAR measured after injecting oligomycin. As shown, overexpressing either GLUT1 or GLUT4 in HEK293 cells elevates glycolytic capacity (*see* Fig. 6.1).
2. Treatment with phloretin or compound **20** (a GLUT4 selective inhibitor) effectively decreases glycolytic capacity in cells overexpressing GLUT1 or GLUT4, respectively. The glycolytic capacity of A549 lung cancer cells is also suppressed upon treatment with phloretin.

4 Notes

1. If other cell lines are used, cell seeding number and concentration of oligomycin need to be optimized. Cell seeding numbers are typically between 5,000 and 100,000 per well and the concentration of oligomycin is usually between 0.5 and 5 μM. The optimal cell seeding number should meet these criteria: minimum ECAR of 20 mpH/min, should lie within the linear portion of ECAR/cell seeding number graph.

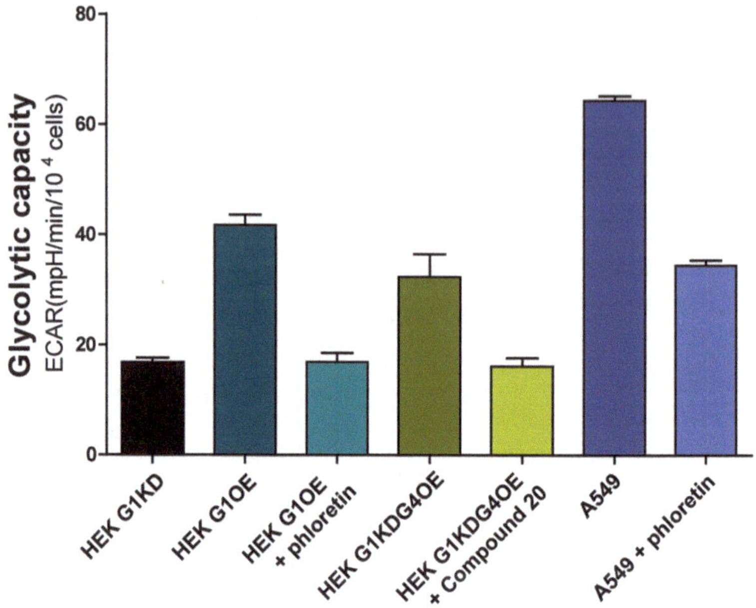

Fig. 6.1 Glycolytic capacity of HEK293 cells and A549 cells treated with GLUT inhibitors. HEK G1KD: HEK293 cells with GLUT1 knockdown. HEK G1OE: HEK293 cells overexpressing human GLUT1. HEK G1KDG4OE: HEK293 cells overexpressing human GLUT4 with GLUT1 knockdown. The final concentration of GLUT1 inhibitor phloretin is 100 μM and 20 μM for GLUT4 inhibitor Compound **20**

2. Compounds may change the pH of the solution. Make sure that injected compounds are at pH 7.35 +/− 0.05 before loading.
3. Add medium gently to avoid cell detachment. Do not touch the bottom of the plate. Look at cells under microscope to ensure that cells are not washed away.
4. Load vehicle or compound in the ports being used for background measurements. During loading, the hydrated XF sensor cartridge must remain in the utility plate. Do not lift, angle, or tap the cartridge.
5. For more details on setting up the assay, please refer to Agilent Seahorse Bioscience website.

References

1. Metallo CM, Vander Heiden MG (2013) Understanding metabolic regulation and its influence on cell physiology. Mol Cell 49(3):388–398. https://doi.org/10.1016/j.molcel.2013.01.018
2. Vander Heiden MG, Cantley LC, Thompson CB (2009) Understanding the Warburg effect: the metabolic requirements of cell proliferation. Science 324(5930):1029–1033. https://doi.org/10.1126/science.1160809
3. Manolescu AR, Witkowska K, Kinnaird A, Cessford T, Cheeseman C (2007) Facilitated hexose transporters: new perspectives on form and function. Physiology (Bethesda)

22:234–240. https://doi.org/10.1152/physiol.00011.2007

4. Thorens B, Mueckler M (2010) Glucose transporters in the 21st century. Am J Phys Endocrinol Metab 298(2):E141–E145. https://doi.org/10.1152/ajpendo.00712.2009
5. Deng D, Xu C, Sun P, Wu J, Yan C, Hu M, Yan N (2014) Crystal structure of the human glucose transporter GLUT1. Nature 510(7503):121–125. https://doi.org/10.1038/nature13306
6. Bryant NJ, Govers R, James DE (2002) Regulated transport of the glucose transporter GLUT4. Nat Rev Mol Cell Biol 3(4):267–277. https://doi.org/10.1038/nrm782
7. Macheda ML, Rogers S, Best JD (2005) Molecular and cellular regulation of glucose transporter (GLUT) proteins in cancer. J Cell Physiol 202(3):654–662. https://doi.org/10.1002/jcp.20166
8. McBrayer SK, Cheng JC, Singhal S, Krett NL, Rosen ST, Shanmugam M (2012) Multiple myeloma exhibits novel dependence on GLUT4, GLUT8, and GLUT11: implications for glucose transporter-directed therapy. Blood 119(20):4686–4697. https://doi.org/10.1182/blood-2011-09-377846
9. Divakaruni AS, Paradyse A, Ferrick DA, Murphy AN, Jastroch M (2014) Analysis and interpretation of microplate-based oxygen consumption and pH data. Methods Enzymol 547:309–354. https://doi.org/10.1016/B978-0-12-801415-8.00016-3
10. Mookerjee SA, Goncalves RL, Gerencser AA, Nicholls DG, Brand MD (2015) The contributions of respiration and glycolysis to extracellular acid production. Biochim Biophys Acta 1847(2):171–181. https://doi.org/10.1016/j.bbabio.2014.10.005
11. Afzal I, Cunningham P, Naftalin RJ (2002) Interactions of ATP, oestradiol, genistein and the anti-oestrogens, faslodex (ICI 182780) and tamoxifen, with the human erythrocyte glucose transporter, GLUT1. Biochem J 365(Pt 3):707–719. https://doi.org/10.1042/BJ20011624
12. Mishra RK, Wei C, Hresko RC, Bajpai R, Heitmeier M, Matulis SM, Nooka AK, Rosen ST, Hruz PW, Schiltz GE, Shanmugam M (2015) In silico modeling-based identification of glucose transporter 4 (GLUT4)-selective inhibitors for cancer therapy. J Biol Chem 290(23):14441–14453. https://doi.org/10.1074/jbc.M114.628826
13. Wei C, Bajpai R, Sharma H, Heitmeier M, Jain AD, Matulis SM, Nooka AK, Mishra RK, Hruz PW, Schiltz GE, Shanmugam M (2017) Development of GLUT4-selective antagonists for multiple myeloma therapy. Eur J Med Chem 20;139:573–586. https://doi.org/10.1016/j.ejmech.2017.08.029

Chapter 7

Glucose Transport Activity Measured in Giant Vesicles

Jesper S. Hansen and Karin Lindkvist-Petersson

Abstract

Incorporation of membrane proteins and internal reporter systems directly into giant vesicles, during their formation from a hydrogel surface, has emerged as a promising new concept in membrane protein characterization. Here, we provide the detailed protocol for a glucose transporter activity assay based on giant vesicles containing a fluorescent enzyme-linked reporter system internally. This assay is applicable for the functional analysis of a variety of hexose-transporting proteins. We furthermore believe that it can aid in the development of drugs targeting hexose transporters.

Key words Glucose transporters, GLUT1, Glucose transport, Membrane protein, Activity assay, Fluorescence microscopy, Resorufin, Giant vesicles

1 Introduction

Glucose transporters (GLUTs) are energy-independent solute carriers that mediate bidirectional diffusion of monosaccharides across the biological cell membrane [1]. Besides their central role in maintaining glucose homeostasis, overexpression and altered activity of GLUTs have been associated with various cancer types (reviewed in [2, 3]) and also play a role in the pathogenesis of diabetes [4–6]. Still, this protein class is underrepresented as drug targets in treatment of disease [7]. This is largely attributed to general lack of functional assays suitable for drug screening [7]. Such activity assays do not necessarily need to retain the biological activity that is characteristic of the protein in its natural environment, but rather enable investigation of protein activity in a reliable manner in vitro.

Current techniques for screening the activity of membrane proteins use cultured cells, or artificial lipid bilayers as simplified models of biological membranes. The use of artificial lipid bilayers in early-stage drug development can potentially save both time and cost by eliminating the need to grow and maintain cells in culture [7].

Karin Lindkvist-Petersson and Jesper S. Hansen (eds.), *Glucose Transport: Methods and Protocols*, Methods in Molecular Biology, vol. 1713, https://doi.org/10.1007/978-1-4939-7507-5_7,

Previously, artificial lipid bilayers with incorporated proteins have been troublesome to prepare, primarily due to challenges with detergent removal and exchange between amphiphiles [8]. Furthermore, it is often not straightforward to incorporate such bilayers into devices capable of measuring membrane protein activity. Activity measurements of membrane proteins transporting uncharged soluble molecules present a particular challenge, because the movement of such solutes is not straightforwardly measured. Methods that could overcome the present challenges would represent an important step toward a better understanding of membrane proteins and the identification of potential new drugs for the treatment of disease.

An emerging new concept in reconstituting membrane proteins into artificial lipid bilayers is the swelling of giant vesicles from a membrane protein-containing hydrogel substrate [9–14]. This approach enables direct reconstitution of membrane proteins during the lipid bilayer formation process (*see* Fig. 1a). In this protocol, purified membrane proteins are dissolved in a molten agarose hydrogel, which is then spread onto a solid substrate and then partially dehydrated. Lipids of the desired membrane composition are spread as a thin film on top of the partially dehydrated protein-containing agarose hydrogel. Subsequent rehydration by addition of aqueous buffer causes spontaneous swelling of the lipid-protein matrix and formation of protein-incorporated giant vesicles (*see* Fig. 1b) [13]. The formed protein-incorporated giant vesicles can then be harvested for biochemical or biophysical analyses (*see* Fig. 1c).

A key advantage with this approach is that the final detergent concentration in the rehydrated system is significantly lower than the critical micelle concentration (CMC) of the detergent used to solubilize the membrane protein. This means that protein assays can be performed without the need to carry out further detergent removal steps.

Secondly, the protocol reliably produces protein-incorporated giant vesicles with good yields and at a variety of ionic strength buffers and lipid compositions [15–17].

In our experience, another clear advantage with this approach is the ease of which protein incorporated lipid bilayers are produced. The protocol is, unlike most other giant vesicle formation methods, straightforward to set up in any laboratory without requiring extensive knowledge of lipid biophysics or membrane protein-handling skills. It only requires low picomolar quantities of purified membrane proteins [13], but the protocol can even be used directly on membrane preparations from cell lines expressing high levels of protein, which thereby circumvents the requirement of protein purification [10, 11].

We have used this system to build a functioning artificial cell model that is able to take up glucose and process it (*see* Fig. 2) [12].

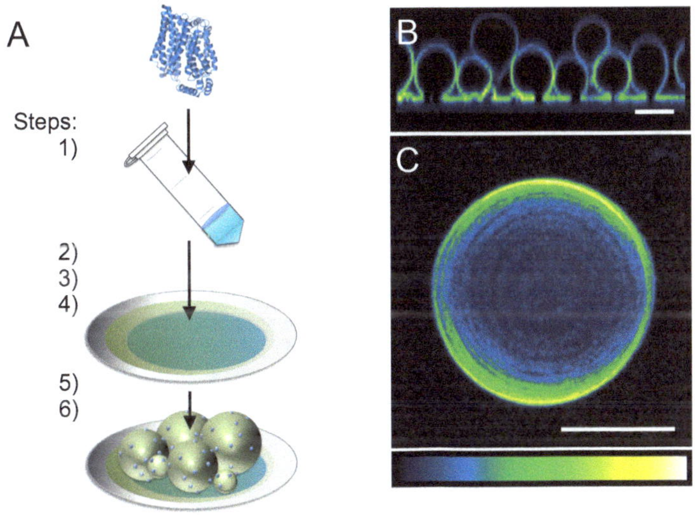

1) Dilute GLUT proteins into molten agarose.
2) Spread a thin layer of GLUT-agarose gel on a coverslip.
3) Partially dehydrate the gel.
4) Deposit lipids on top.
5) Add aqueous buffer and allow the giant vesicles to form.
6) Harvest the formed giant vesicles.

Fig. 1 Reconstitution of GLUT1 into giant vesicles. (**a**) Simplified scheme of the approach for hydrogel-assisted formation of GLUT-incorporated giant vesicles. (**b**, **c**) False-color fluorescence micrographs showing (**b**) Rehydrating a DPhPC lipid-coated partially dried hydrogel film containing pure ATTO565-labeled GLUT1, and (**c**) Harvested GLUT1-incorporated giant vesicle. Scale bars are 10 μm; normalized min–max fluorescence intensity is shown by color calibration bar

By formation of giant vesicles reconstituted with the facilitative glucose transporter GLUT1 and a glucose oxidase and hydrogen peroxidase-linked fluorescent reporter internally, we have demonstrated that facilitated hexose transport can be coupled to an interior multienzymatic reaction yielding a quantifiable fluorescent signal when glucose is internalized [12]. Here, we provide a detailed protocol for setting up this activity assay for GLUT1. In principle, the glucose transporter can be freely exchanged to other membrane proteins and the internal reporter system can be replaced to measure the uptake of a variety of other solutes. Thus, this protocol is by no means restricted to hexose-transporting proteins.

2 Materials

Prepare all aqueous buffers and stock solutions using ultrapure water (18.2 MΩ.cm). Analytical grade reagents are used throughout and without further purification. Diligently follow all safety regulations for handling chemicals and solvents.

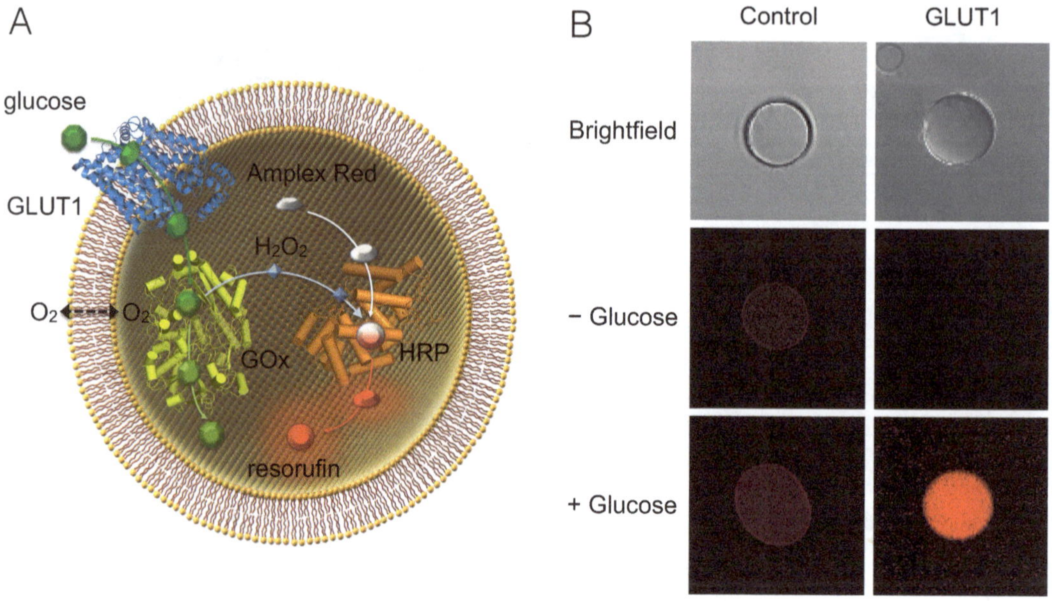

Fig. 2 Assay principle and expected results. (**a**) Illustration showing the principle of the assay. Facilitated glucose transport (GLUT1; *blue*) first leads to conversion of the glucose (*green*) to gluconolactone (*lighter green*) and hydrogen peroxide (*blue diamond*) by GOx (*yellow*). Subsequently, the nonfluorescent Amplex Red (*grey*) is converted to fluorescent resorufin (*red glowing*) in the presence of HRP (*orange*) and hydrogen peroxide. (**b**) Functional incorporation of GLUT1 and reconstitution of active enzymes HRP and GOx elicit fluorescent resorufin inside vesicles following addition of glucose (final conc. 1 mM). In contrast, giant vesicles without GLUT1 (Control) do not develop any notable fluorescence following glucose addition. Adapted from ref. [12]

2.1 Giant Vesicle Chamber Preparation Materials

1. Ultrasonic bath.
2. Handheld corona treater (Electro-Technic Inc.) or benchtop plasma cleaner (Harrick Plasma).
3. Wash-N-Dry™ coverslip rack (Sigma-Aldrich).
4. Glass coverslips: round 25 mm diameter, type 1.0 thickness.
5. Glass beaker: 100 mL.
6. Sterile filters: 0.22 μm syringe filters.
7. Disposable plastic syringes: 2, 10, and 20 mL.
8. Cell chambers: AttoFluor cell chambers for microscopy (Thermo Fisher Scientific) or Sykes-Moore chambers (Bellco Glass).
9. Sodium hydroxide (NaOH) solution: 1 M NaOH. To make a 1 M NaOH solution, dissolve 4.0 g of sodium hydroxide in 100 mL ultrapure water. Notice that NaOH is corrosive and has to be handled accordingly.
10. Glass-bottom dishes with chamber micro-insert wells, e.g., 2- to 4-well inserts (Ibidi) (*see* **Note 1**).
11. Biotin-PEG-Silane solution: 25 mg/mL in ethanol. We use biotin-PEG-silane with a molecular weight of 3400 kDa

(Laysan Bio Inc.). Prepare the solution of biotin-PEG-silane by dissolving 25 mg biotin-PEG-silane powder in 1 mL 96 % EtOH.

12. Avidin solution: 1 mg/mL in ultrapure water. Prepare the avidin solution by dissolving 1 mg lyophilized avidin in 1 mL ultrapure water.
13. Bovine serum albumin (BSA) solution: 1 mg/mL in ultrapure water. We use essentially fatty acid-free BSA. Filter the solution through a 0.22 μm syringe filter.

2.2 Buffers, Lipids, and Reagents

1. Glassware: Volumetric flasks of 10 and 50 mL; glass beakers with 100 mL capacity; 2 mL glass amber vials with PTFE screw top.
2. Magnetic stirrer.
3. Heat block.
4. Plastic tubes: Standard 1.5 mL Eppendorf tubes and 15 mL Falcon tubes.
5. Hamilton syringe: 100 μL capacity.
6. Purified GLUT proteins (*see* **Note 2**).
7. Phosphate-buffered saline (PBS), 10× concentrate: 1.37 M NaCl, 27 mM KCl, 100 mM Na_2HPO_4, 18 mM KH_2PO_4, pH 7.4.
8. Raffinose-PBS buffer solution: 100 mM raffinose in PBS. First prepare a 0.2 M raffinose stock solution (*see* **Note 3**). Dissolve 5.95 g D-(+)-raffinose pentahydrate in ultrapure water and adjust to a final volume of 50 mL using volumetric flasks. Filter the solution through 0.22 μm filters. Then make the raffinose-PBS buffer solution, which consists of 100 mM raffinose in 1× PBS, by adding 5 mL 0.2 M raffinose stock solution and 1 mL 10× PBS to a 15 mL Falcon tube and adjust to a final volume of 10 mL with 4 mL ultrapure water. Degas the buffer solution by bubbling it through with N_2 gas or use ultrasonic bath and keep on ice until use (*see* **Note 4**).
9. KCl-PBS buffer solution: 67.8 mM KCl in PBS. Start by preparing 1 M KCl stock solution. Dissolve 3.73 g KCl in ultrapure water. Adjust the solution to a final volume of 50 mL using volumetric flasks. Filter the solution through 0.22 μm filters. To prepare a solution of 67.8 mM KCl in 1× PBS, add 0.678 mL of the 1 M KCl and 1 ml of the 10× PBS solutions to a 15 mL Falcon tube and adjust to a final volume of 10 mL with ultrapure water (8.322 mL). The KCl-PBS solution is osmotically balanced to the raffinose-PBS solution (*see* **Note 5**). Degas the buffer solution by bubbling it through with N_2 gas or use ultrasonic bath and keep on ice until use.
10. Lipid solution: 10 mM 1,2-diphytanoyl-*sn*-glycero-3-phosphocholine (DPhPC), 0.1 mM 1,2-dipalmitoyl-*sn*-glycero-3-

phosphoethanolamine-N-(cap biotinyl) (DPPE-biotin) in chloroform (*see* **Note 6**). Prepare 10 mg/mL DPhPC and 1 mg/mL DPPE-biotin stock solutions in $CHCl_3$, respectively (*see* **Note 7**). To a 2 mL glass amber vial, mix 84.6 μL DPhPC with 10.5 μL DPPE-biotin. Evaporate the solvent using a gentle stream of nitrogen gas. Dissolve the dried lipids in 100 μL $CHCl_3$. It can be stored at −20 °C for several weeks. The lipid solution should be equilibrated to room temperature prior to its use.

11. Molten agarose solution: 1 % w/v agarose type IX ultra-low gelling temperature in ultrapure water. Prepare the agarose solution by dissolving 5 mg agarose powder in 500 μL ultrapure water in a 1.5 mL Eppendorf tube. Mix vigorously by pipetting up and down several times to bring the agarose in suspension. Place the Eppendorf tube with the agarose suspension in a dry block heater at 80 °C for ~10 min. Mix the agarose suspension frequently until the agarose is completely melted. Cool the agarose solution to 25 °C and maintain it at this temperature until use. Agarose type IX ultra-low gelling temperature has a gel temperature ≤ 18 °C and will stay molten at or above this temperature. Please also note that special storage conditions, in our experience, apply for dry powder type IX ultra-low gelling temperature agarose (*see* **Note 8**).
12. Glucose solution: 1 M glucose stock solution. Dissolve 0.18 g D-(+)-glucose in 1 mL KCl-PBS buffer solution.
13. Amplex Red glucose/glucose oxidase assay kit: Dissolve and handle all assay components according to the manufacturer's instructions. Accordingly, the following stock solutions are prepared: 10 mM Amplex® Red (i.e., 10-acetyl-3,7-dihydroxyphenoxazine), 10 U/mL horseradish peroxidase (HRP), and 100 U/mL glucose oxidase (GOx) (*see* **Note 9**).
14. Rehydration buffer (with glucose reporter system): 50 μM Amplex® Red reagent, 0.2 U/mL HRP, and 2 U/mL GOx in raffinose-PBS buffer solution. Immediately before use, mix 5 μL of 10 mM Amplex® Red reagent stock solution, 20 μL of 10 U/mL HRP stock solution, and 20 μL of 100 U/mL glucose oxidase stock solution with 955 μL of raffinose-PBS buffer solution. Keep the solution on ice in the dark.
15. GLUT inhibitors, i.e., cytochalasin B, WZB-117, or other relevant test compounds.

2.3 Assay Tools and Data Analysis Software

1. Inverted fluorescence microscope: Epifluorescence, spinning disc, or confocal laser scanning microscope all work for this assay.
2. Appropriate fluorescence excitation light source around 561 nm. It can be a xenon lamp, LED excitation source, or any appropriate laser line.

3. Filter set appropriate for resorufin fluorescence, i.e., rhodamine filter set.
4. Detection appropriate for fluorescence: CCD camera or PMT detector.
5. Objectives: 20× and 40×.
6. ImageJ open-source software for data analysis.

3 Methods

3.1 Chamber Preparations for Giant Vesicles

1. Clean glass coverslips: Gently place the coverslips in the ceramic coverslip rack. Submerge the rack with the coverslips into a 100 mL glass beaker containing a 1 M NaOH solution. Place the beaker with the coverslips into the ultrasonic bath, and clean for 30 min. Thoroughly rinse the coverslips in ultrapure water. Submerge the coverslips in a beaker with ultrapure water, subsequently place the beaker back into the ultrasonic bath, and clean the coverslips for another 30 min. Take out the coverslip rack and place it onto a dry paper towel and air-dry the coverslips using a gentle air stream, or by placing the cleaned coverslips in an oven set to 60 °C.
2. Clean the cell chambers: Clean the chambers to be used by placing them in a glass beaker containing 1 M NaOH and clean them using an ultrasonic bath for 30 min. Rinse the chambers three times with ultrapure water and dry them.
3. Biotinylation of glass-bottom dishes with micro-insert wells: Add 0.1 mL of the biotin-PEG-silane solution to the micro-inserts of the glass-bottom dishes and incubate for 60 min. Rinse carefully with ultrapure water (*see* **Note 10**).
4. Precondition the micro-insert wells of the petri dishes with avidin: Add 0.1 mL of the avidin solution to the chamber micro-inserts of the glass-bottom dishes and incubate for 30 min. Rinse with ultrapure water.
5. Albumin surface treatment of glass-bottom dishes with micro-insert wells: Add 0.1 mL of the BSA solution to the chamber micro-inserts and incubate for minimum 60 min (*see* **Note 11**). It is a good idea to leave the BSA solution in the chamber inserts until immediately before use, and then gently wash the chamber inserts one time with ultrapure water.

3.2 Preparation of GLUT-Containing Agarose Hydrogel for Giant Vesicle Formation

1. Place two clean glass coverslips on a clean surface. One will be used for depositing the molten agarose and the other will be used to evenly spread the agarose on the depositing glass surface. If a handheld corona treater tool is used proceed to the following step. If a benchtop plasma etcher is used, **steps 1** and **2** can be performed in reverse order (*see* **Note 12**).

2. Plasma etch treat the coverslips with the handheld corona treater to render the glass surface hydrophilic (*see* **Note 13**). Set the plasma etch tool to medium intensity using the screw knob on the top end of the plasma treater. Slowly move the plasma treater over the glass coverslips to ensure even etching of the glass surfaces. Perform plasma etching for 1–2 min. The treatment will last several minutes.

3. Dilute purified detergent-stabilized GLUT proteins into the molten agarose solution. Typically, the protein concentration we use ranges from 1 to 10 mg/mL and the protein is diluted 5–10 times into the molten agarose solution (*see* **Note 14**). For example, transfer 40 μL of molten agarose to a new 1.5 mL Eppendorf tube and add 10 μL of GLUT protein solution to the molten agarose suspension. Gently mix by pipetting. Do not vortex.

4. Deposit 10 μL of the still molten protein–agarose solution onto the middle of a clean and plasma etch-treated coverslip (*see* Fig. 3a).

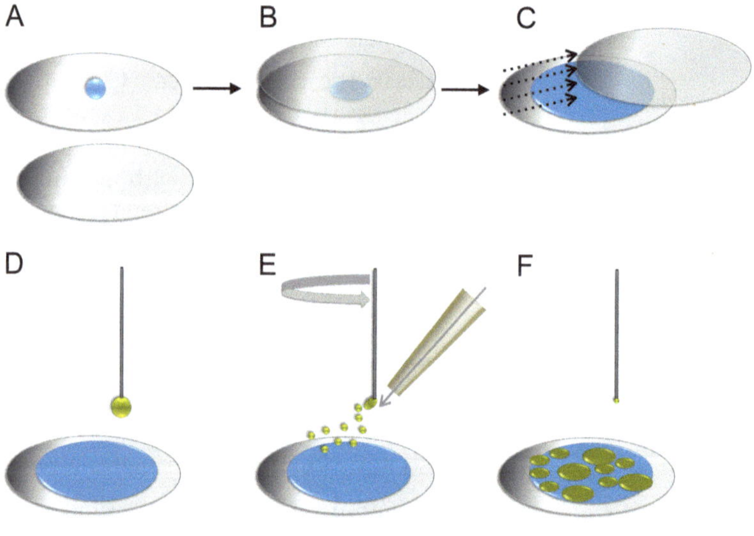

Fig. 3 Deposition of molten GLUT1-agarose and lipids onto a glass surface. Illustration of the protocol steps for deposition of protein–agarose gel and lipids: (**a**) Place two, plasma etch-treated, coverslips on a clean surface. Add 10 μL of molten GLUT1-agarose to the center of one coverslip. (**b**) Invert the other coverslip, such that the clean hydrophilic surfaces of the two coverslips are facing each other. (**c**) In a swift motion, slide the top coverslip across the bottom one. This creates a thin and evenly distributed layer of the GLUT1-agarose gel on the coverslip. Partially dehydrate the gel. (**d**) With a Hamilton pipette let a small drop of lipid solution hang from the syringe needle. (**e**) With a stream of N_2 gas (or air) gently blow this drop onto the partially dried protein–agarose surface as a mist. (**f**) Repeat this procedure until the syringe is emptied. Ensure to evaporate all solvent before adding aqueous solution to form the giant vesicles

5. Take the other plasma etch-treated coverslip and place upside down, such that the two treated hydrophilic glass surfaces face each other (*see* Fig. 3b).
6. Spread the protein–agarose solution across the surface evenly by swiping the top coverslip across (*see* Fig. 3c).
7. Place the protein–agarose-deposited coverslip in the cell chamber. Do not screw on the top-half of the chamber.
8. Allow the thin protein–agarose film to gel by cooling it to 4 °C in the refrigerator for ~5–10 min.
9. Partially dehydrate the thin protein–agarose gel by applying a stream of air across. It is important that the coverslip appears completely dry to the naked eye. The giant vesicles will not form properly if the surface still looks wet. High water content will remain in the gel, even when dehydrated in this manner (*see* **Note 15**).
10. Fill a Hamilton syringe with 10 μL of the lipid solution.
11. A thin layer of lipid is pipetted onto the partially dried protein–agarose gel in a N_2 gas stream to immediately evaporate the solvent. Let a small drop of lipid hang from the Hamilton syringe needle held vertically (*see* Fig. 3d). With a stream of N_2 gas (or air), gently blow this drop onto the dried protein–agarose surface as a mist (*see* **Note 16** and *see* Fig. 3e). Repeat this procedure until the syringe is emptied (*see* Fig. 3f).
12. Dry the surface in a stream of N_2 gas (or air). Again, it is important that the lipid-gel surface appears completely dry to the naked eye.
13. Fully assemble the cell chamber, i.e., screw on the top half of the chamber assembly.

3.3 Formation of GLUT-Containing Giant Vesicles with Internal Reporter System

1. Rehydrate the lipid-hydrogel film with the rehydration buffer containing the glucose reporter system. Typically, a volume of 400 μL is added to the dehydrated lipid-hydrogel film. This results in the spontaneous swelling of the lipid-hydrogel layer and formation of giant vesicles.
2. Cover the chambers to protect from light, i.e., with aluminum foil (*see* **Note 17**).
3. Allow the giant vesicle formation process to proceed for ≥30 min at 4 °C in a refrigerator.
4. Harvest the resulting giant vesicles by aspirating the rehydration medium and diluting it fivefold into KCl-PBS buffer solution in a 15 mL Falcon tube. The sugar-salt density gradient causes the giant vesicles to settle on the bottom of the Falcon tube. Use wide bore or cut pipette tips in this step.

5. Allow the vesicles to settle in the 15 mL Falcon tube placed on ice.
6. Transfer 100 μL of the vesicle suspension from the bottom of the Falcon tube to the micro-inserts of the glass-bottom petri dishes preconditioned with biotin-PEG-silane, avidin, and BSA. Again, use wide bore or cut pipette tips in this step.
7. Allow the vesicles to settle on the coverslip surface before imaging.

3.4 Running the GLUT Glucose Transport Activity Assay in Giant Vesicles

1. Wash the vesicles attached to the bottom of the petri dish with 2.5 chamber volumes of the KCl-PBS buffer solution. Do so by aspirating 50 μL volume from the petri dish micro-insert containing the vesicles and replenish with 50 μL of KCl-PBS buffer solution. Repeat this step until 2.5 chamber volumes have been exchanged.
2. Acquire a background micrograph (fluorescence and brightfield) of the vesicles on the bottom of the petri dish before addition of glucose. Select only those vesicles displaying faint fluorescent background and a clear phase contrast for GLUT activity measurements (*see* Fig. 4 and *see* **Note 18**).
3. If inhibitors are used, add selected chemical compound at desired concentrations and incubate for 5 min prior to the addition of glucose (*see* **Note 19**). For GLUT activity measurements without inhibitor present proceed directly to the next step.
4. Make 10 and 40 mM glucose-working solutions by dilution of the 1 M glucose stock solution with KCl-PBS buffer.
5. Add glucose to the giant vesicle-containing chamber. Typically, D-glucose is added to a final concentration of 1–4 mM.
6. Acquire fluorescence and brightfield micrographs after addition of glucose. For endpoint measurements, incubate the vesicles with glucose for 5 min and then acquire images. For time-lapse experiments, acquire micrographs of the same vesicle(s) for 10 min with a 30-s delay cycle. Initiate data collection one micrograph before addition of glucose, which will function as reference point.

3.5 Data Analysis

1. Open the micrographs in ImageJ. Use the Bio-Formats Import tool and select split channels to separate the fluorescence channel from the brightfield image.
2. Threshold the fluorescence image, using the Auto Threshold function.
3. Use the particle analysis tool to automatically obtain the fluorescence mean intensities of the giant vesicles contained in the micrograph (*see* Fig. 4).

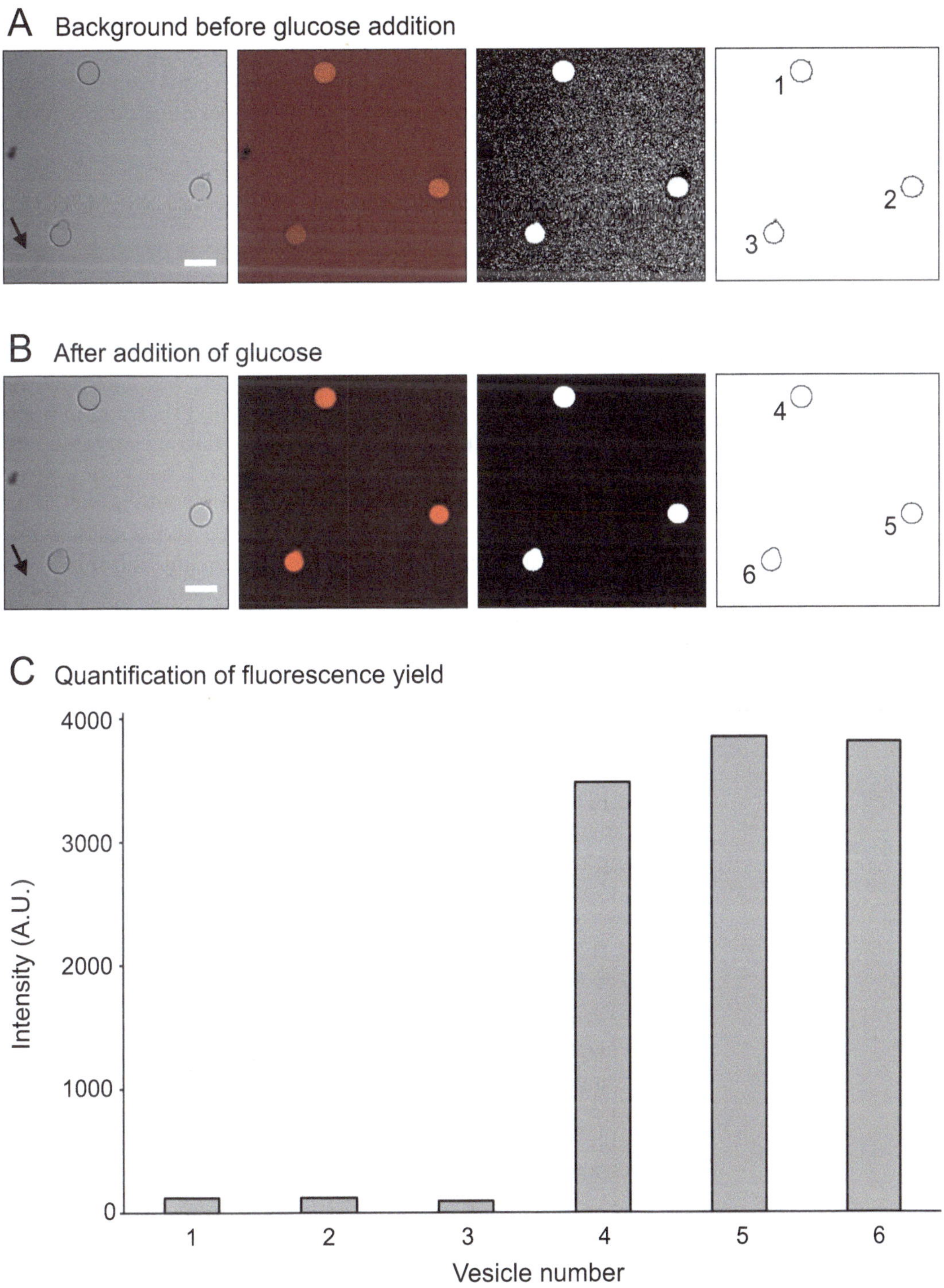

Fig. 4 Example of experimental data quantification. (**a**) Shows how processing of the background, before addition of glucose, typically appears. The fluorescence intensity has been auto adjusted in order to clearly identify vesicles with faint interior fluorescence prior to the addition of glucose. (**b**) Shows the corresponding result after glucose addition. (**c**) Expected outcome after quantification of the fluorescence intensities. *Black arrows* in (**a**) and (**b**) show the appearance of a morphologically inadequate giant vesicle ("ghost vesicle") that should be excluded from the experiment beforehand

4. Save the output Results as ASCII text file.
5. Open the Results file in suitable data analysis software and perform the desired data analysis, e.g., convert raw data to relative fluorescence, perform calculations, and make diagrams, curves, etc.

4 Notes

1. It is possible to make your own chamber micro-insert wells. Materials: Polydimethylsiloxane elastomer (PDMS) and curing agent (Dow Corning, Sylgard 184); 4″ silicon wafer (Ted Pella); Harris Uni-core 8 mm hole-puncher (Ted Pella); Harris cutting mat (Ted Pella), and aluminum foil. Procedure in brief: Wrap a sheet of aluminum foil around the silicon wafer to create a reservoir. Pour PDMS elastomer (35 g) and then curing agent (10:1 weight ratio elastomer:curing agent) into a weighing boat. Mix the two components thoroughly with a spatula. Put the PDMS in a desiccator and apply vacuum until all air bubbles are removed. Pour the PDMS into the silicon wafer wrapped in aluminum foil. Place the PDMS-filled silicon wafer into the desiccator again to remove any residual air bubbles. Cure the elastomer overnight in an oven at 60 °C. When the PDMS is solidified, punch holes with the hole puncher and cut out small squares around the holes with a scalpel. These comprise the custom-made chamber wells. Place a clean glass coverslip and the PDMS on a clean table with the bonding side up. Plasma treat etch both the glass coverslip and the PDMS piece for 30 s using the handheld plasma etch tool. Gently place the PDMS on top of the glass slide with the bonding sides together. Bond overnight in an oven at 60 °C. The prepared coverslips, containg the micro-insert wells, are then placed in AttoFluor or Sykes-Moore cell chambers for imaging.
2. GLUT proteins, purified, from natural sources or as membrane preparations, are required for this protocol. For details about cloning, expression, and purification of glucose transporters please *see* Chapters 1, 2, and 3 in this book. Alternatives to purified protein are membrane preps from transfected overexpression cell lines, as well as small-scale cell-free expression.
3. The sugar density gradient created with osmotically balanced sucrose and glucose buffers is typically used to settle giant vesicles on the coverslip surface for imaging. Sucrose, presumably due to the presence of glucose impurities, triggers the glucose assay reaction producing fluorescent resorufin. In our hands, raffinose does not trigger the glucose assay reaction and is therefore used in combination with an osmotically balanced KCl solution to settle the giant vesicles on the bottom.

4. Proper degassing of buffers is essential. The 10-acetyl-3,7-dihydroxyphenoxazine reagent (also sold under brand names such as Amplex Red, ADHP, A 6550, and Amplifluᵀᴹ Red) is very sensitive to oxidation by air, which creates fluorescent resorufin.
5. The buffer system described, consisting of the raffinose-PBS and KCl-PBS solutions, has been osmotically balanced. It is therefore not necessary to measure buffer osmolarity, provided that the described buffers are used. If other buffers are desired, use an osmometer to measure the buffer osmolarity in order to osmotically balance different buffers.
6. The DPPE-biotin in the lipid mixture creates anchor moieties on the giant vesicles for adhering the vesicles to the glass surface through biotin-avidin-biotin conjugation.
7. Notice that $CHCl_3$ is a harmful volatile solvent. Make sure to perform this procedure with appropriate protection and ventilation, i.e., in a fume hood. Only use Hamilton syringes for solvent transfers.
8. Agarose type IX ultra-low gelling temperature should be stored dry in a vacuum desiccator. Inappropriate storage of the agarose causes leaky giant vesicle membranes in our experience, and may thus lead to deceptive results.
9. 10-Acetyl-3,7-dihydroxyphenoxazine (e.g., Amplex Red) reagent, and GOx and HRP enzymes, can be purchased separately. In this case, prepare stock solutions with concentrations as indicated.
10. Biotinylation of glass surfaces with biotin-PEG-silane creates an anchor for adhering the giant vesicles to the glass coverslip surface through biotin-avidin-biotin conjugation.
11. BSA acts as a cushion for settling giant vesicles. Protein-incorporated giant vesicles are very brittle and will burst on direct contact with the glass surface, if the surface is not treated.
12. Surface treatments using benchtop plasma cleaners last approx. 1 h compared to a few minutes for the handheld model. It is therefore not necessary to plasma treat the coverslip surfaces immediately before adding the molten agarose.
13. The plasma treatment of glass surfaces is applied to make a clean hydrophilic surface. If not applied, it will not be possible to slide a coverslip across another to make an even hydrogel coating on the glass surface.
14. Ensure that the pure GLUT proteins are in a suitable detergent buffer, and with a detergent concentration as close to the detergent's CMC as possible. Typically, we use 1–1.5× detergent CMC in buffers for reconstitution purposes. We recommend the use of dialyzable detergents for reconstituting

membrane proteins into giant vesicles, albeit giant vesicles can be formed with some nondialyzable detergents as well.

15. Work by Horger *et al.* shows that a residual water content of ≥15 % by weight remains in the hydrogel after dehydration in the described manner [15]. If the hydrogel is not sufficiently gelled and dehydrated it will result in low giant vesicle yield and leaky giant vesicle membranes.
16. A stream of N_2 gas (or air) is applied to gently blow lipids, dropwise, onto the dehydrated protein–agarose surface as a mist. The mist should not be too fine or too coarse. An air paintbrush typically creates too fine of a mist, which leads to a low yield of giant vesicles. If a larger (> 1 μL) lipid drop accidentally comes in contact with the protein–agarose gel, promptly blow it across the surface and onto the side of the chamber. This will generally not affect the overall outcome, and thus proceed with the protocol.
17. 10-Acetyl-3,7-dihydroxyphenoxazine (e.g., Amplex Red) is extremely sensitive to light. Thus, protect the reagent from light.
18. Select only vesicles displaying faint fluorescence inside and a clear phase contrast for the activity measurements. This is indicative of non-leaky GLUT-containing vesicles with all required assay components reconstituted. You will find giant vesicles that do not live up to these criteria. We commonly call them "ghost vesicles" as they, due to their weak phase contrast in the microscope, appear as shadows against the diffraction index of the aqueous buffer solution.
19. The assay is suitable for screening and characterization of GLUT inhibitors [12]. Indicated inhibitors only serve as suggestions.

Acknowledgements

We thank Karin Elbing for the production of GLUT1 proteins, and Serena Fortunato for carefully reading the manuscript. The Swedish Research Council (2011-2891), the Cancer Foundation (2010/1171 and 2014/575), and Novo Nordisk Foundation (9807) supported this work.

References

1. Zhao FQ, Keating AF (2007) Functional properties and genomics of glucose transporters. Curr Genomics 8(2):113–128
2. Barron CC, Bilan PJ, Tsakiridis T, Tsiani E (2016) Facilitative glucose transporters: implications for cancer detection, prognosis and treatment. Metabolism 65(2):124–139. https://doi.org/10.1016/j.metabol.2015.10.007
3. Szablewski L (2013) Expression of glucose transporters in cancers. Biochim Biophys Acta

1835(2):164–169. https://doi.org/10.1016/j.bbcan.2012.12.004

4. Hajiaghaalipour F, Khalilpourfarshbafi M, Arya A (2015) Modulation of glucose transporter protein by dietary flavonoids in type 2 diabetes mellitus. Int J Biol Sci 11(5):508–524. https://doi.org/10.7150/ijbs.11241
5. Orci L, Ravazzola M, Baetens D, Inman L, Amherdt M, Peterson RG, Newgard CB, Johnson JH, Unger RH (1990) Evidence that down-regulation of beta-cell glucose transporters in non-insulin-dependent diabetes may be the cause of diabetic hyperglycemia. Proc Natl Acad Sci U S A 87(24):9953–9957
6. Zisman A, Peroni OD, Abel ED, Michael MD, Mauvais-Jarvis F, Lowell BB, Wojtaszewski JF, Hirshman MF, Virkamaki A, Goodyear LJ, Kahn CR, Kahn BB (2000) Targeted disruption of the glucose transporter 4 selectively in muscle causes insulin resistance and glucose intolerance. Nat Med 6(8):924–928. https://doi.org/10.1038/78693
7. Rask-Andersen M, Masuram S, Fredriksson R, Schioth HB (2013) Solute carriers as drug targets: current use, clinical trials and prospective. Mol Asp Med 34(2–3):702–710. https://doi.org/10.1016/j.mam.2012.07.015
8. Hansen JS, Vararattanavech A, Vissing T, Torres J, Emneus J, Helix-Nielsen C (2011) Formation of giant protein vesicles by a lipid cosolvent method. Chembiochem 12(18): 2856–2862. https://doi.org/10.1002/cbic.201100537
9. Gutierrez MG, Jalali-Yazdi F, Peruzzi J, Riche CT, Roberts RW, Malmstadt N (2016) G protein-coupled receptors incorporated into rehydrated diblock copolymer vesicles retain functionality. Small 12(38):5256–5260. https://doi.org/10.1002/smll.201601540
10. Gutierrez MG, Malmstadt N (2014) Human serotonin receptor 5-HT(1A) preferentially segregates to the liquid disordered phase in synthetic lipid bilayers. J Am Chem Soc 136(39):13530–13533. https://doi.org/10.1021/ja507221m
11. Gutierrez MG, Mansfield KS, Malmstadt N (2016) The functional activity of the human serotonin 5-HT1A receptor is controlled by lipid bilayer composition. Biophys J 110(11): 2486–2495. https://doi.org/10.1016/j.bpj.2016.04.042
12. Hansen JS, Elbing K, Thompson JR, Malmstadt N, Lindkvist-Petersson K (2015) Glucose transport machinery reconstituted in cell models. Chem Commun 51(12):2316–2319. https://doi.org/10.1039/c4cc08838g
13. Hansen JS, Thompson JR, Helix-Nielsen C, Malmstadt N (2013) Lipid directed intrinsic membrane protein segregation. J Am Chem Soc 135(46):17294–17297. https://doi.org/10.1021/ja409708e
14. Horger KS, Liu H, Rao DK, Shukla S, Sept D, Ambudkar SV, Mayer M (2015) Hydrogel-assisted functional reconstitution of human P-glycoprotein (ABCB1) in giant liposomes. Biochim Biophys Acta 1848(2):643–653. https://doi.org/10.1016/j.bbamem.2014.10.023
15. Horger KS, Estes DJ, Capone R, Mayer M (2009) Films of agarose enable rapid formation of giant liposomes in solutions of physiologic ionic strength. J Am Chem Soc 131(5):1810–1819. https://doi.org/10.1021/ja805625u
16. Lopez Mora N, Hansen JS, Gao Y, Ronald AA, Kieltyka R, Malmstadt N, Kros A (2014) Preparation of size tunable giant vesicles from cross-linked dextran(ethylene glycol) hydrogels. Chem Commun 50(16):1953–1955. https://doi.org/10.1039/c3cc49144g
17. Peruzzi J, Gutierrez MG, Mansfield K, Malmstadt N (2016) Dynamics of hydrogel-assisted giant unilamellar vesicle formation from unsaturated lipid systems. Langmuir 32(48):12702–12709. https://doi.org/10.1021/acs.langmuir.6b01889

Chapter 8

Design, Synthesis, and Evaluation of GLUT Inhibitors

Carlotta Granchi, Tiziano Tuccinardi, and Filippo Minutolo

Abstract

The Warburg effect describes how most cancer cells exhibit higher-than-normal glucose consumption, not only under hypoxic conditions, but also when normal oxygen levels are present. Although glucose transporter 1 (GLUT1) has been found to play a key role in the cellular uptake of glucose, especially in cancer cells, where it is generally overexpressed, it has not been given consideration as a suitable target for the development of anticancer drugs. In this chapter, an example of molecular design and realization of novel GLUT1 inhibitors, including in silico modeling, chemical synthesis, and biological characterization, is provided. This process started with the identification of a focused series of oxime derivatives, originally designed as estrogen receptor (ER) ligands, which were structurally optimized in order to direct their activity towards GLUT1 and to minimize their binding to the ERs, leading to the production of efficient and selective inhibitors of glucose uptake in cancer cells.

Key words Glucose transporters, Warburg effect, Inhibitors, Cancer, Molecular design, Oximes, 2-NBDG

1 Introduction

The management of glucose in living cells is a crucial process that starts with its uptake inside the cytoplasm through specific transporters. Cancer cells generally overexpress glucose transporters (GLUTs), in order to guarantee a massive influx of glucose to support their high proliferation rates [1]. This is generally explained as the Warburg effect, consisting of a metabolic switch from OXPHOS to glycolysis that tumor cells display even under normoxic conditions [2]. This switch provides cancer cells with increased amount of energetic and anabolic supplies, in order to support their extraordinarily remarkable growth and invasiveness. This justifies the increased expression of GLUTs, which is also exploited by the clinically utilized diagnostic PET imaging, which measures the increased glucose uptake in cancer masses by means of a radioactive glucose analogue (FDG) [3].

Karin Lindkvist-Petersson and Jesper S. Hansen (eds.), *Glucose Transport: Methods and Protocols*,
Methods in Molecular Biology, vol. 1713, https://doi.org/10.1007/978-1-4939-7507-5_8,

Of the several GLUTs present in human cells, GLUT1 is the isoform that is most frequently found to be overexpressed in human tumors, where it is also considered as a negative prognostic factor in terms of clinical outcome, and its expression is directly regulated by hypoxia-inducible factor 1 (HIF-1) [4]. Therefore, the ability to modulate the cellular uptake of glucose by means of inhibitors of GLUTs, in particular of GLUT1, is currently being considered as a promising approach for the development of innovative anticancer therapies, which has so far led to the discovery of several chemical classes of GLUT1 inhibitors [5]. Herein, we report the process of design and development we have followed for the production of GLUT1 inhibitors belonging to the chemical class of salicylketoximes [6, 7].

2 Design and Preliminary Biological Assays

There are several ways to start with the design and development of new biologically active molecules. We started by the observation that most of the GLUT inhibitors reported in literature have some common structural features [5], consisting of central scaffolds of various types (aromatic, heteroaromatic, olefinic, etc.), which are all characterized by the presence of peripheral hydroxyl groups, mostly phenolic (*see* Fig. 1). This observation highlighted an evident similarity with the structure of some estrogen receptor (ER) ligands that we had previously designed and synthesized as selective ERβ ligands [8–10]. In particular, we had previously developed salicylaldoxime and salicylketoxime derivatives (*see* Fig. 1), which possess a hydroxy-substituted six-membered pseudo-ring, formed by an intramolecular hydrogen bond between the nitrogen atom of the oxime portion and the phenolic group of the salicylic moiety, which is supposed to mimic the phenolic ring of 17β-estradiol present in natural estrogens, in order to establish a highly energetic hydrogen bond network with the receptor.

Interestingly, other compounds that bind nuclear receptors, such as some thiazolidinedione derivatives, which were initially developed as peroxisome proliferator-activated receptor γ (PPAR γ) agonists, proved to exert part of their action thanks to their ability to block glucose entry by inhibition of GLUT1 [11]. Furthermore, some flavonoids such as phloretin, which showed a well-established inhibition of GLUT1, also proved to bind to ERs [12]. This interplay between glucose transport by GLUT1 and estrogen receptor modulation in cells [13, 14] prompted us to discover if some of our ER ligands could also be active as GLUT1 inhibitors [6, 7].

In addition to this approach based on structure similarities, we also aimed at building a model of the human GLUT1 by using a homology modeling approach. The deposited structure of XylE

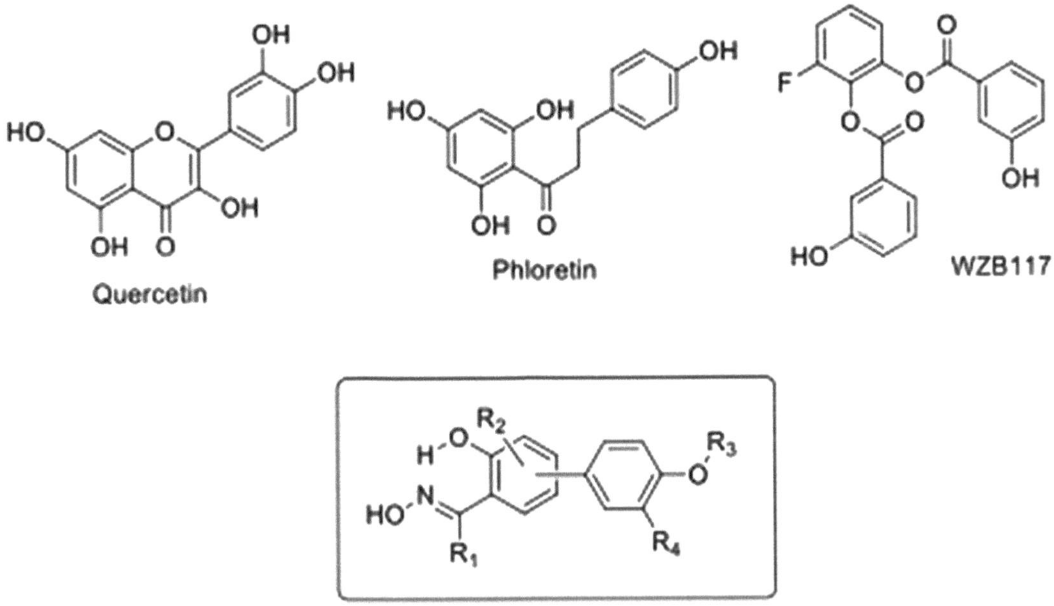

Fig. 1 Comparison of some known GLUT1 inhibitors and the general structure of oxime derivatives

(4gbz PDB code [15]), an *Escherichia coli* homologue of GLUT1-4, showed the best sequence similarity (29 % of identities) and was thus used as a template for developing the model. The human GLUT1 was constructed directly from the coordinates of the corresponding amino acids in XylE. Starting from this protein, ten structures were generated by means of the "very slow MD annealing" refinement method, as implemented in Modeller program [16], and the best receptor model was chosen on the basis of the discrete optimized protein energy (DOPE) assess method. The human GLUT1 model was then complexed with a D-glucose moiety and refined by means of 10 ns of molecular dynamics (MD) simulation in a fully hydrated phospholipid bilayer environment made up of palmitoyloleoylphosphatidylcholine (POPC) molecules solvated by TIP3 water molecules. The minimized average structure of the last 6 ns of the MD simulation was used for the identification of possible binding sites by using the FLAPsite pocket detection algorithm included in the FLAP software [17]. As shown in Fig. 2, this analysis highlighted the presence of one large extracellular cavity (yellow), one pocket in the transmembrane domain (green), and a third cavity in the intracellular region of the receptor (magenta). On the basis of further docking calculations and MD simulations the intracellular cavity seemed to be the most suitable for interacting with GLUT1 inhibitors possessing the aryl-substituted salicylketoxime scaffold, as described below.

A representative series of our oxime derivatives was screened in a standard glucose uptake assay to evaluate their ability to

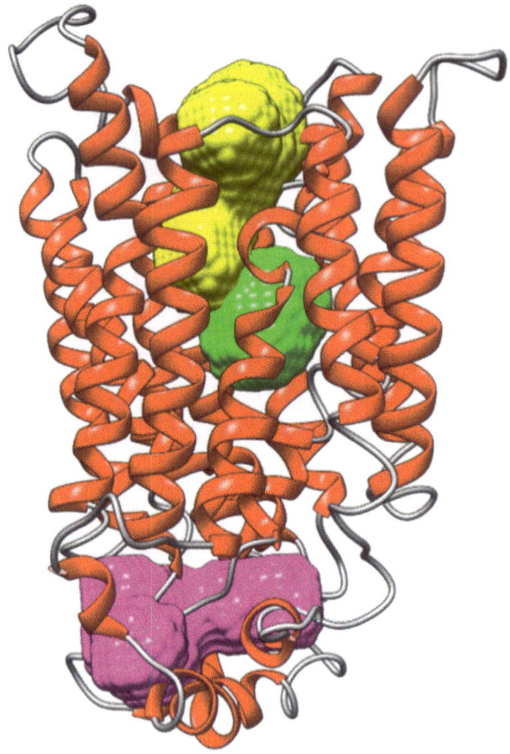

Fig. 2 Analysis of the *h*GLUT-1 model cavities

inhibit glucose transport through GLUT1 and then their antiproliferative activity was tested by a cytotoxicity assay in human non-small-cell lung cancer H1299 cells. The glucose uptake assay consists of measuring the cell uptake of 2-deoxy-D-[^{3}H] glucose after incubation of the cells for 15 min, with the compounds at a final concentration of 30 μM. Glucose uptake was initiated by the addition of 37 MBq/L 2-deoxy-D-[^{3}H] glucose and 1 mM regular glucose as final concentrations. Glucose uptake was terminated by washing the cells with cold phosphate-buffered saline (PBS). Then cells were lysed with 0.2 M NaOH and the radioactivity retained by the cell lysates was measured [18]. A series of oxime derivatives were tested (*see* Fig. 3) and some of them (**1c**, **1f**, and **3a–c**) demonstrated to efficiently reduce the glucose uptake (*see* Fig. 4), displaying activities that are comparable or even better than those of the two reference GLUT1 inhibitors used in this assay, such as phloretin and WZB117. Most of the tested compounds possess a 5-monoaryl-substituted salicylaldoxime scaffold (compounds **1a–g**) with the exception of compound **1b**, which bears an additional aryl ring in position 3 of the central scaffold, compounds **2a–c**, which are 4-aryl-substituted salicylaldoximes, and compounds **3a–c** are 4-aryl-substituted salicylketoximes bearing a chlorine atom in position 3 of the central ring.

Subsequently, these compounds were subjected to an antiproliferative assay in the same cancer cell line. Cell growth and

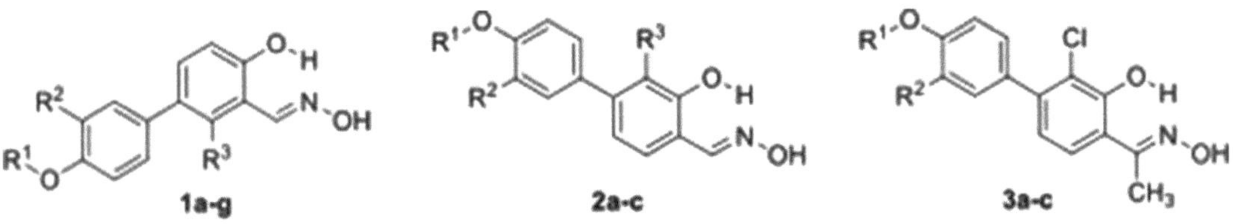

a: $R^1,R^2,R^3 = H$
b: $R^1,R^2 = H, R^3 = 4\text{-OH-Ph}$
c: $R^1,R^2 = H, R^3 = Cl$
d: $R^1 = H, R^2 = F, R^3 = Cl$
e: $R^1,R^3 = H, R^2 = F$
f: $R^1 = CH_3, R^2 = F, R^3 = H$
g: $R^1 = CH_3, R^2,R^3 = H$

a: $R^1,R^2 = H, R^3 = CN$
b: $R^1 = CH_3, R^2 = H, R^3 = CH_3$
c: $R^1,R^2 = H, R^3 = CH_3$

a: $R^1 = H, R^2 = F$
b: $R^1 = CH_3, R^2 = F$
c: $R^1,R^2 = H$

Fig. 3 Structures of the first oxime derivatives screened for glucose uptake inhibition: 5-aryl-substituted salicylaldoximes **1a–g**, 4-aryl-substituted salicylaldoximes **2a–c**, and 4-aryl-substituted salicylketoximes **3a–c**

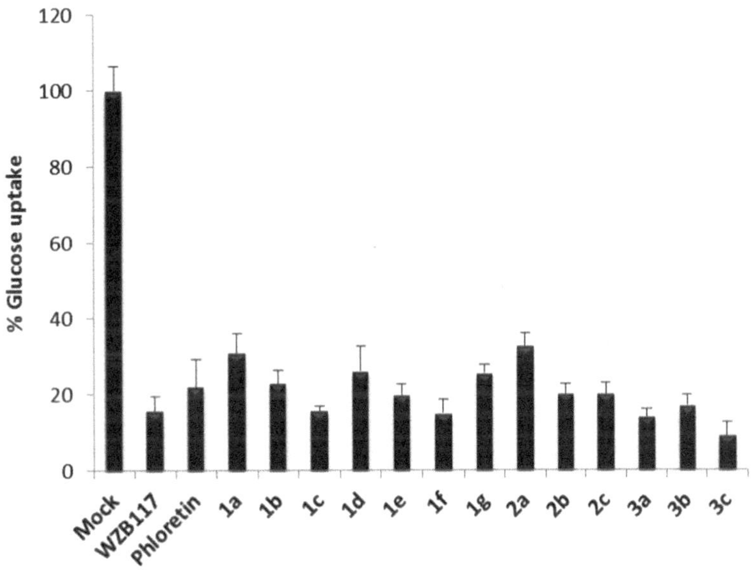

Fig. 4 Inhibition percentage of glucose uptake in H1299 cancer cells by compounds **1a–g, 2a–c, 3a–c**

proliferation were assessed using the MTT proliferation assay. Briefly, H1299 cancer cells were treated with a 30 μM concentration of the compounds. Cells were cultured for 48 h, and then 10 μL of MTT reagent was added to each well, and then incubated for 3 h. After the incubation, the culture medium was removed, 100 μL crystal dissolving solution was added to each well, the absorbance of the solution was measured at 570 nm, and data were analyzed [19]. Some compounds such as **1e–1g**, **3b**, and **3c** showed a good inhibition of cell viability, by reducing the percentage

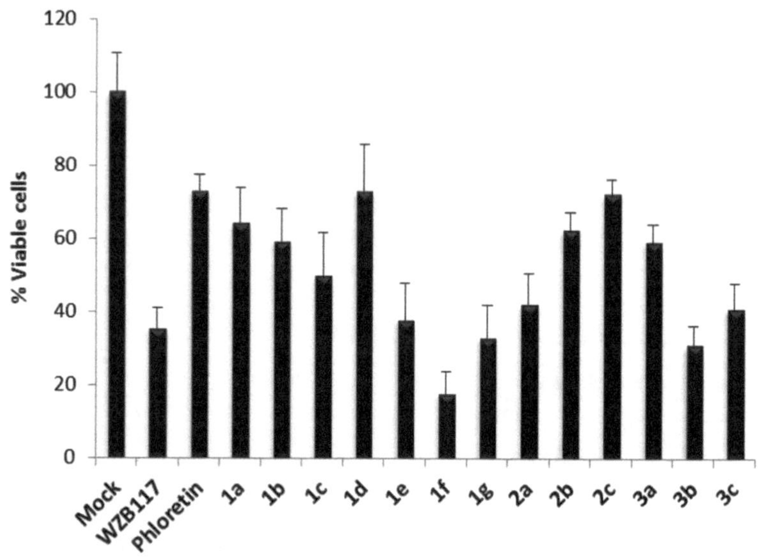

Fig. 5 Inhibition percentage of cell viability in H1299 cancer cells by compounds **1a–g**, **2a–c**, **3a–c**

of viable cells with a potency that is similar to or, in some cases, even better than that of WZB117 (*see* Fig. 5).

Considering the results obtained in both of these preliminary assays, we observed that aldoximes **1e–g** and ketoximes **3b–c** display the best combined activities of inhibition of glucose uptake and cell viability. Nevertheless, in order to further develop these compounds as GLUT1 inhibitors we also needed to consider their activity on the estrogen receptors for which they were initially designed. In fact, the possible binding of these molecules to the ERs could complicate their therapeutic profile. ERα and ERβ binding affinities of these oxime derivatives were previously determined by a radiometric competitive binding assay [10], and the relative binding affinity (RBA %) values for the compounds are referred to that of estradiol, which is set at 100 %. For most of these compounds (**1f**, **1g**, **3b**, and **3c**) the binding affinities for ERs were negligible or very low (from 0.001 to 0.123 %). On the contrary, compound **1e** displayed a certain binding affinity for ERβ (RBA = 0.97 %, K_d = 51 nM) (*see* Table 1). Therefore, only the compounds that showed a negligible affinity for the ERs were considered eligible to further determine their IC_{50} values in glucose uptake and antiproliferative assays, in order to be compared with those of the two reference inhibitors phloretin and WZB117 (*see* Table 1). Compounds **1f**, **1g**, **3b**, and **3c** were tested at different concentrations (0, 5, 10, 30, and 60 μM) to measure their IC_{50} values: IC_{50} values of the oxime derivatives were always lower than those of phloretin in both the assays, whereas they were either lower or similar if compared to those of WZB117.

Table 1
Inhibitory activities on glucose uptake and cell growth (H1299) and relative binding affinities (RBA) for the estrogen receptors α and β of compounds *1e–1g* and *3b–c*. Phloretin and WZB117 are reported as reference compounds for glucose uptake and cell viability assays

	IC_{50} (μM)		RBA (%)[a]	
Compounds	**Glucose uptake**	**Cell viability**	***h*ERα**	***h*ERβ**
1e	n.d.[b]	n.d.	0.021 ± 0.001	0.970 ± 0.110
1f	8.5 ± 2.0	14.1 ± 4.8	0.007 ± 0.001	0.013 ± 0.001
1g	23.4 ± 5.1	20.4 ± 5.4	0.003 ± 0.001	0.011 ± 0.003
3b	15.5 ± 3.8	39.6 ± 11.8	< 0.001	0.002 ± 0.001
3c	10.6 ± 2.8	34.8 ± 5.8	0.012 ± 0.004	0.123 ± 0.030
Phloretin	21.4 ± 5.1	54.0 ± 14.6	0.206[c]	n.d.
WZB117	10.9 ± 3.6	20.4 ± 4.8	n.d.	n.d.

[a]Determined by a competitive radiometric binding assay with [^{3}H]estradiol, estradiol is set as 100 % [10]; [b]not determined; [c]determined in rat uterine estrogen receptor [12]

At this point, we decided to analyze the possible binding mode of these compounds in the *h*GLUT1 model we had developed (*see* Fig. 2), by means of molecular modeling studies. By using homology modeling techniques, molecular dynamic simulations, pocket detection algorithms, and docking calculations, we hypothesized that the compounds actually interact with the cavity present in the intracellular region of the receptor (magenta, Fig. 6), where they establish profitable lipophilic and H-bond interactions. For example, Fig. 6 shows the possible interactions of compound **3c** inside our *h*GLUT1 model. The hydroxyl group forms a H-bond with the backbone nitrogen of F460 and *p*-hydroxyphenyl ring shows a lipophilic interaction with the same residue. The ketoximic central scaffold displays an arginine π-stacking interaction with R212 and the chlorine atom is inserted into a polar pocket delimited by S148 and R232. The oxime oxygen atom forms a H-bond with the backbone of Q397, whereas the hydroxyl group in position 2 participates to a H-bond network between the backbone oxygen of E146 and the backbone nitrogen of R212.

3 Development of 4-Aryl-Substituted Salicylketoximes

The preliminary results that we obtained with our collection of compounds originally designed as ER ligands validated our initial similarity-based hypothesis that the chemical series of aryl-substituted salicylaldoxime and salicylketoxime derivatives could be further

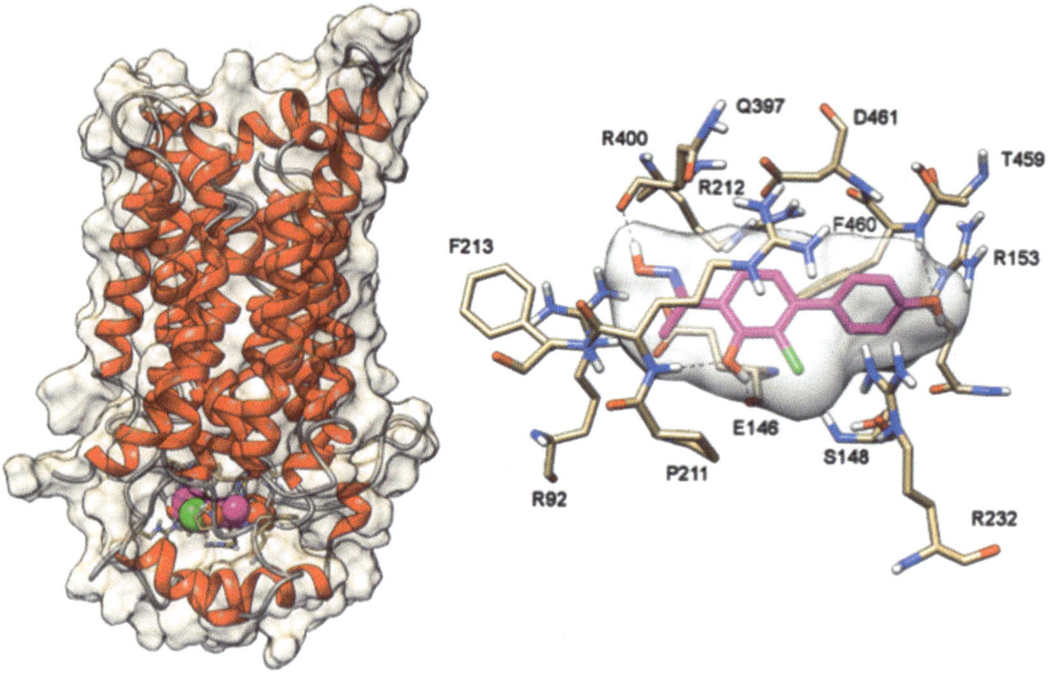

Fig. 6 Binding pose and main interactions of compound **3c** docked into the *h*GLUT1 model

developed to design and synthesize more potent GLUT1 inhibitors devoid of any undesired interaction with the ERs. In particular, we decided to stop working with aryl-substituted salicylaldoximes [10] and 5-aryl-salicylketoximes [19], since they had generally demonstrated to efficiently bind to ERs, and to expand, instead, only the subclass of 4-aryl-salicylketoximes, because we had never found any significant ER binding affinity for any of the compounds belonging to this subclass (*see* Fig. 7).

Thus, the second step of our work was to synthesize a focused series of salicylketoximes (compounds **4a–m**, Fig. 8) by varying only the substituents present in *meta* or *para* position of the aryl ring in position 4 while maintaining the central ketoximic scaffold.

Ketoximes **4a–m** were synthesized following a common general synthetic scheme (*see* Fig. 9). The synthesis starts from 3-bromo-phenol: the hydroxyl group of this precursor was acylated with sodium hydride and diethylcarbamoyl chloride to get carbamate **6**, which was subjected to an *ortho* lithiation step to insert with high regioselectivity a chlorine atom in position 2 of the phenyl ring. Hydrolysis gave phenol **8**, and then this intermediate was acetylated to produce compound **9**. A Fries rearrangement step with aluminum trichloride gave compound **10** in which the acetyl group was transposed into position *ortho* to the phenol group. Biaryl compounds **11** were obtained by a Pd-catalyzed cross-coupling reaction of **10** with properly substituted arylboronic acids. Biaryl compounds bearing a free hydroxyl group, such

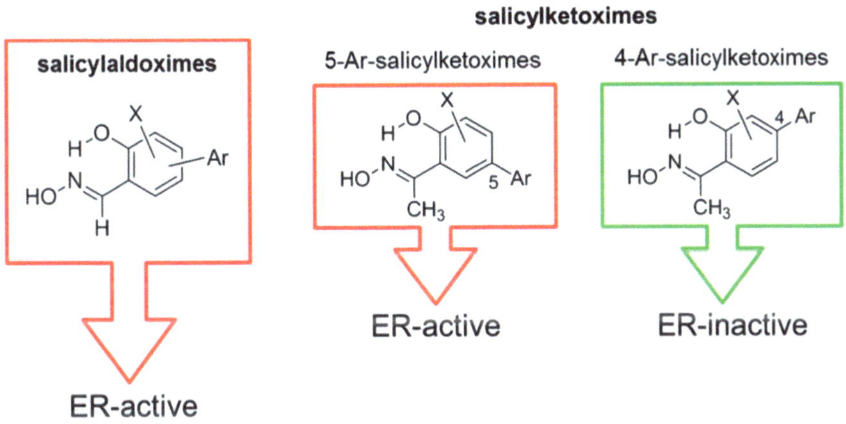

Fig. 7 Selection of the chemical class of 4-aryl-salicylketoximes as potential GLUT1 inhibitors, because they are devoid of any significant binding affinities for ERs

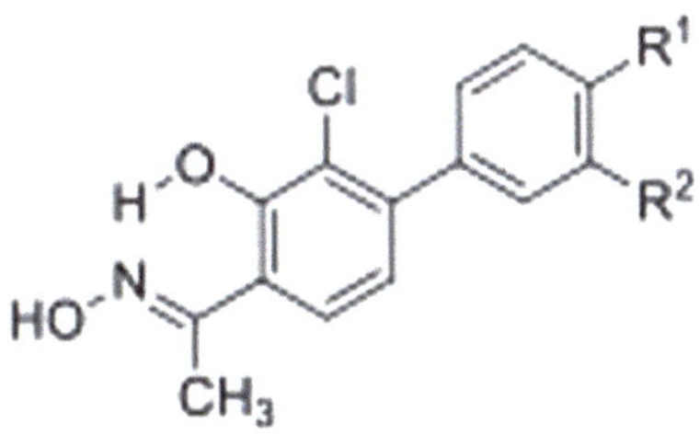

4a-m

a : R^1 = F; R^2 = H.
b : R^1 = CF_3; R^2 = H.
c : R^1 = H; R^2 = OH.
d : R^1 = OH; R^2 = Cl.
e : R^1 = OCF_3; R^2 = H.
f : R^1 = OCH_3; R^2 = H.
g : R^1 = H; R^2 = OCF_3.
h : R^1 = H; R^2 = OCH_3.
i : R^1 = CN; R^2 = H.
j : R^1 = H; R^2 = CN.
k : R^1 = OCH_3; R^2 = Cl.
l : R^1 = OCH_3; R^2 = CH_3.
m : R^1 = OH; R^2 = CH_3.

Fig. 8 Newly synthesized 4-aryl-substituted salicylketoximes **4a–m**

as **11c**, **d**, **m**, were obtained after deprotection of the methoxy groups by treatment with boron tribromide. Finally, all the resulting acetophenone derivatives **11** were directly transformed into the corresponding ketoximes by reaction with hydroxylamine hydrochloride. These 4-aryl-substituted ketoximes were exclusively obtained as diastereomers of *E* configuration in their oxime portion, since a highly energetic intramolecular hydrogen bond between the phenolic OH group and the nitrogen atom of the oxime portion strongly stabilizes this configuration. ^{1}H and ^{13}C NMR analysis confirmed this configuration: the ^{1}H NMR chemical

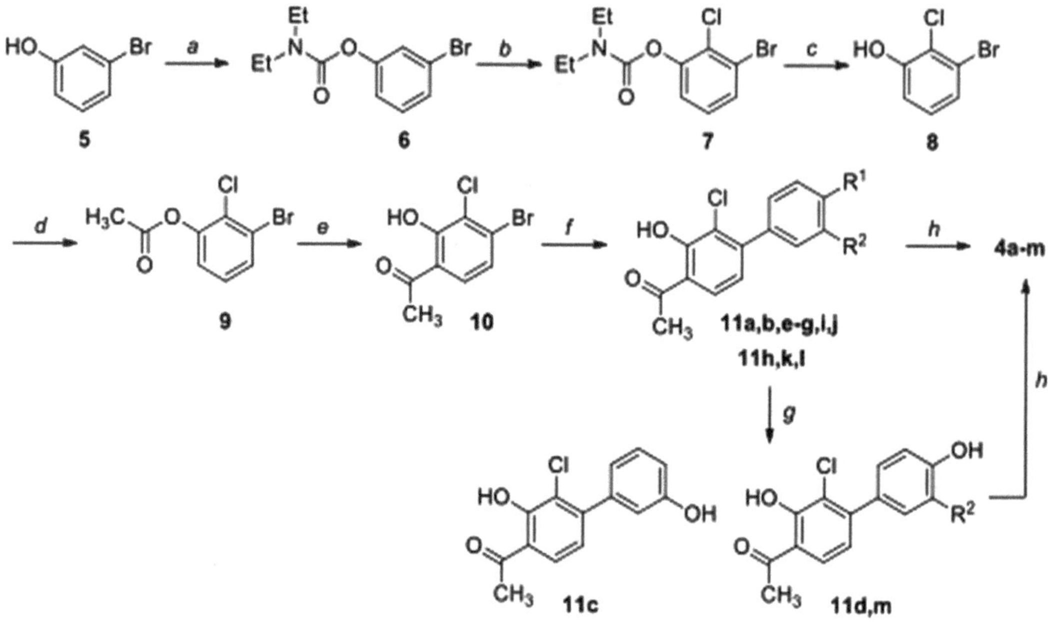

Fig. 9 Synthesis of salicylketoximes **4a–m**. *Reagents and conditions*: (*a*) NaH, *N,N*-diethylcarbamoylchloride, THF, RT, 3 h; (*b*) (*1*) *n*BuLi, (*i*Pr)$_2$NH, THF, 0 to −78 °C, 1 h; (*2*) C_2Cl_6, THF, 30 min; (*c*) NaOH, EtOH, reflux, overnight; (*d*) NaOH, acetyl chloride, TBAHS, dioxane, 1 h, RT; (*e*) $AlCl_3$, 130 °C, 3 h; (*f*) ArB(OH)$_2$, Pd(OAc)$_2$, PPh_3, aq. 2M Na_2CO_3, 1:1 toluene/EtOH, 100 °C, 16 h; or ArB(OH)$_2$, Pd(OAc)$_2$, PPh_3, solid K_2CO, toluene, 100 °C, 24 h; (*g*) BBr_3, CH_2Cl_2, −78 to 0 °C, 1 h; (*h*) $NH_2OH{\cdot}HCl$, EtOH-H_2O, 50 °C, 16 h

shift (δ) values of the ketoximic methyl protons ($2.40 \leq \delta \leq 2.43$ ppm) and the ^{13}C NMR δ values of the ketoximic methyl carbon atom ($10.91 \leq \delta \leq 10.96$ ppm) are in agreement with those reported for the *E* isomers of similar methylketoxime derivatives [20].

The newly synthesized ketoximes were screened for inhibition of glucose transport through GLUT1 by a glucose uptake assay and were also subjected to an antiproliferative assay in H1299 cancer cells, similarly to the previous group of oxime derivatives. Among the new compounds derivatives **4c**, **4d**, and **4m** produced a significant reduction of glucose uptake, reducing it below 20 % at a 30 μM concentration (grey bars, *see* Fig. 10). These potent compounds also showed good cytotoxic activities (black bars, *see* Fig. 10), in particular compound **4c**. There is a general parallelism between GLUT1 inhibitory potency and antiproliferative effect, with some exceptions such as compound **4d** because it displays the highest potency as GLUT1 inhibitor, but a relatively low cytotoxicity.

Further studies on the three most potent inhibitors of glucose uptake at various concentrations allowed the determination of IC_{50} values (*see* Table 2). We can observe that compounds **4c** and **4m** have IC_{50} values of about 11–16 μM as glucose uptake inhibitors.

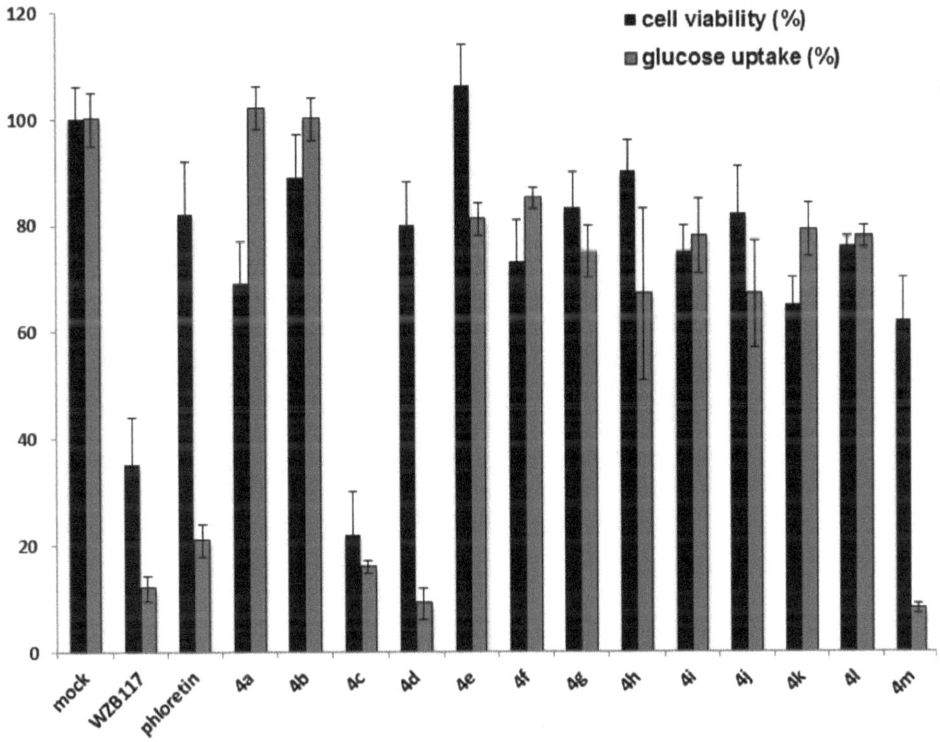

Fig. 10 Inhibition percentage of cell viability (*black* bars) and of glucose uptake (*grey* bars) in H1299 cancer cells by compounds **4a–m**

Table 2
Inhibitory activities on glucose uptake and cell growth (H1299) and relative binding affinities (RBA) for the estrogen receptors α and β of compounds 4c, 4d, and 4m

	IC_{50} (μM)		RBA (%)[a]	
Compounds	Glucose uptake	Cell viability	*h*ERα	*h*ERβ
4c	15.7 ± 2.0	17.8 ± 2.8	< 0.001	0.003 ± 0.001
4d	7.0 ± 2.0	45.8 ± 7.4	0.008 ± 0.002	0.010 ± 0.002
4m	11.1 ± 2.6	30.3 ± 5.9	0.018 ± 0.004	0.034 ± 0.004

[a]Determined by a competitive radiometric binding assay with [^{3}H]estradiol, estradiol is set as 100 % [21, 22]

These are comparable or even slightly better than those of reference GLUT1 inhibitors (*see* Table 1). Furthermore, **4d** proved to be the most potent inhibitor with an IC_{50} of 7 μM. In the cytotoxicity assay, the IC_{50} values of these three ketoximes are in the range of 18–46 μM, with compound **4c** being the most potent, showing a stronger antiproliferative effect than that of reference GLUT1 inhibitors.

In order to assess possible cross-activities of these compounds (**4c**, **d**, **m**) on the estrogen receptors, their binding affinities for the two estrogen receptor subtypes α and β were tested and, fortunately, turned out to be negligible, thus confirming the selectivity of these 4-aryl-substituted salicylketoximes for GLUT1 over ERs (*see* Table 2). A structure-activity relationship (SAR) analysis revealed the importance of a free phenolic group that is present in *para* (**4d**, **4m**) or *meta* (**4c**) position of the 4-phenyl substituent for an efficient GLUT1 inhibition. The activity was maintained with the insertion of a *meta*-chloro (**4d**) or *meta*-methyl group (**4m**), whereas the presence of different substituents such as fluorine, trifluoromethyl, trifluoromethoxy, methoxy, and cyano led to less active compounds (*see* Figs. 8 and 10).

In order to confirm the mechanism of action of these GLUT1 inhibitors, a fluorescent glucose analogue 2-[*N*-(7-nitrobenz-2-oxa-1,3-diazol-4-yl)amino]-2-deoxy-D-glucose (2-NBDG) was used. 2-NBDG is a suitable probe for the detection of glucose taken up by cells, and it is used to visualize and quantify the inhibition of glucose uptake in cancer cells by fluorescence microscopy. Non-small-cell lung cancer A549 cells were treated with 30 μM of compounds or control inhibitor phloretin in a glucose-free medium; after 30 min cells were treated with 2-NBDG (50 μM) and the Hoechst 33,342 dye (1 μg mL^{-1}) which is used to stain DNA. After 15 min, the cells were washed and the intracellular accumulation of fluorescent 2-NBDG (excitation wavelength at 488 nm) was observed using a confocal laser scanning microscope. We decided to analyze in this assay only salicylketoximes **3b**, **4c**, **4d**, and **4m**, which are the most potent GLUT1 inhibitors of this oxime-based series of compounds, and are devoid of any significant binding to the ERs. Therefore, compound **3c**, which showed a certain activity on ERβ (RBA = 0.123 %), was excluded from this study. In these experiments a reduction of green fluorescence of the cytosol maintaining the same nuclear blue fluorescence was particularly evident in treated cells when compared to control (only vehicle), and this effect is due to the reduction of 2-NBDG uptake by cancer cells, which is more evident than that caused by phloretin, a reference GLUT1 inhibitor. The mean fluorescence intensities of about 100 cells were measured to quantify these effects (*see* Fig. 11): these data confirmed that all the ketoximes exert a more potent effect than that of phloretin (residual fluorescence ranging from 33 to 53 %). In particular, the most active compound is **4m**, which is able to reduce the cellular uptake of 2-NBDG of about 70 %, whereas phloretin reduced the uptake only of 30 %.

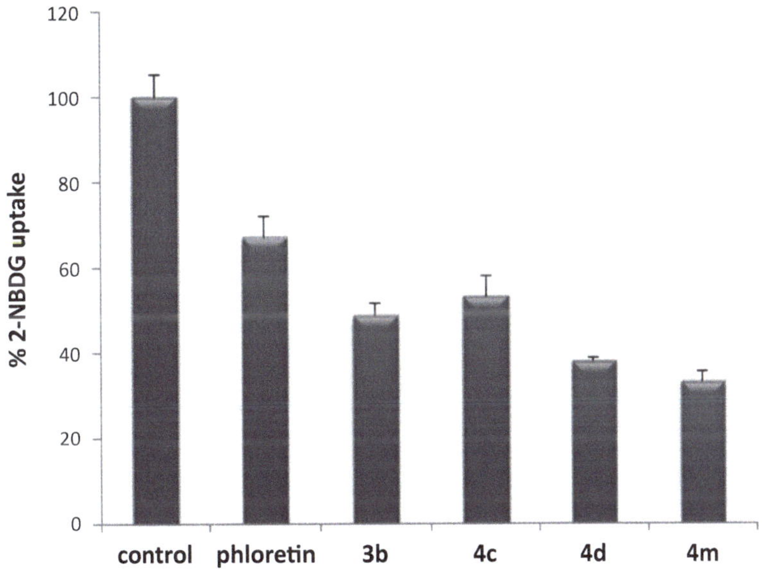

Fig. 11 2-NBDG uptake percentage in A549 cancer cells in the presence of compounds **3b**, **4c**, **4d**, and **4m** and phloretin

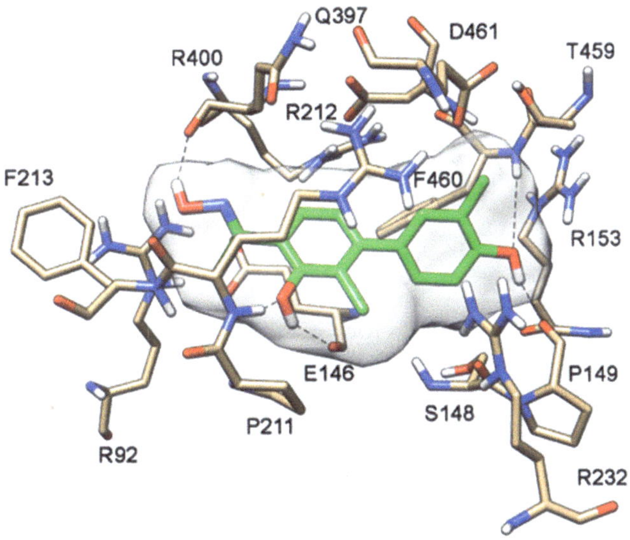

Fig. 12 Docking analysis of compound **4m** into the *h*GLUT1 model

In order to analyze the binding mode of these compounds, the most promising GLUT1 inhibitors were docked into our *h*GLUT1 homology model. Figure 12 shows the main interactions of compound **4m**. Similarly to **3c**, the oxime hydroxy group forms an H-bond with the carbonyl backbone of Q397. The phenolic hydroxyl in position 2 participates to two H-bonds with the

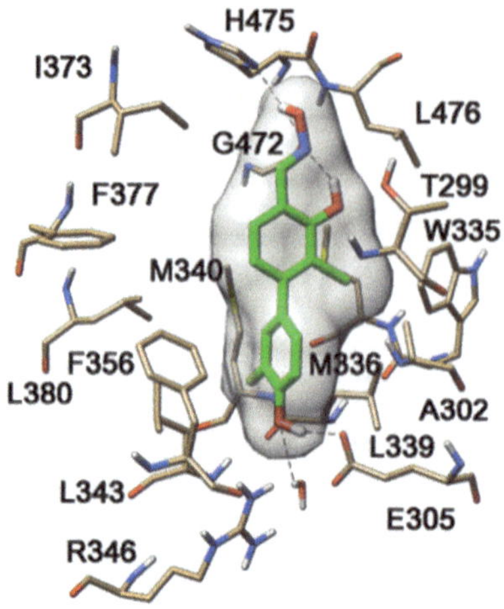

Fig. 13 Docking analysis of compound **4m** into ERβ

carbonyl backbone of E146 and the nitrogen backbone of R212, respectively. The ketoximic central scaffold shows a π-arginine stacking interaction with R212, whereas the chlorine atom present in the central scaffold is inserted into a polar pocket mainly delimited by S148 and R232. The aryl substituent in position 4 of the central ring showed a lipophilic interaction with F460, the *p*-hydroxy groups participated to a network of H-bonds with the nitrogen backbone of F460 and the carbonyl backbone of P149, and, finally, the methyl substituent displays a good lipophilic interaction with the methyl group of T459.

Most importantly, when compared to compound **3c**, the presence of the methyl substituent in **4m** determined a decrease of its affinity for ERβ (*see* Tables 1 and 2), thus increasing its specificity for GLUT1. In order to explain the differential ERβ binding affinity of **4m** and **3c**, a molecular modeling analysis was carried out and the docking of compound **4m** into the ligand-binding cavity of ERβ is shown in Fig. 13. The resulting binding pose displays the following features: the pseudocycle/oxime system forms H-bonds with H475 and the backbone carbonyl of G472; the chlorine atom is inserted into the lipophilic pocket delimited by A302, W335, M336, and L339; and the phenol-type *para*-OH group is involved in a H-bond network that includes E305, R346, and a water molecule. All these interactions are similar to those found for compound **3c**. However, the presence of an additional methyl substituent in compound **4m** determines a lower polarization degree of the phenol OH group (lower acidity), thus reducing the strength of its interactions with the E305/R346/water system in the ERβ-binding site, when compared to **3c**.

4 Conclusions

A new chemical class of GLUT1 inhibitors was designed starting from an initial structural similarity approach, where some common pharmacophoric features were identified between previously reported GLUT1 inhibitors and an in-house-developed class of ERβ ligands. The molecular design was aided by the development of an in silico model of GLUT1, which indicates a preferential binding pocket for the oxime class within this transporter, which is located close to the intracellular portion of the protein. An analysis of the most relevant interactions between these first inhibitors and the protein residues was useful in the rationalization of the experimentally found results, which were determined by means of a glucose uptake assay in cancer cells overexpressing GLUT1 (H1299), together with a cytotoxicity test in the same cell line. Then, keeping in mind that any undesired interaction with ERβ had to be avoided, a specific structural selection of the chemical subclass of 4-aryl-substituted salicylketoxime derivatives led to the production of novel molecules that proved to significantly interfere with glucose uptake and viability in cancer cells, with negligible interactions with the ERs. The synthesis of these new oxime derivatives was straightforward and produced a relevant number of analogues in a relatively short time, so that a satisfactory structure-activity relationship analysis could be done. A visualization of the efficient inhibition of the uptake of a fluorescent glucose analogue (2-NBDG) operated by some selected oxime derivatives further confirmed their mechanism of action.

References

1. Macheda ML, Rogers S, Best JD (2005) Molecular and cellular regulation of glucose transporter (GLUT) proteins in cancer. J Cell Physiol 202(3):654–662. https://doi.org/10.1002/jcp.20166
2. Warburg O (1956) On the origin of cancer cells. Science 123(3191):309–314
3. Smith TA (2001) The rate-limiting step for tumor [18F]fluoro-2-deoxy-D-glucose (FDG) incorporation. Nucl Med Biol 28(1):1–4
4. Denko NC (2008) Hypoxia, HIF1 and glucose metabolism in the solid tumour. Nat Rev Cancer 8(9):705–713. https://doi.org/10.1038/nrc2468
5. Granchi C, Fortunato S, Minutolo F (2016) Anticancer agents interacting with membrane glucose transporters. Medchemcomm 7(9):1716–1729. https://doi.org/10.1039/C6MD00287K
6. Granchi C, Qian Y, Lee HY, Paterni I, Pasero C, Iegre J, Carlson KE, Tuccinardi T, Chen X, Katzenellenbogen JA, Hergenrother PJ, Minutolo F (2015) Salicylketoximes that target glucose transporter 1 restrict energy supply to lung cancer cells. ChemMedChem 10(11):1892–1900. https://doi.org/10.1002/cmdc.201500320
7. Tuccinardi T, Granchi C, Iegre J, Paterni I, Bertini S, Macchia M, Martinelli A, Qian Y, Chen X, Minutolo F (2013) Oxime-based inhibitors of glucose transporter 1 displaying antiproliferative effects in cancer cells. Bioorg Med Chem Lett 23(24):6923–6927. https://doi.org/10.1016/j.bmcl.2013.09.037

8. Bertini S, De Cupertinis A, Granchi C, Bargagli B, Tuccinardi T, Martinelli A, Macchia M, Gunther JR, Carlson KE, Katzenellenbogen JA, Minutolo F (2011) Selective and potent agonists for estrogen receptor beta derived from molecular refinements of salicylaldoximes. Eur J Med Chem 46(6):2453–2462. https://doi.org/10.1016/j.ejmech.2011.03.030
9. Minutolo F, Bellini R, Bertini S, Carboni I, Lapucci A, Pistolesi L, Prota G, Rapposelli S, Solati F, Tuccinardi T, Martinelli A, Stossi F, Carlson KE, Katzenellenbogen BS, Katzenellenbogen JA, Macchia M (2008) Monoaryl-substituted salicylaldoximes as ligands for estrogen receptor beta. J Med Chem 51(5):1344–1351. https://doi.org/10.1021/jm701396g
10. Minutolo F, Bertini S, Granchi C, Marchitiello T, Prota G, Rapposelli S, Tuccinardi T, Martinelli A, Gunther JR, Carlson KE, Katzenellenbogen JA, Macchia M (2009) Structural evolutions of salicylaldoximes as selective agonists for estrogen receptor beta. J Med Chem 52(3):858–867. https://doi.org/10.1021/jm801458t
11. Wang D, Chu PC, Yang CN, Yan R, Chuang YC, Kulp SK, Chen CS (2012) Development of a novel class of glucose transporter inhibitors. J Med Chem 55(8):3827–3836. https://doi.org/10.1021/jm300015m
12. Hillerns PI, Zu Y, Fu YJ, Wink M (2005) Binding of phytoestrogens to rat uterine estrogen receptors and human sex hormone-binding globulins. Z Naturforsch C 60(7–8):649–656
13. Frolova A, Flessner L, Chi M, Kim ST, Foyouzi-Yousefi N, Moley KH (2009) Facilitative glucose transporter type 1 is differentially regulated by progesterone and estrogen in murine and human endometrial stromal cells. Endocrinology 150(3):1512–1520. https://doi.org/10.1210/en.2008-1081
14. Takizawa I, Lawrence MG, Balanathan P, Rebello R, Pearson HB, Garg E, Pedersen J, Pouliot N, Nadon R, Watt MJ, Taylor RA, Humbert P, Topisirovic I, Larsson O, Risbridger GP, Furic L (2015) Estrogen receptor alpha drives proliferation in PTEN-deficient prostate carcinoma by stimulating survival signaling, MYC expression and altering glucose sensitivity. Oncotarget 6(2):604–616. https://doi.org/10.18632/oncotarget.2820
15. Sun L, Zeng X, Yan C, Sun X, Gong X, Rao Y, Yan N (2012) Crystal structure of a bacterial homologue of glucose transporters GLUT1-4. Nature 490(7420):361–366. https://doi.org/10.1038/nature11524
16. Fiser A, Do RK, Sali A (2000) Modeling of loops in protein structures. Protein Sci 9(9):1753–1773. https://doi.org/10.1110/ps.9.9.1753
17. Cross S, Baroni M, Carosati E, Benedetti P, Clementi S (2010) FLAP: GRID molecular interaction fields in virtual screening. Validation using the DUD data set. J Chem Inf Model 50(8):1442–1450. https://doi.org/10.1021/ci100221g
18. Liu Y, Zhang W, Cao Y, Liu Y, Bergmeier S, Chen X (2010) Small compound inhibitors of basal glucose transport inhibit cell proliferation and induce apoptosis in cancer cells via glucose-deprivation-like mechanisms. Cancer Lett 298(2):176–185. https://doi.org/10.1016/j.canlet.2010.07.002
19. Berridge MV, Herst PM, Tan AS (2005) Tetrazolium dyes as tools in cell biology: new insights into their cellular reduction. Biotechnol Annu Rev 11:127–152. https://doi.org/10.1016/S1387-2656(05)11004-7
20. Paterni I, Bertini S, Granchi C, Tuccinardi T, Macchia M, Martinelli A, Caligiuri I, Toffoli G, Rizzolio F, Carlson KE, Katzenellenbogen BS, Katzenellenbogen JA, Minutolo F (2015) Highly selective salicylketoxime-based estrogen receptor beta agonists display antiproliferative activities in a glioma model. J Med Chem 58(3):1184–1194. https://doi.org/10.1021/jm501829f
21. Carlson KE, Choi I, Gee A, Katzenellenbogen BS, Katzenellenbogen JA (1997) Altered ligand binding properties and enhanced stability of a constitutively active estrogen receptor: evidence that an open pocket conformation is required for ligand interaction. Biochemistry 36(48):14897–14905. https://doi.org/10.1021/bi971746l
22. Katzenellenbogen JA, Johnson HJ Jr, Myers HN (1973) Photoaffinity labels for estrogen binding proteins of rat uterus. Biochemistry 12(21):4085–4092

Chapter 9

Applying Microfluidic Systems to Study Effects of Glucose at Single-Cell Level

Niek Welkenhuysen, Caroline B. Adiels, Mattias Goksör, and Stefan Hohmann

Abstract

Microfluidic systems in combination with microscopy (e.g., fluorescence) can be a powerful tool to study, at single-cell level, the behavior and morphology of biological cells after uptake of glucose. Here, we briefly discuss the advantages of using microfluidic systems. We further describe how microfluidic systems are fabricated and how they are utilized. Finally, we discuss how the large amount of data can be analyzed in a "semi-automatic" manner using custom-made software. In summary, we provide a guide to how to use microfluidic systems in single-cell studies.

Key words Microfluidic system, Optical tweezers, Fluorescence microscopy, Microbiology, Single cell analysis, Glucose uptake, *Saccharomyces cerevisiae*

1 Introduction

The development of microfluidic systems has been seminal for the development of single-cell analysis techniques. Microfluidic systems offer the experimentalist the advantage of incubating cells in a controlled and constant environment (e.g., growth media). It further offers the possibility to rapidly and reliably switch between media with different compositions (e.g., growth medium containing high versus low glucose concentrations). Therefore, microfluidic techniques have become important tools to study the effect of changing conditions on biological cells. In our studies, we use proteins tagged with fluorescent reporters to study the activation of certain glucose signaling pathways in yeast *Saccharomyces cerevisiae*. By changing the extracellular glucose concentration, downstream pathway activation results in altered intracellular protein localization, which in turn can be tracked by imaging [1]. In a typical experiment two tagged proteins are used: Nrd1 fused to the red fluorescent protein mCherry and Mig1 bound to GFP. Nrd1 is

Karin Lindkvist-Petersson and Jesper S. Hansen (eds.), *Glucose Transport: Methods and Protocols*, Methods in Molecular Biology, vol. 1713, https://doi.org/10.1007/978-1-4939-7507-5_9,

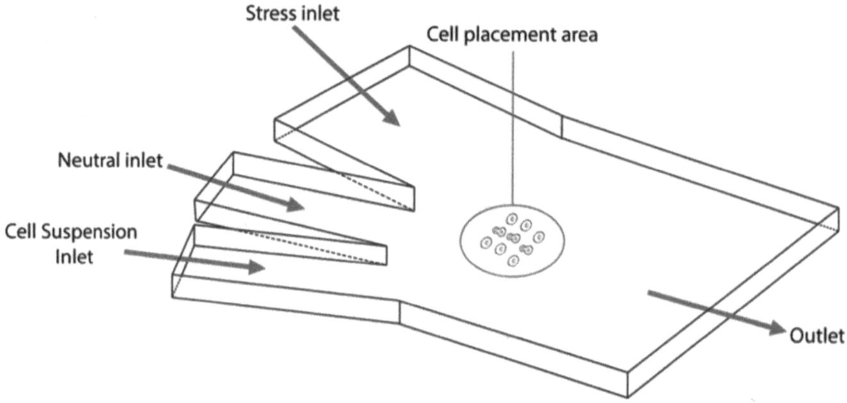

Fig. 1 Three-inlet channel microfluidic device. Cells are injected through the cell suspension inlet, trapped by the optical tweezers, and placed in the cell placement area. Which environment the cells will experience will depend on the flow velocities in the different inlets. The cell suspension and neutral inlet contain the media in which the cells are grown in, and the stress inlet contains the altered media in which the cell environment will be switched to (i.e., altered glucose concentration)

always localized in the nucleus [2] (*see* Fig. 1), while Mig1 is a transcription factor that translocates to the nucleus when glucose is taken up into the cell. Mig1 is hence employed as readout for the activity of the glucose repression-signaling pathway [3].

In the microfluidic experiments explained herein, glucose uptake is controlled by either steering the glucose concentration in the media or using yeast strain with glucose uptake capacities altered compared with the wild-type cell [4, 5]. Microfluidic systems to study the influence of the uptake of glucose on cellular processes and signaling have already been applied in research on the yeast *Saccharomyces cerevisiae* as well as on mammalian cells [1, 6, 7]. In our experiments, we employ a three-inlet channel device, in which cells are introduced into one channel and the two remaining channels are used to switch between media (*see* Fig. 2). Cells are trapped within the device by optical tweezers and immobilized by adhering to the concanavalin A-coated glass surface of the channel. Our microfluidic setup can change glucose concentration in less than 2 s, a requirement when studying rapidly activated pathways such as glucose signaling [8].

Other types of microfluidic setup and methods of immobilization are possible, but almost all have in common that they represent capillary systems in the micrometer scales. Due to the small dimensions, the flow behavior is in the laminar regime; hence a sharp concentration gradient will be created in the junction of the three inlets. As a result, microfluidic devices offer a superior performance due to spatial and temporal resolution. The volumes typically processed are in the nano- and pico-liter range; hence the sample and reagent consumption is small, which in turn lowers the experimental costs. Furthermore, the efficiency is often increased compared to traditional biological methodology and on-chip treatments otherwise impossible

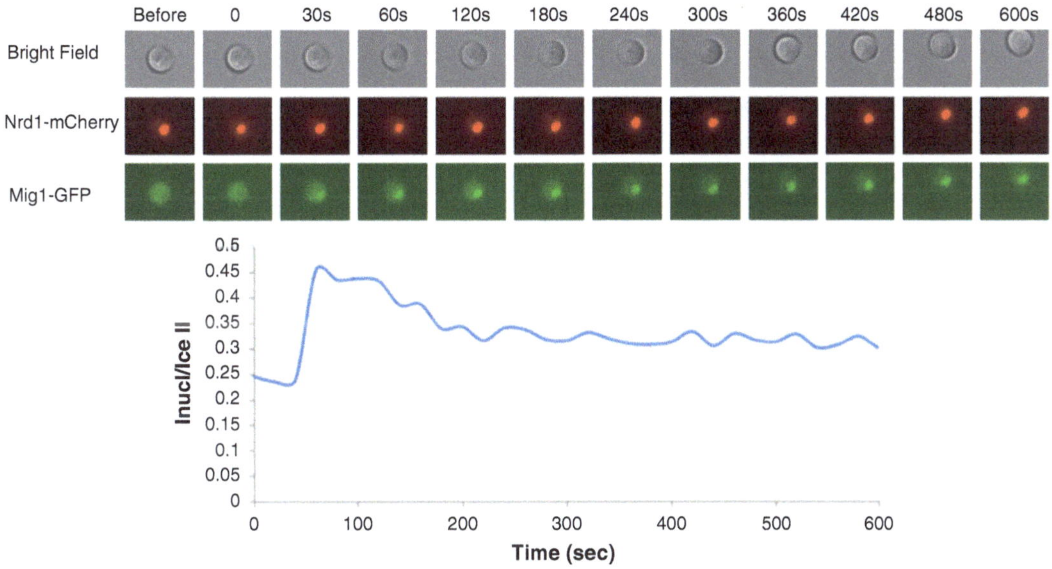

Fig. 2 Typical single-cell data. The *upper panel* shows time-lapse images of one cell from the transmission, Nrd1-mCherry and Mig1-GFP channel. The *lower panel* shows the fluorescence trace of the single cell over time, given as the ratio of Mig1 intensity in the nucleus to whole cell. A normal experiment tracks approximately 30–40 cells

to perform can be executed [9]. A very common biocompatible material used in the fabrication of microfluidic systems is the soft polymer polydimethylsiloxane (PDMS). This polymer is very suitable for optical studies of biological systems since it is transparent, permeable to gasses such as oxygen (but not fluids), and chemically inert, and has widely controllable mechanical characteristics [10].

The process of using microfluidic systems in combination with fluorescence microscopy can be described as three distinct steps: (i) microfluidic system production, (ii) utilization, and, finally, (iii) analysis of the acquired data. Microfluidic system fabrication and utilization will be discussed in a more comprehensive manner. For data analysis we will restrict ourselves to suggestions of software and algorithms that can be utilized for a "semi-automatic" analysis of large data sets of images. The growth conditions and the media we describe apply to experiments on the yeast *Saccharomyces cerevisiae*. Other single-cell organisms and cells from cell cultures might need other growth media and conditions and therefore have to be adapted to the biological model used.

2 Materials

2.1 Microfluidic Device Fabrication

1. Air pistol attached to nitrogen source (gas).
2. Chromium mask (Ångström Microstructure Laboratory).
3. Dissection microscope.

4. Desiccator.
5. Hot plate.
6. Isopropanol, 99.5 %.
7. Photoresist: SU-8 2025.
8. Plasma chamber: PDC-32G.
9. Plastic Pasteur pipette.
10. Plastic syringes: 1 mL.
11. Shaking incubator.
12. Silicon rubber polydimethylsiloxane (PDMS) Sylgard 184 2-component kit.
13. Silicon wafer, 3″ (Si-Mat Silicon Materials).
14. Spin processor (Laurell Technologies).
15. SU–8 developer.
16. Syringe needles, Sterican 27-gauge 0.40 × 20 mm.
17. Microbore PTFE Tubing, 0.012″ID × 0.030″OD.
18. UV exposure, 140,001 UV-light source-vacuum: This is a simplified and cheaper substitute for a mask aligner.

2.2 Microfluidic Utilization

1. Air pistol attached to nitrogen source (gas).
2. Concanavalin A solution: 1 mg/mL concanavalin A, 10 mM Tris–HCl, 100 mM NaCl and is adjusted to pH 8.0 by adding HCl (*see* **Note 1**).
3. Hamilton glass syringes, 250 μL (*see* **Note 2**): The volume of the glass needles depends on the temporal length of the experiment and working flow of the pumps.
4. Microbore PTFE Tubing, 0.012″ID × 0.030″OD.
5. Optical density meter.
6. Plastic syringes, 1 mL.
7. Sterile medical needles, syringe needle gauge depends on the size of the tubing.
8. Syringe pumps.
9. Tweezers or forceps.
10. YNB without amino acids: For 1 L YNB; dissolve 1.7 g yeast nitrogen base and 5 g $(NH_4)_2SO_4$ in 800 mL ddH_2O. Adjust pH value to 5.8, and autoclave appropriately. After autoclaving the solution, let it cool down and bring to the appropriate carbon source concentration from a sterile stock solution (for example: 100 mL of a 20 % w/v glucose/dextrose stock solution to have a final concentration of 2 %) and add the required amount of amino acids to the solution. Finally, dilute to a final volume to 1 L with sterile ddH_2O (*see* **Note 3**).

11. For the imaging, an epifluorescence microscope is required (*see* **Note 4**). Images are acquired using a 14-bit dynamic range EM-CCD camera. A fluorescence light source is used together with a GFP filter cube (472/30 nm exciter, 520/35 nm emitter, and 495LP dichroic mirror) and a mCherry filter cube (560/40 nm exciter, 630/75 nm emitter, and 585LP dichroic mirror, ET-Texas Red). The exposure time for the transmission is 27 ms. As mentioned before, for our glucose uptake experiments we often follow the localization of two proteins tagged with either mCherry or GFP as fluorescent marker, and the exposure times used for these fluorescent signals were 150 ms and 300 ms, respectively.
12. For the trapping of the single cells we apply optical tweezers. An extension of the 1070 nm ytterbium fiber laser (YLD-5-LP-IPG laser) was used (*see* **Note 5**).

3 Methods

3.1 Microfluidic Device Fabrication

In this paragraph, the fabrication method for the microfluidic devices is described in detail. The completed device is shown in Fig. 3 and consists of PDMS covalently bonded to a thin cover glass. To fabricate a microfluidic device, a master corresponding to the designed pattern of interest is created by a technique called soft photolithography. The lithographic master is used as a mold for the PDMS. To facilitate a tight junction between the finished PDMS device and tubing, the access holes are not drilled, but formed already at the molding stage. This is to assure that no cracks are generated that could cause leakage to occur. By using a guider in the molding process, a lid with pre-drilled holes can be used in which flattened needles are inserted (*see* Fig. 4). After polymerization, the PDMS is peeled off the master, access holes are cleaned, and thereafter the device is sealed to a cover glass in a plasma chamber. The master is reused several times while the resulting PDMS device is only used once and then discarded.

1. Create the channel design of interest using a CAD program or a similar one (*see* **Note 6**).
2. The corresponding chromium mask is printed by e-beam lithography. The places where the channels are supposed to be are left opaque, while the rest is black.
3. Gently clean a silicon wafer with nitrogen gas (*see* **Note 7**). Carefully place the wafer in a spin coater and spin coat an appropriate amount of SU-8 2025 on top of it. Make sure to perform this step in a fume hood. The photo resist is a photopolymer that changes its properties when exposed to light; SU-8 2025 is a negative photoresist, which cross-links when

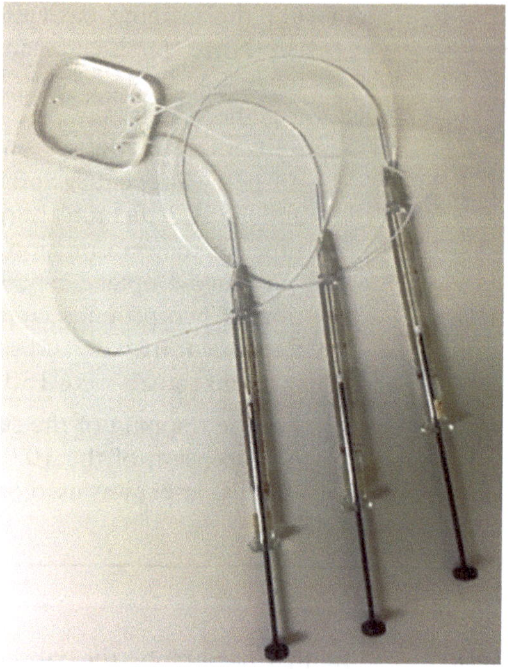

Fig. 3 The ready-to-use microfluidic device. The transparent microfluidic device covalently bonded to a coverslip and connected to three glass syringes through plastic tubing

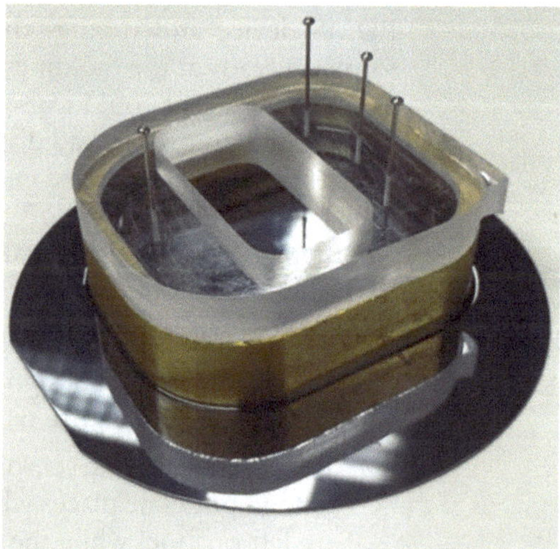

Fig. 4 Image of the molding process. The lid with pre-drilled holes for the needles (*top* transparent part) fits into the guider (metallic parts). Four flattened needles are inserted into the lid all the way down to the wafer, corresponding to the inlet and outlet holes of the microfluidic device

exposed to UV. Depending on which thickness of the photo resist (i.e., the height of you channels) is desired, different viscosities and spin speeds can be used (see data sheet of SU-8). For our purpose, the 27 μm thick layer is chosen, which can be achieved by using the following program; accelerate from 0 to 500 rpm for 5 s, and then at 3750 rpm for 30 s (acceleration 118 rpm/s).

4. To dissolve the solvents and densify the photo resist, progressively heat the master on a hot plate (prebake). Again, baking times will have to be adjusted for each respective thickness. We use 2–3 min at 65 °C and 5–6 min at 95 °C.
5. Expose the SU-8 spun wafer with the chromium mask on top to collimated light at 350–400 nm, preferably using a mask aligner or similarly. The appropriate exposure times can be found in the SU-8 data sheet. In our setup, we have found 10 min sufficient for achieving stable masters.
6. To initiate polymerization of the exposed photo resist, post-bake as before first at 65 °C and then at 95 °C according to the SU-8 data sheet. For 27 μm high channels, 1 min and 5–6 min, respectively, have proven sufficient.
7. Make sure to let the master cool slightly before immersing it into the developer solution, preferably under shaking. After exposure to light and developer, the photo resist is highly cross-linked and highly resistant to chemicals, hence the name photoresist. This step should be performed in a fume hood. Be careful not to overexpose, since this will lead to unstable reliefs. The actual exposure times are to be found in the SU-8 data sheet. We let the master soak for 4–5 min to dissolve all remnant photo resist not polymerized.
8. Rinse the master with isopropanol the remove the photoresist that has not been exposed to light. If a white precipitate is formed, continue developing a few more minutes. If not, gently blow-dry the master with nitrogen gas. Check the master for impurities and imperfections using a, for instance, low-magnification dissection microscope. If approved, the master is now finished and the second part of the device production can be initiated.
9. Pour PDMS base and curing agent in a 10:1 weight ratio in a plastic beaker and stir.
10. Degas the mixture in a desiccator until gas bubbles disappear (*see* **Note 8**).
11. Place the guider, if at hand (custom-made equipment), on top of the channel inlets (*see* **Note 9**).

12. Pour the PDMS onto the master using a plastic Pasteur pipette. Since no adapters are used for the tubing, the resulting device requires a minimum thickness off about 6–8 mm. Insert flattened needles into the lid placed on the guider and incubate at 95 °C for 1 h.
13. Pull out the PDMS from the guider and lift it off the master (*see* **Note 10**). Trim the access holes with some tweezers under magnification, rinse with isopropanol and dry using nitrogen gas. Then place it upside down (channels facing the air) on a clean coverslip.
14. Treat the PDMS and yet another clean coverslip with air plasma for 30 s in the plasma chamber (*see* **Note 11**).

3.2 Microfluidic Usage

In this paragraph, we describe the utilization of the microfluidics system. Briefly, the PDMS system is connected to low-resistance glass syringes, by tubing, which is directly fastened into the inlet holes of the device. A syringe pump generates flow through the channels. The outlet hole is just left as it is, and any flow-through material will collect as a droplet on the surface of the PDMS device. The channels within this particular device are shaped as a trident and are 27 μm high, spanning from 100 to 400 μm in width, and their lengths are on the cm scale (*see* Fig. 1).

1. On the day before the experiment start a pre-culture at 30 °C; for a fast-growing culture (for example, growth on glucose) start culture with an optical density (OD) of 0.002–0.005. For a culture that is not expected to grow fast (for example, growth on ethanol) start with an OD that is between 0.05 and 0.08. The OD on the day for the experiment should be between 0.5 and 1 for the fast-growing culture and for slow-growing cultures 0.3 and 0.7 (*see* **Note 12**).
2. Cut the appropriate amount of polytetrafluoroethylene tubes to the correct length depending on the distance between the flow pumps and the microscope (*see* **Note 13**). To attach the tubes to the syringes the tubes are pulled over a needle (*see* **Note 14**).
3. Fill a plastic syringe with concanavalin A solution. A tube is attached to the syringe and inserted into the outlet of the channel. Fill the system with concanavalin A solution until drops appear on the inlet channels of the system. Pretreat the microfluidic device with the concanavalin A solution for at least 30 min, as this is the required time to saturate the glass surface of the channel.
4. Clean the other tubes by flushing them with a small amount of the culture media to remove chemical, organic, and inorganic contamination (*see* **Note 15**).
5. The glass syringes are cleaned by loading and unloading three times with deionized water, once with ethanol or isopropanol

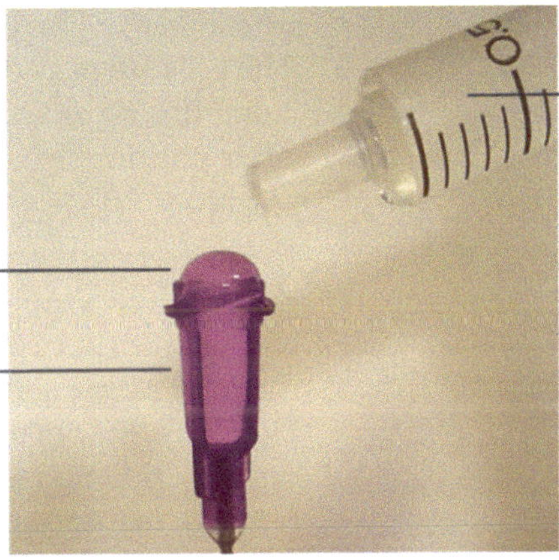

Fig. 5 Attaching the needle to the syringe. To avoid bubbles in the syringe, prefill the base of the needle with the liquid from the syringe. This should avoid air getting trapped in the base of the needle

and finally again with deionized water. Blow the syringes dry with nitrogen.

6. Pre-charge both the tubes and syringes with the culture media. Make sure that no air bubbles are left in the tubes or in the syringes (*see* **Note 16**).
7. Attach the tubes to the syringes. Prefill the base of the needle with the culture media such that no air is trapped in between the tube and the syringe (*see* Fig. 5).
8. Plug the tubing into the device inlets. Put a small drop of media on the channel inlet and gently put pressure on the syringe while inserting the tube in the device inlet in order to minimize the chance of air getting trapped in the system (*see* **Note 16**). Start with the "neutral" media, then the "stress" media, and finally the cell suspension (*see* **Notes 17** and **18**).
9. Put the microfluidic device on the microscopic stage (*see* **Note 19**) and load the glass syringes into the pump.
10. A syringe pump drives the flow. Install the syringes in the pressure pump and initiate the flow. The flow is set, in our case, to 320 nL/min for the cell inlet and the neutral inlet, while the inlet for the stress media is 160 nL/min. The flow in the "stress" media is much lower than the flow in the cell and neutral input in order not to contaminate the area in the microfluidics system with the cells and hence preexpose the cells to non-intended environment.
11. In our case, the cells are placed in the cell area with the help of optical tweezers (*see* **Note 5**). The laser is turned on at 0.9 A and 0.4 W. Cells are kept trapped in the optical tweezers as short as possible since the laser might cause stress or damage

to the cells. The cell placement is performed automatically. The cells are placed in 6 × 6 cell array with a distance of 10 μm along the *x*-axis and 8 μm along the *y*-axis.

After positioning of the cell, the image acquisition is initiated (*see* **Note 20**).

12. In our setup the imaging acquisition, the exact flow rates, through the syringe pumps, and the position of the microscope stage (in case optical trapping is used) are controlled by the microscope software OpenLab (Improvision Inc.) and the OpenLab Automator extension of the OpenLab software. In our experiments, we acquired images at different time points (at 30 s before the shift, at the shift, at 60 s, every 60 s for 420 s, every 120 s for 360 s, and every 180 s for 360 s after the shift); this sampling rate would not have been achieved without automatization of the whole process.

3.3 Data Analysis

Time-lapse single-cell studies with microfluidics provide a large amount of data. Manual analysis of this data is time ineffective, inaccurate, and often biased. Therefore, it is strongly advised to utilize specialized analysis software for this.

Most important is that the algorithm is able to identify the contours of the cell. This allows identifying cell morphology parameters such as size and shape in transmission images and also allows analyzing protein abundance and localization if combined with fluorescence images. In computing programs such as Matlab® or ImageJ it is also possible to write segmentation and analysis scripts. A set of instructions on how to segment cells can be easily found online. However, own scripts are often inadequate and much slower than the available segmentation software. There are several noncommercial software programs and algorithms available for semiautomated image analysis (*see* **Note 21**).

4 Notes

1. We find it optimal to add the concanavalin A fresh to the buffer solution.
2. Other types of glass needles can be used but it is important that the material is glass due to technical reasons.
3. In order to avoid contaminants, such as fibers from clothing, which can obstruct the tubes or the microfluidics device filter all media and solutions.
4. We use a Leica DMI 6000B inverted epifluorescence microscope containing a motorized xy-stage.

5. For the trapping of the single cells we apply optical tweezers; other strategies of trapping and immobilizing single cells are also available and can be used [8, 11].
6. The flow in the "stress" media will be much lower than the flow in the cell and neutral input in order not to contaminate the area in the microfluidics system where the cells are (i.e., cell placement area) and hence pre-expose the cells to non-intended environment (*see* Fig. 1). This area needs to be evaluated and estimated via flow simulations using, e.g., COMSOL Multiphysics to know exactly the distribution of the diffusion gradients [8].
7. Be careful not to use too high pressure from the nitrogen pistol. Around 1 bar is sufficient.
8. Slowly release the air valve of the desiccator to avoid the PDMS holder to fall over.
9. Include markings on the CAD design where to place the guider during this step.
10. This can be a bit tricky. Use a scalpel and tweezers. Carefully start bending the PDMS off the master in one corner and work around the whole guider until the PDMS is completely relieved from the silicon surface.
11. Make sure to use the correct plasma by adjusting the color to deep pinkish purple.
12. If the cell culture has not grown sufficiently and the cell density is too low, it is possible to spin the culture briefly at low speed. A loose pellet of cells will be created in the bottom and the cell suspension can be concentrated by removing supernatant.
13. When cutting the tubes, use a sharp scalpel knife. Avoid too much pressure on the tube to cut it, as this can deform the tube and affect the flow.
14. Using tweezers makes it easier to pull the tube over the needle.
15. If your solutions permit it, filtering them highly decreases the probability attaining clogged channels.
16. Remove all bubbles as they can block tubes or the microfluidic device, and also remove cells that possibly have adhered inside the microfluidic system. Air bubbles adhered in the needle can be removed by tapping on the needle with the exit pointing upwards. When the air is in the tip of the needle, eject it. For bubbles trapped in the tubing and in microfluidic channels, keep flushing until the bubble disappears with the flow of the injected fluid.
17. Some types of biological cells tend to clump. Vortex or sonicate cells briefly just before loading into the microfluidics device. Cells that are clumped together might clog up the microfluidics device or make the analysis of the images hard.

18. Start the experiment as fast as possible after loading the cells into the syringe. The cells might sediment in the syringe if no pressure is applied for a longer time and the cell suspension flow will cease containing cells.
19. To keep the microfluidic device stable on the microscope stage it can be taped with transparent tape or glued to the microscope stage with nail polish. Residual glue or nail polish can be easily removed from the microscopic stage with acetone.
20. A highly dynamic and efficient microfluidic system is achieved by automating the whole system (the operation of the flow pump and the microscope) through one software.
21. For the cell analysis program the best segmentation is done with transmission images that are slightly defocused. Examples of such programs are CellStat [12], CellStress [13], CellX [14], and Cell-ID [15]. CellStress uses the contours computed by CellStat and can both measure protein intensity in cell compartments and formation of protein aggregation.

Acknowledgements

We wish to acknowledge the Science Faculty of the University of Gothenburg for financing and supporting the PDMS Microfluidic Fabrication Facility.

References

1. Bendrioua L, Smedh M, Almquist J, Cvijovic M, Jirstrand M, Goksor M, Adiels CB, Hohmann S (2014) Yeast AMP-activated protein kinase monitors glucose concentration changes and absolute glucose levels. J Biol Chem 289(18):12863–12875. https://doi.org/10.1074/jbc.M114.547976
2. Steinmetz EJ, Brow DA (1998) Control of pre-mRNA accumulation by the essential yeast protein Nrd1 requires high-affinity transcript binding and a domain implicated in RNA polymerase II association. Proc Natl Acad Sci U S A 95(12):6699–6704
3. Treitel MA, Kuchin S, Carlson M (1998) Snf1 protein kinase regulates phosphorylation of the Mig1 repressor in Saccharomyces Cerevisiae. Mol Cell Biol 18(11):6273–6280
4. Elbing K, Stahlberg A, Hohmann S, Gustafsson L (2004) Transcriptional responses to glucose at different glycolytic rates in Saccharomyces cerevisiae. Eur J Biochem 271(23–24): 4855–4864. https://doi.org/10.1111/j.1432-1033.2004.04451.x
5. Elbing K, Larsson C, Bill RM, Albers E, Snoep JL, Boles E, Hohmann S, Gustafsson L (2004) Role of hexose transport in control of glycolytic flux in Saccharomyces cerevisiae. Appl Environ Microbiol 70(9):5323–5330. https://doi.org/10.1128/aem.70.9.5323-5330.2004
6. Lin Y, Sohn CH, Dalal CK, Cai L, Elowitz MB (2015) Combinatorial gene regulation by modulation of relative pulse timing. Nature 527(7576):54–58. https://doi.org/10.1038/nature15710
7. Zambon A, Zoso A, Luni C, Frommer WB, Elvassore N (2014) Determination of glucose flux in live myoblasts by microfluidic nanosensing and mathematical modeling. Integr Biol (Camb) 6(3):277–288. https://doi.org/10.1039/c3ib40204e
8. Eriksson E, Sott K, Lundqvist F, Sveningsson M, Scrimgeour J, Hanstorp D, Goksor M, Graneli A (2010) A microfluidic device for reversible environmental changes around single cells using optical tweezers for cell selection

and positioning. Lab Chip 10(5):617–625. https://doi.org/10.1039/b913587a

9. Whitesides GM (2006) The origins and the future of microfluidics. Nature 442(7101):368–373. https://doi.org/10.1038/nature05058
10. Cademartiri L, Ozin GA (2009) Concepts of nanochemistry. Wiley, Weinheim
11. Eriksson E, Engstrom D, Scrimgeour J, Goksor M (2009) Automated focusing of nuclei for time lapse experiments on single cells using holographic optical tweezers. Opt Express 17(7):5585–5594
12. Kvarnstrom M, Logg K, Diez A, Bodvard K, Kall M (2008) Image analysis algorithms for cell contour recognition in budding yeast. Opt Express 16(17):12943–12957
13. Smedh M, Adiels CB, Sott K, Goksör M (2010) CellStress—open source image analysis program for single-cell analysis. Proc SPIE Int Soc Opt Eng 7762. https://doi.org/10.1117/12.860403
14. Dimopoulos S, Mayer CE, Rudolf F, Stelling J (2014) Accurate cell segmentation in microscopy images using membrane patterns. Bioinformatics 30(18):2644–2651. https://doi.org/10.1093/bioinformatics/btu302
15. Chernomoretz A, Bush A, Yu R, Gordon A, Colman-Lerner A (2008) Using Cell-ID 1.4 with R for microscope-based cytometry. Current protocols in molecular biology/edited by Frederick M Ausubel [et al] Chapter 14:Unit 14.18. doi:https://doi.org/10.1002/0471142727.mb1418s84

Chapter 10

A Growth-Based Screening System for Hexose Transporters in Yeast

Eckhard Boles and Mislav Oreb

Abstract

As the simplest eukaryotic model system, the unicellular yeast *Saccharomyces cerevisiae* is ideally suited for quick and simple functional studies as well as for high-throughput screening. We generated a strain deficient for all endogenous hexose transporters, which has been successfully used to clone, characterize, and engineer carbohydrate transporters from different source organisms. Here we present basic protocols for handling this strain and characterizing sugar transporters heterologously expressed in it.

Key words Sugar transport, EBY.VW4000 strain, Sugar uptake assay, Yeast-screening system

1 Introduction

Characterization of substrate specificities, kinetics, or pharmacological properties of glucose transporters ideally requires an expression system without endogenous glucose uptake activity. Moreover, for screening purposes a growth-based system is desirable as it provides advantages such as high speed, low costs, and easy handling. The yeast *Saccharomyces cerevisiae* represents such a powerful system. Its physiology has been extensively studied, and its molecular and genetic tools are well established. Although yeast cells normally have a multitude of endogenous glucose transporters [1], we have constructed a strain in which all genes encoding glucose transporters and other transporters with glucose uptake activity have been deleted [2]. The strain is called the hexose transporter-null (*hxt*^{0}) strain, EBY.VW4000. This strain is no longer able to grow on media with glucose, fructose, or mannose as the sole carbon source and grows only very poorly with galactose. It does not show any residual glucose consumption or glucose uptake activity. Nevertheless, it is still able to grow normally with maltose or ethanol as carbon sources, which is important for its propagation and cultivation.

Karin Lindkvist-Petersson and Jesper S. Hansen (eds.), *Glucose Transport: Methods and Protocols*, Methods in Molecular Biology, vol. 1713, https://doi.org/10.1007/978-1-4939-7507-5_10,

The hxt^0 strain has already been used extensively to screen, clone, and characterize hexose transporters from a variety of other microorganisms, animal cells, humans, parasites, and plants [3–10]. Moreover, derivatives of this strain have been engineered with new metabolic pathways enabling the characterization and engineering of transporters for related carbohydrates such as xylose [11–13], arabinose [14], polyols [15], sucrose [7], or N-acetylglucosamine [16] by simple growth tests and uptake assays using radiolabeled substrates. Furthermore, by engineering the hxt^0 strain for xylose utilization with a simultaneous block of glucose phosphorylation, we were able to screen for xylose transport activity in the presence of glucose, a feature that has not been found in nature before [17].

Even though the heterologous sugar transporters are sometimes not correctly expressed in yeast cells or are not targeted to the plasma membrane, the powerful tools of yeast genetics can be used to bypass these problems. For instance, the human glucose transporters GLUT1 and GLUT4 initially did not support growth on glucose when expressed in the hxt^0 strain [18, 19]. However, after prolonged incubation on a glucose-containing medium, suppressor colonies appeared that obviously had acquired mutations enabling the human GLUT transporters to support growth of the hxt^0 strain on glucose [3]. Genetic analysis showed that this was due to mutations either in the GLUT transporters themselves or in the yeast genome. For example, for functional expression of GLUT1 in the yeast plasma membrane, a C-terminal truncation of the Efr3 (equal to Fgy1) phosphoinositide 4-kinase assembly protein turned out to be necessary. Moreover, for the relocation of human GLUT4 from the yeast endoplasmatic reticulum to the plasma membrane, perturbations in the ergosterol biosynthetic pathway (e.g., *erg4/fgy4* disruption) were required [3; our unpublished results].

Here, we present and comment on the methods that are necessary to screen for and to characterize heterologous glucose transporters in the yeast hxt^0 strain.

2 Materials

The hxt^0 strain EBY.VW4000, a suitable yeast expression vector (*see* **Note 1**) and a plasmid encoding a yeast glucose transporter as a positive control, can be obtained on request from the authors.

Prepare all media and buffers using ultrapure water and analytical grade reagents.

2.1 Maintenance, Transformation, and Cultivation of the hxt^0 Strain

1. YPM medium: 10 g/L Yeast extract, 20 g/L bactopeptone, 10 g/L maltose. It is recommended to separately autoclave the YP base and a 100 g/L maltose solution and mix them (9:1) after cooling down to 60 °C or a lower temperature (*see* **Note 2**).

2. Synthetic complete (SC) selective medium (*see* **Note 3**): 1.7 g/L Yeast nitrogen base, 5 g/L ammonium-sulfate, 11.2 mg/L adenine, 38.4 mg/L arginine, 57.6 mg/L isoleucine, 57.6 mg/L lysine, 38.4 mg/L methionine, 48.0 mg/L phenylalanine, 57.6 mg/L threonine, 14.4 mg/L tyrosine, 57.6 mg/L valine. 26.5 mg/L Histidine, 57.6 mg/L leucine, 19.0 mg/L tryptophan, and/or 19.2 mg/L uracil are added depending on the selective marker of the plasmids used. For example, uracil is omitted for the selection of cells transformed with a *URA3*-based plasmid. The pH value is adjusted with KOH buffer to 6.3 after mixing all components (*see* **Note 4**). 10 g/L Maltose, 20 g/L glucose, or other carbohydrates of interest are added from separately autoclaved (e.g., 10× concentrated) stock solutions (see above and *see* **Note 2**). For agar plates, add 20 g/L agar.
3. Master-mix for transformation: Sufficient for n + 1 samples; 240 μL of 50 % w/v PEG 3350, 36 μL of 1 M lithium acetate, 50 μL of 2 mg/mL salmon sperm single stranded carrier DNA (*see* **Note 5**).
4. Sterile velvet cloths and a pad for replica plating of 10 cm agar plates.

2.2 Plasmid Construction

1. Commercial kits for high-fidelity PCR, DNA purification, and plasmid preparation.
2. Acid-washed glass beads (0.45 mm).
3. Reagents and media for *E. coli* transformation according to standard molecular biology methods.

2.3 Sugar Uptake Assays

1. A Büchner funnel with a valve, a Büchner flask, a Woulff's bottle, stand clamps.
2. An electrical vacuum pump.
3. A heating device for 1.5 mL tubes or a water bath.
4. 50–300 mCi/mmol, 0.1 mCi/mL ^{14}C-labeled sugar.
5. Unlabeled sugar solution (*see* **Note 6**).
6. Cooled (4 °C) centrifuges for 50 mL tubes.
7. Phosphate buffer: Prepare 1 M KH_2PO_4 stock solution (adjust with KOH to pH 6.5 before filling up to final volume), autoclave. Dilute to a 0.1 M working solution (control the pH and adjust with KOH to pH 6.5 before filling up to final volume).
8. Quenching buffer: 500 mM Sugar (unlabeled), 0.1 M K_iPO_4, pH 6.5.
9. PVDF membrane filters, 0.22 μm pore size, 47 mm diameter.
10. Scintillation counter vials (e.g., 20 mL LDPE vials).

11. Scintillation cocktail (e.g., plus LSC universal cocktail).
12. A freeze-dryer
13. Pipettes (10 μL, 100 μL, 10 mL range), tweezers.

3 Methods

For general yeast genetic and molecular biology methods *see* **Note 7**.

3.1 Preparation of a PCR Cassette and Linearized Vector

We recommend cloning the transporter of interest into a yeast expression vector by homologous recombination (HR) cloning (*see* **Note 8**). However, alternative methods such as classical restriction/ligation or Gibson assembly can also be used.

1. Generate a PCR product of the open reading frame encoding the transporter of interest flanked by 30–40 bps overhangs identical to the promoter and terminator regions of the vector (*see* **Note 8**). Purify the PCR fragment, e.g., using a commercial kit.
2. Linearize a yeast expression vector with two restriction enzymes cutting between the promoter and terminator sequences (*see* **Note 9**).
3. Purify the linearized vector, e.g., using a commercial kit.

3.2 *S. cerevisiae* Transformation

The protocol can be up-scaled where necessary, e.g., for screening mutagenized plasmid libraries (*see* **Note 10**).

1. Inoculate a 10 mL YPM pre-culture of the *hxt*^{0} strain grown on an agar plate and incubate overnight at 30 °C on a rotary shaker at 180 rpm.
2. Determine the cell density of the pre-culture by measuring the optical density at 600 nm (dilutions should be made above an OD_{600nm} of 0.5).
3. Inoculate 50 mL of pre-warmed YPM medium in a 250 mL flask with an OD_{600nm} of about 0.2 and grow the cells at 30 °C and 180 rpm until they have undergone two to three divisions (OD_{600nm} of about 0.8–1.5 should be reached).
4. Collect the cells by centrifugation at 3,000 × *g* for 5 min at 20 °C in 50 mL tubes.
5. To remove the residual medium, resuspend the pellet in 25 mL of sterile distilled water, and centrifuge again at 3,000 × *g* for 5 min at 20 °C.
6. Prepare the master mix for transformation, depending on the number of transformations (n).
7. Remove the supernatant from **step 5** and resuspend the cells in 1 mL of sterile distilled water. This suspension is sufficient

for up to ten transformation samples. Prepare the required number of 100 μL aliquots in 1.5 mL reaction tubes and centrifuge the cells at 3,000 × *g* for 1 min at 20 °C. Discard the supernatant.

8. Overlay the pellets in 326 μL of the transformation mixture.
9. Add 0.1–0.3 μg linearized plasmid, 0.1–0.3 μg of the PCR product, and sterile water to a final volume of 360 μL.
10. Resuspend the cells by vortexing or pipetting. Foaming of the mixture should be avoided.
11. Heat-shock the cells at 42 °C for 30 min without shaking.
12. Centrifuge the transformation tubes at 3,000 × *g* for 30 s and remove the supernatant with a pipette. Resuspend the cells in 150 μL of sterile water.
13. Plate the suspension on agar plates containing selective SC medium supplemented with 1 % w/v maltose (*see* **Note 11**).
14. Incubate the plates at 30 °C for 2–3 days (*see* **Note 12**).

3.3 Complementation Analysis

1. After obtaining transformants on selective media with maltose, make replica plates of the transformants onto the desired hexose-containing agar plates (e.g., glucose, mannose, fructose, galactose; suggested final concentrations: 2 g/L and 20 g/L) (*see* **Note 13**). A velvet cloth can be replica plated successively at least on five further agar plates.
2. Incubate the replicates for up to 1 week at 30 °C. A comparison with the negative (empty plasmid) and positive (*HXT* plasmid) control transformants is crucial for assessing the success of the complementation: if the colony number on hexose-containing media approaches that of the positive control, the transporter of interest is likely functional in the *hxt*^{0} strain (*see* **Note 14**).
3. Regardless of the procedure resulting in hexose-growing colonies, these should be replica plated once more on hexose medium to confirm the results. For further experiments (e.g., re-isolation of the plasmids) take colonies only from the corresponding maltose plates and grow them in selective maltose medium. This is absolutely necessary in order to avoid the selection of secondary-site mutations accelerating growth on hexose media (*see* **Note 11**).

3.4 Plasmid Preparation and Analysis

1. Isolate the plasmids from yeast cells by alkaline lysis protocols as used for *E. coli* cells but, in addition, yeast cell walls must be broken mechanically. For that purpose, 2/3 volumes of glass beads (diameter 0.45 mm) are added to yeast cells after they have been suspended in the common three buffers (resuspen-

sion, lysis, and neutralization buffers) for alkaline lysis. The cells are broken by vigorous vortexing for 5 min (*see* **Note 15**).

2. After centrifugation of cell debris and glass beads (13,000 × *g*, 5 min, 20 °C), the supernatant is subjected to DNA precipitation using isopropanol by standard biochemical protocols. Since plasmid preparations from yeast cells yield only very low amounts of DNA that are not sufficient for further analysis, re-isolated plasmid DNA must first be transformed into *E. coli* (*see* **Note 16**) and isolated in larger quantities from the bacterial cells by standard procedures, e.g., using commercial kits.
3. The amplified plasmids should be transformed back into the hxt^0 strain or into the plasmid-cured mutant strains, plated first on maltose medium, and should confer growth to all transformants after replica plating on the corresponding hexose-media plates.

3.5 Assays for Uptake of Radiolabeled Sugars

Most steps of the uptake assay must be performed quickly and it is therefore necessary to prepare all devices and materials in advance. The working space should be large enough to accommodate all equipment and reagents within the reach of the experimenter. Despite the short incubation time of the cells with the radiolabeled sugar, yeast will produce small amounts of radioactive CO_2 during the experiment. It is therefore recommended to perform the experiment under a fume hood.

Before the experiment, mount a Büchner funnel onto a Büchner flask. Connect the Büchner flask with a pump via a Woulff's bottle. Secure the equipment with a stand clamp. For smooth handling, practicing the procedure described below without radioactive solutions is strongly recommended.

1. Grow the hxt^0 cells expressing the transporter of interest until OD_{600nm} of about 1–2 (the culture volume depends on the number of aliquots needed; *see* **steps 6** and **8** below). It is important that the cells are in the exponential growth phase.
2. Prepare the required radiolabeled sugar solutions, which should have concentration threefold higher than the concentration to be investigated (*see* **Note 17**). For each uptake sample, 50 μL of the threefold concentrated sugar solution is required.
3. Harvest the cells by centrifugation, e.g., in 50 mL disposable tubes (3,000 × *g*, 5 min at 20 °C). Weigh the empty tube in advance for later determination of the cell fresh weight (**step 5**).
4. Resuspend the cells in ice-cold phosphate buffer and centrifuge the suspension (3,000 × *g* for 5 min at 4 °C). Discard the supernatant and repeat this washing step. It is important to remove the liquid with a pipette completely after the second washing step.

5. Weigh the tube with the cell pellet and calculate the fresh cell weight (by subtracting the weight of the empty tube determined in **step 3**).
6. Resuspend the pellet in ice-cold phosphate buffer to reach 60 mg/mL (corresponds roughly to 40 OD_{600nm} Units; this should help estimating the culture volume required in **step 1**; also take **step 8** into account).
7. Prepare cell suspension aliquots of 110 μL and keep them on ice.
8. Keep residual cell suspension (not less than 10 mL) for dry-weight determination (freeze at −80 °C).
9. Fill two scintillation counter vials per uptake sample with 5 ml scintillation liquid. These can be prepared in advance, but note that the liquid is light sensitive.
10. Set up the uptake reaction (*see* **Note 18**): Pre-warm one 110 μL aliquot of the cell suspension and one 50 μL aliquot of the threefold concentrated sugar solution for 5 min at 30 °C.
11. Insert a filter into the Büchner funnel and overlay it with 10 mL of ice-cold quenching buffer (pump on, valve closed).
12. Pipet 100 μL from the cell suspension aliquot into the tube containing 50 μL of the threefold concentrated sugar solution. Mix quickly by up-and-down pipetting three times. Start counting time beginning from the first contact of both solutions.
13. Incubate the mixture for 5–20 s (*see* **Note 19**).
14. Pipet 100 μL of the mixture into the Büchner funnel containing the quenching buffer and filtrate (open the valve of the Büchner funnel).
15. Wash the filter two times with 10 mL of ice-cold quenching buffer. The valve should be closed while overlaying the filter with the buffer. This ensures that residual cells adsorbed on the wall of the funnel are collected.
16. Using tweezers, transfer the filter into a prepared scintillation counter vial. Shake the vial vigorously to strip the cells from the filter.
17. Add 10 μL of the residual mixture from **step 12** directly into another scintillation counter tube. This sample is required to determine the total radioactive counts in the uptake mixture.
18. In order to obtain accurate values for the uptake efficiency, the unspecific binding of the radiolabeled sugar to the cell surface and the filter must also be determined (below referred to as cpm_B) for each transformant variant. For the control, ***skip* steps 12–14** and pipet 67 μL of the cell suspension and 33 μL of the sugar solution sequentially (do not premix) into the quenching buffer, and ***go to* steps 15** and **16**.

19. Insert the vials containing the filters from **step 16**, vials with 10 μL of the cell/sugar mixture from **step 17**, and control vials from **step 18** into a scintillation counter and determine the counts per minute (cpm) in a window of 10 min.

3.6 Cell Dry Weight Determination

The cell dry weight is most precisely determined by freeze-drying.

1. A defined volume (not less than 10 mL) of the cell suspension (from Subheading 3.5, **step 8**) should be frozen at −80 °C in a 50 mL tube (determine the weight of the empty tube in advance). Cap the tube with perforated aluminum foil and freeze-dry the cells for 16 h or longer in a freeze-dryer.
2. To determine the cell dry weight, subtract the weight of the empty tube from the total weight of the freeze-dried sample. If no freeze-dryer device is available, filtering and subsequent heating can perform the cell drying. Weigh a nitrocellulose filter (0.45 μm) and insert it into a Büchner funnel (the same setup as for the uptake assay can be used). Filter the cell suspension (not less than 10 mL) and wash two times with distilled water. Heat the filter in the microwave at 140 W for 15 min. Additionally, dry the filter overnight in a desiccator. Weigh the dry filter and subtract the empty filter weight to obtain cell dry weight.

3.7 Calculations

The uptake velocity is usually expressed as nmol sugar imported per min per mg cell dry weight.

1. To calculate the amount of intracellular sugar (S_i), the counts on the filter (cpm_F, *see* **step 16**, Subheading 3.5) are corrected for background binding (cpm_B, *see* **step 18**, Subheading 3.5) and the proportion of the imported sugar is calculated relative to the total sugar in the mixture (cpm_M, *see* **step 17** in Subheading 3.5) (*see* **Note 20**):

$$S_i\,[\mathbf{nmol}] = \frac{\mathrm{cpm_F} - \mathrm{cpm_B}}{\mathrm{cpm_M} \times 10} \times \text{sugar concentration}\left[\frac{\text{nmol}}{\mu\text{L}}\right] \times \text{reference volume}\,[100\,\mu\text{L}]$$

2. The uptake velocity (v) is calculated taking into account the incubation time (t, *see* **step 13**, Subheading 3.5) and normalized to the cell dry weight (cdw) (*see* **Note 21**):

$$v\left[\frac{\mathbf{nmol}}{\mathbf{min \times mg}}\right] = \frac{S_i\,[\text{nmol}]}{t\,[\text{min}] \times \text{cdw}\,[\text{mg}]}$$

3. For determination of K_M values, v is plotted against the sugar concentrations and fitted to the Michaelis-Menten equation, preferably by nonlinear regression. Linearization approaches such as Eadie-Hofstee or Lineweaver-Burk plots may be used as alternatives.

4 Notes

1. Do not use vectors comprising sugar-regulated promoters, such as *GAL1*, as they are usually subject to catabolite repression in the presence of glucose.
2. In this way, a reaction between the sugar and amino acids (Maillard reaction), which occurs at high temperatures, is prevented.
3. Other formulations of synthetic media are found in the literature. In principle, they can also be used, but the variant presented here is routinely used in our laboratory.
4. Avoid using NaOH for any media and solutions to be used for EBY.VW4000; the strain is slightly sensitive to sodium.
5. Prepare as described by Gietz and Schiestl [20].
6. The concentration depends on the concentration range to be investigated and on the solubility of the respective sugar. For glucose, we usually prepare a 2 M stock solution, which can be used for the preparation of the quenching solution and dilution of the radiolabeled sugar to the desired specific activity.
7. General yeast genetic and molecular biology methods have been described, e.g., by Guthrie and Fink [21].
8. We recommend cloning the transporter of interest into a yeast expression vector by homologous recombination (HR) cloning [22].
9. The restriction sites do not necessarily have to be directly adjacent to the integration sites in the promoter/terminator regions, as the overhanging sequence will be removed during homologous recombination in vivo.
10. We have also successfully transformed frozen competent cells of EBY.VW4000. See [23].
11. It is important to perform the transformation and propagation of the transformants on maltose. Immediate plating on hexoses will result in considerably lower numbers of transformants and could result in suppressor mutations in the EBY. VW4000 genome. Mixture of sugars (e.g., maltose/glucose) should never be used due to their possible interferences in the catabolite signaling and as the yeast maltose transporter is inhibited by glucose.

12. This protocol should typically yield about 100–1000 colonies per plate. It is recommended to include the following controls into the transformation procedure: (1) one transformation sample without any DNA (to control the competent cells and reagents for contaminations; no colonies should appear); (2) one sample with only the linearized plasmid (to control the HR-independent re-circularization of the plasmid; not more than 5–10 % of the colony number of the transformation including the PCR product should appear); (3) one sample with the diluted empty circular plasmid (to control the competence of the cells and as a negative control for later functional analyses; about 100–1000 colonies should appear); (4) one tube with a yeast *HXT* control plasmid (as a positive control for functional analyses; about 100–1000 colonies should appear).
13. We observed that the *hxt*^{0} strain expressing heterologous transporters grows better on low hexose concentrations. This applies especially to glucose. The reason is not known in detail, but we speculate that a discrepancy between the high extracellular hexose concentration (as sensed by hexose sensors) and a low intracellular concentration (due to inefficient transport through a heterologous protein) may negatively affect the regulation of the central carbon metabolism. Moreover, colony sizes are not indicative of the preferred substrate specificities and performances of the transporters.
14. A certain proportion of the colonies observed on maltose will not grow on hexoses due to an unsuccessful assembly of the PCR product into the vector. In order to confirm the identity of the cloned transporter and propagate the plasmid, a few of the positive colonies should be subcultured for plasmid preparation. A small number of colonies on hexose medium (less than 1 % of colonies observed on maltose) are usually found even in the negative control transformed with the empty plasmid. This is due to spontaneous suppressor mutations in the genome of EBY.VW4000 (e.g., due to mutations in the *SNF3* glucose sensor gene inducing a normally silent still unknown glucose transporter). If a comparable number of colonies appears with the transporter of interest, the cloned transporter is likely not functional in the *hxt*^{0} strain. In some cases, an intermediate colony number is found; this is indicative of mutations occurring either in the transporter sequence or in the genome, which render the transporter functional in yeast. For instance, mammalian glucose transporter GLUT1 was functional in the *hxt*^{0} strain only when the transporter gene contained specific PCR-born missense substitutions or when the host strain acquired an additional genomic mutation, called the *fgy1*-allele [3]. Both type of mutations occur randomly and are selected for on hexose-containing media. If necessary, the mutation frequency can be increased by error-prone PCR of the transporter gene (as

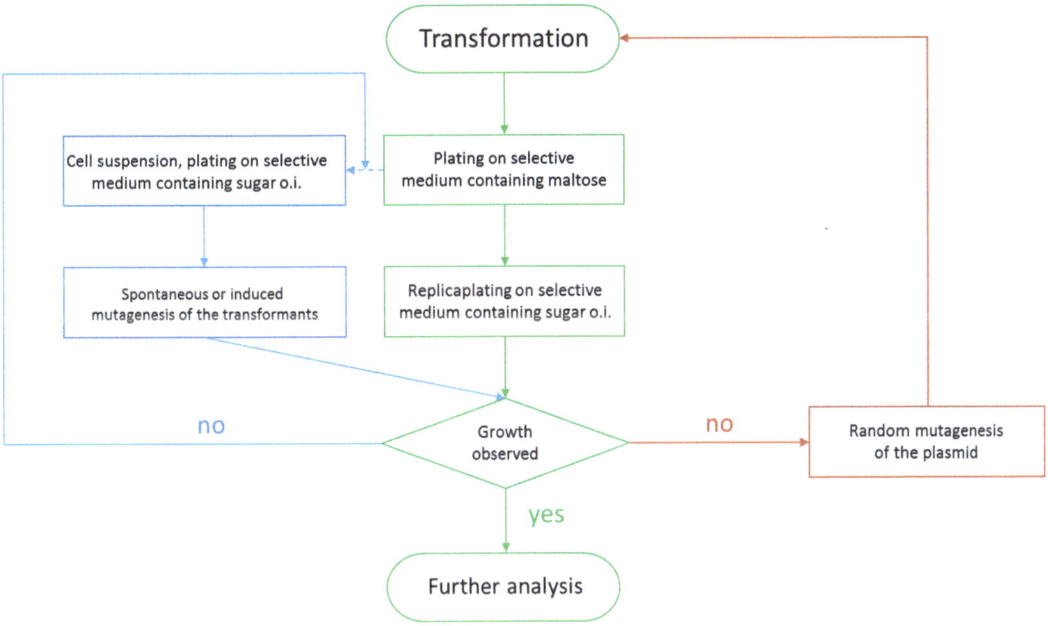

Fig. 1 Workflow for the complementation of the *hxt⁰* strain by the transporter of interest. After co-transformation of the *hxt⁰* strain with the linearized vector and the PCR product encoding the transporter of interest, the transformants are plated on selective media containing maltose. The cells, which have successfully integrated the PCR product into the plasmid, will yield colonies. They are subsequently replica plated on selective medium containing the sugar of interest (e.g., glucose). If a significant number of colonies grow on this medium, further analyses can be performed, but only after isolation, amplification in *E. coli*, sequence analysis, and re-transformation of verified plasmids into the *hxt⁰* strain. If the complementation is not immediately successful (*green lines*), the ORF of the transporter can be subjected to random mutagenesis (e.g., error-prone PCR as described in [17]) and the transformation procedure is repeated (*red lines*). Alternatively or additionally, transformants are collected from the maltose medium, resuspended in water, and plated on selective media containing the sugar of interest. The cells are subsequently subjected to UV irradiation to mutagenize the plasmid and the genome of the *hxt⁰* strain (*blue lines*). The detailed procedure was described by Boles [24]. o.i. = of interest

described, e.g., by ref. 17) and/or UV mutagenesis of the transformants (*see* Fig. 1; the approach has been described by refs [3, 24].).

15. The cells are broken by vigorous vortexing for 5 min according to Boles and Zimmermann [25].
16. Isolated plasmid DNA must first be transformed into *E. coli* [26].
17. The total sugar concentration in the uptake mixture is dependent on the purpose of the experiment and on the properties of the transporter. For determination of the Michaelis constant (K_M value), the concentrations should be varied within one to three orders of magnitude. Typically, seven to ten different concentrations should be measured at least in triplicates, whereby saturation of the transporter should be reached (if possible). The solutions are prepared as a threefold concentrated stock containing a mixture of

radiolabeled and unlabeled sugar, since two volumes of the cell suspension are mixed with one volume of the sugar solution. Thereby, the specific activity can be varied within a wide range. The lower limit we employed in our laboratory was in the range of 8 nCi/μmol sugar. It is advised to determine the sugar concentration range and the optimum specific activity in a pilot experiment.

18. Control uptake reactions using cells transformed with the empty vector and the *HXT* plasmid should be performed in parallel with each set of experiments.
19. The time is dependent on the uptake efficiency of the respective transporter and should be determined in a pilot experiment.
20. The cpm_M should be multiplied by factor 10 to account for the different volumes used in **steps 14** and **17** in Subheading 3.5. The sugar concentration refers to the final concentration after mixing the cell suspension with the threefold concentrated sugar solution (**step 12**, Subheading 3.5). The reference volume is 100 μL (the volume of the filtered cell/sugar mixture from **step 14**, Subheading 3.5).
21. The *cdw* must relate to the actual cell mass on the filter (i.e., it must correspond to 67 μL of the cell suspension from **step 6**, Subheading 3.5).

Acknowledgement

We thank Dr. Joanna Tripp for helpful comments on the manuscript.

References

1. Boles E, Hollenberg CP (1997) The molecular genetics of hexose transport in yeasts. FEMS Microbiol Rev 21(1):85–111. https://doi.org/10.1111/j.1574-6976.1997.tb00346.x
2. Wieczorke R, Krampe S, Weierstall T et al (1999) Concurrent knock-out of at least 20 transporter genes is required to block uptake of hexoses in Saccharomyces cerevisiae. FEBS Lett 464(3):123–128. https://doi.org/10.1016/S0014-5793(99)01698-1
3. Wieczorke R, Dlugai S, Krampe S et al (2003) Characterisation of mammalian GLUT glucose transporters in a heterologous yeast expression system. Cell Physiol Biochem 13(3):123–134. https://doi.org/10.1159/000071863
4. Vignault C, Vachaud M, Cakir B et al (2005) VvHT1 encodes a monosaccharide transporter expressed in the conducting complex of the grape berry phloem. J Exp Bot 56(415):1409–1418. https://doi.org/10.1093/jxb/eri142
5. Schussler A, Martin H, Cohen D et al (2006) Characterization of a carbohydrate transporter from symbiotic glomeromycotan fungi. Nature 444(7121):933–936. https://doi.org/10.1038/nature05364
6. Price DRG, Tibbles K, Shigenobu S et al (2010) Sugar transporters of the major facilitator superfamily in aphids; from gene prediction to functional characterization. Insect Mol Biol 19(2):97–112. https://doi.org/10.1111/j.1365-2583.2009.00918.x

7. Chen L, Hou B, Lalonde S et al (2010) Sugar transporters for intercellular exchange and nutrition of pathogens. Nature 468(7323): 527–532. https://doi.org/10.1038/nature09606
8. Young E, Poucher A, Comer A et al (2011) Functional survey for heterologous sugar transport proteins, using Saccharomyces cerevisiae as a host. Appl Environ Microbiol 77(10):3311–3319. https://doi.org/10.1128/AEM.02651-10
9. Coelho MA, Goncalves C, Sampaio JP et al (2013) Extensive intra-kingdom horizontal gene transfer converging on a fungal fructose transporter gene. PLoS Genet 9(6):e1003587. https://doi.org/10.1371/journal.pgen.1003587
10. Xuan YH, Hu YB, Chen L et al (2013) Functional role of oligomerization for bacterial and plant SWEET sugar transporter family. Proc Natl Acad Sci U S A 110(39):E3685–E3694. https://doi.org/10.1073/pnas.1311244110
11. Young EM, Comer AD, Huang H et al (2012) A molecular transporter engineering approach to improving xylose catabolism in Saccharomyces cerevisiae. Metab Eng 14(4):401–411. https://doi.org/10.1016/j.ymben.2012.03.004
12. Subtil T, Boles E (2012) Competition between pentoses and glucose during uptake and catabolism in recombinant Saccharomyces cerevisiae. Biotechnol Biofuels 5(1):14. https://doi.org/10.1186/1754-6834-5-14
13. Wang C, Bao X, Li Y et al (2015) Cloning and characterization of heterologous transporters in Saccharomyces cerevisiae and identification of important amino acids for xylose utilization. Metab Eng 30:79–88. https://doi.org/10.1016/j.ymben.2015.04.007
14. Subtil T, Boles E (2011) Improving L-arabinose utilization of pentose fermenting Saccharomyces cerevisiae cells by heterologous expression of L-arabinose transporting sugar transporters. Biotechnol Biofuels 4:38. https://doi.org/10.1186/1754-6834-4-38
15. Jordan P, Choe J, Boles E et al (2016) Hxt13, Hxt15, Hxt16 and Hxt17 from Saccharomyces cerevisiae represent a novel type of polyol transporters. Sci Rep 6:23502. https://doi.org/10.1038/srep23502
16. Scarcelli JJ, Colussi PA, Fabre A et al (2012) Uptake of radiolabeled GlcNAc into Saccharomyces cerevisiae via native hexose transporters and its in vivo incorporation into GPI precursors in cells expressing heterologous GlcNAc kinase. FEMS Yeast Res 12(3):305–316. https://doi.org/10.1111/j.1567-1364.2011.00778.x
17. Farwick A, Bruder S, Schadeweg V et al (2014) Engineering of yeast hexose transporters to transport D-xylose without inhibition by D-glucose. Proc Natl Acad Sci U S A 111(14):5159–5164. https://doi.org/10.1073/pnas.1323464111
18. Kasahara T, Kasahara M (1996) Expression of the rat GLUT1 glucose transporter in the yeast Saccharomyces cerevisiae. Biochem J 315:177–182. https://doi.org/10.1042/bj3150177
19. Kasahara T, Kasahara M (1997) Characterization of rat Glut4 glucose transporter expressed in the yeast Saccharomyces cerevisiae: comparison with Glut1 glucose transporter. Biochim Biophys Acta 1324(1):111–119. https://doi.org/10.1016/S0005-2736(96)00217-9
20. Gietz RD, Schiestl RH (2007) High-efficiency yeast transformation using the LiAc/SS carrier DNA/PEG method. Nat Protoc 2(1):31–34. https://doi.org/10.1038/nprot.2007.13
21. Guthrie C, Fink GR (eds) (2002) Guide to yeast genetics and molecular and cell biology. Methods in enzymology. Acad. Press, Amsterdam
22. Oldenburg KR, Vo KT, Michaelis S et al (1997) Recombination-mediated PCR-directed plasmid construction in vivo in yeast. Nucleic Acids Res 25(2):451–452. https://doi.org/10.1093/nar/25.2.451
23. Gietz RD, Schiestl RH (2007) Frozen competent yeast cells that can be transformed with high efficiency using the LiAc/SS carrier DNA/PEG method. Nat Protoc 2(1):1–4. https://doi.org/10.1038/nprot.2007.17
24. Boles E (2002) Yeast as a model system for studying glucose transport. In: Quick MW, Sibley DR (eds) Transmembrane transporters. Wiley, Hoboken, NJ, pp 19–36
25. Boles E, Zimmermann FK (1993) Saccharomyces cerevisiae phosphoglucose isomerase and fructose bisphosphate aldolase can be replaced functionally by the corresponding enzymes of Escherichia coli and Drosophila melanogaster. Curr Genet 23(3):187–191. https://doi.org/10.1007/BF00351494
26. Nickoloff JA (ed) (1995) Electroporation protocols for microorganisms. Methods in molecular biology, v. 47. Humana Press, Totowa, NJ

Chapter 11

Identification of Insulin-Activated Rab Proteins in Adipose Cells Using Bio-ATB-GTP Photolabeling Technique

Francoise Koumanov and Geoffrey D. Holman

Abstract

We have recently developed a photolabeling method to identify GTP-loaded Rab proteins. The new biotinylated GTP analogue (Bio-ATB-GTP) binds to GTP-binding proteins and after a UV irradiation a covalent bond is formed between the protein and the photoreactive diazirine group on the photolabel. The tagged protein can then be isolated and detected using the classic biotin-streptavidin interaction. In this chapter, we describe the Bio-ATB-GTP photolabel and discuss the advantages of using this photolabeling approach to detect GTP-loaded Rab proteins compared to other existing methodologies. We also describe a step-by-step procedure for detecting the activated state of a Rab protein in primary rat adipocytes.

Key words GLUT4, Adipocytes, Rab, Insulin, Trafficking, Membranes, Bio-ATB-GTP

1 Introduction

Insulin is known to facilitate glucose uptake by stimulating the subcellular trafficking of GLUT4 from its intracellular reservoir compartment to the plasma membrane of target cells that include adipose, heart, and skeletal muscle cells [1]. Recent studies have suggested that exocytosis is the main site of stimulated traffic of the membrane vesicles that contain GLUT4 transporters [2–5]. The early steps in insulin signaling to this process are fairly well resolved; however, the mechanisms by which vesicle traffic is linked to GLUT4 vesicle traffic remain elusive. Among the stimulatory and linking pathways that have been proposed are phosphorylation-induced deactivation of Rab GAPs TBC1D1 and the related TBC1D4 and presumed elevation of the GTP loading of Rab substrates of these GAPs [6–8].

Many of the methods used for studying the activated GTP-loaded state of small GTPases in cells require detailed knowledge of the function and binding of effector proteins. Such effector proteins only bind to the GTP-activated state and therefore binding of

Karin Lindkvist-Petersson and Jesper S. Hansen (eds.), *Glucose Transport: Methods and Protocols*,
Methods in Molecular Biology, vol. 1713, https://doi.org/10.1007/978-1-4939-7507-5_11,

these proteins is taken as a marker of the GTP-loaded state. For example there a numerous assay methods for well-studied proteins such as Ras which binds Raf1 only in its GTP-bound state [9]. Such methods are not available for most Rab proteins as the effector proteins are often unknown. In an attempt to study Rab proteins that are activated by insulin we developed a series of novel biotinylated GTP photolabels [10]. The substitution of diazirine photoreactive moiety was chosen over an azide, because numerous studies have progressively shown that the diazirines offer greater selectivity and a cleaner photochemical reaction [11]. On UV irradiation the diazirines loose nitrogen and produce a carbene that very effectively inserts into target proteins. Azides loose nitrogen and produce a nitrene that inserts into proteins, but can also undergo many side reactions generating intermediates that react electrolytically with targets. Furthermore, diazirines are far more chemically stable, which is important in synthesis of intermediate products and precursor reagents prior to coupling to GTP, and they can be stored for very long periods at 0–4 °C. The biotin moiety is very useful and allows isolation of the photolabeled G-protein using immobilized streptavidin beads (*see* **Subheading 3** below) using well-established and extensively used techniques. The biotin-ATB is synthesized as an activated ester, usually a N-hydroxysuccinimide (*see* Fig. 1), and is attached through to an amine-substituted GTP to form a stable amide linkage. The amine on the GTP is linked via an ethylenediamine spacer to the 2/3-hydroxyl group(s) of the ribose moiety of the GTP. This position, for the bulky biotin-ATB substitution, was chosen as the crystal structure of Rab 3 with bound GppNHp suggested that the ribose hydroxyls are exposed on the surface of the structure while other parts of the nucleotide base and phosphates of the GTP are bound deep within the protein with very little available space around these regions [12]. A previously used azide type of photolabel has also been used to study Rab proteins in insulin target cells [13, 14]. This compound has the azide substitution into the terminal phosphate group of the GTP while the biotin moiety is separately substituted into an oxidized and ring-open form of the ribose (*see* Fig. 2). These comparisons suggest that the diazirine compound offers greater potential for selective labeling and protein identification. Furthermore, the azide-biotin GTP appears to be no longer commercially available.

The use of photolabels for the labeling and subsequent identification of G-proteins that are activated downstream of insulin signaling assumes that these proteins can maintain the activated state form (presumably by binding effectors as described above) and can allow exchange of bound GTP for the biotin-GTP photolabel. The methods described below for facilitating this exchange involve firstly a reduction in magnesium level to facilitate the exchange and then re-addition of magnesium to stabilize the photolabel in the binding site. Whether this manipulation of the magnesium is always required has not been

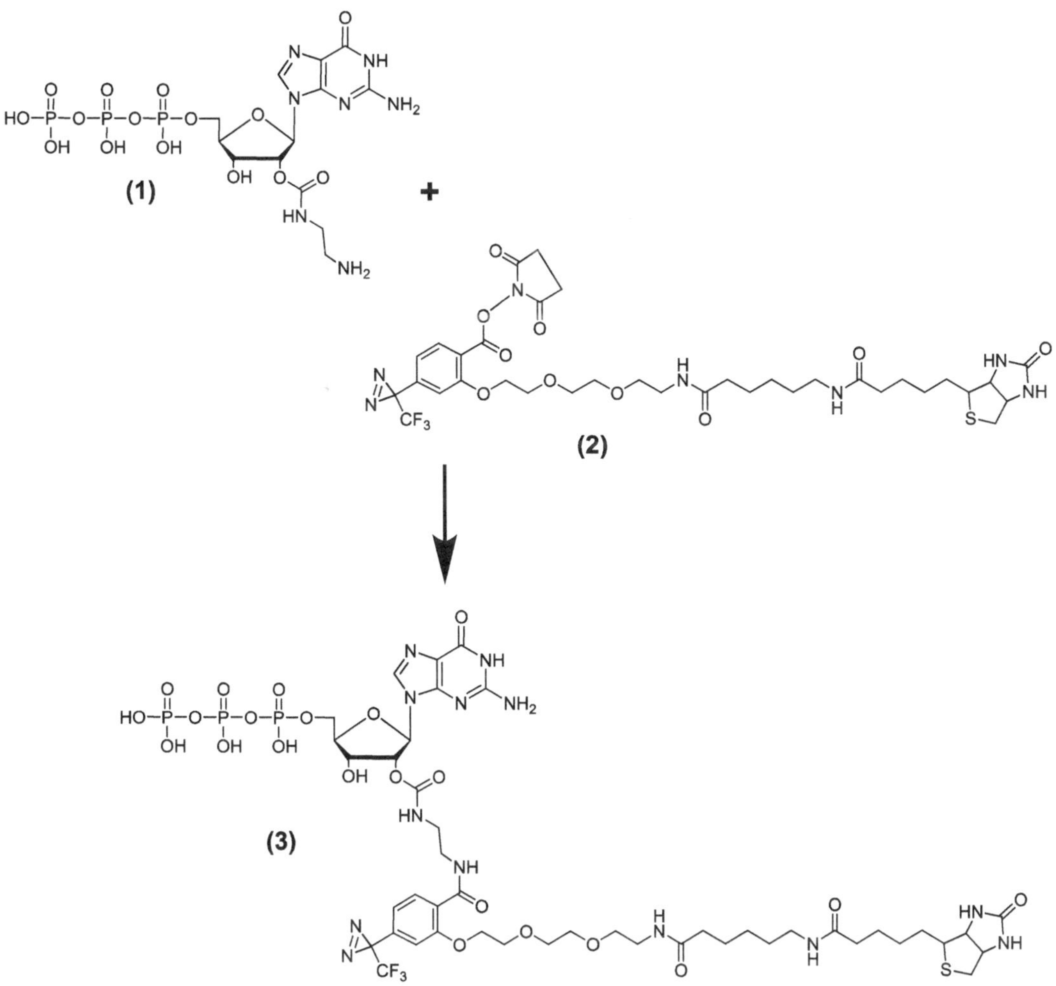

Chemical Formula: $C_{44}H_{63}F_3N_{13}O_{22}P_3S$
Exact Mass: 1307.3096

Fig. 1 Bio-ATB-GTP synthesis and structure. Bio-ATB-GTP full synthesis is described in [2] and the final step is represented here. GTP-ethylene diamine (*1*) and Bio-LC-ATB-NHS (*2*) were combined to generate Bio-ATB-GTP (*3*)

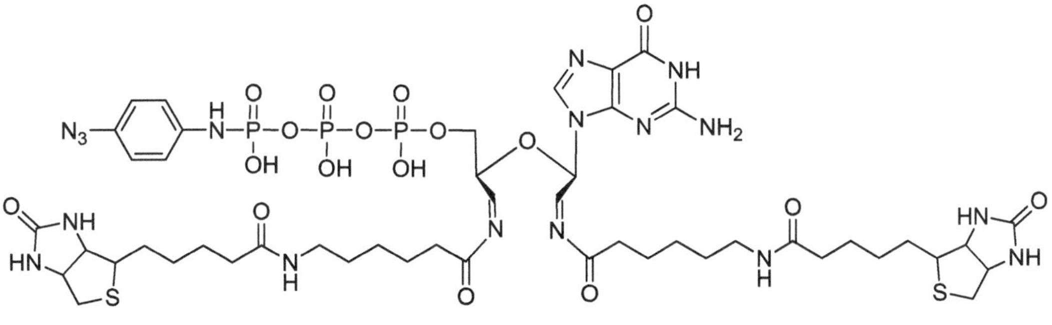

Fig. 2 Azidoanilide-GTP biotin-hydrozone structure. The structure is redrawn from [15]

explored but there are many examples of GTP analogue exchange reactions occurring in permeabilized cells [14] or in cell lysate [15], which do not require such a magnesium manipulation. The mechanistic basis for the proposed retention of Rab proteins in their activated, and possibly effector-bound, states in response to insulin action is unproven, but empirical data suggests that insulin-activated GTP labeling of Rab proteins can occur. It could be that a complex of proteins including GEFs, GAPs, GDIs, and GTP effectors all maintain the activated state and allow GTP-substituted photolabels to bind.

There are several approaches that can be used to determine which Rab proteins are activated by insulin. If antibodies to candidate Rab proteins are available, then a targeted approach can be used. This involves insulin treatment of cells followed by GTP photolabeling of isolated membranes, separation of solubilized proteins by SDS-PAGE, and immunoblotting with the target-specific antibody. We give an example of this approach below (*see* Fig. 3). A more generic, and potentially more powerful, approach involves the same insulin treatment and photolabeling steps, but then solubilized protein is subjected to 2D separation of components firstly by isoelectric focusing and then by SDS-PAGE in a second dimension. Using this method we have recently identified Rab3 as an insulin-activated Rab protein in adipocytes [10]. Numerous proteins are labeled specifically and the protein spots are clearly reduced by GTP competition. Furthermore, insulin action appears to reduce the GTP labeling of many proteins; in fact more proteins appear to have lower rather than higher labeling following insulin treatment (*see* Fig. 4). Clearly, it would be ideal to be able to identify all of the spots on the 2D gels that either increase or decrease in comparing with and without insulin treatment. However, there are currently several technical problems that limit the potential of this generic approach. Firstly, the method used for visualizing the biotin-positive spots involves partial transfer of some of the 2D gel material to a blotting membrane, which is subsequently treated with detecting reagents (usually based on binding streptavidin-peroxidase). This detection is extremely sensitive when compared to protein staining and we have found that some very strongly detected spots have very little protein content, which is often below the protein detection limit. Secondly, the protein distribution on the 2D gels is highly complex with many overlapping spots occurring in some regions of the gels, particularly around the 20 kDa region where one would expect to find small G-proteins. Contaminating proteins are also present on the 2D gels as no selection of solubilized proteins occurs. Many highly abundant scaffolding proteins and metabolic enzyme proteins often occur in regions of the gel which are identified by blotting to also have a potentially interesting G-protein. Fortunately, some Rab proteins (such as the identified Rab 3 and Rab 11) occur in "cleaner" regions of the gel where aligning the streptavidin-peroxidase detecting blot with the protein gel allows the spot to be picked and identified by mass spectroscopy (*see* Fig. 4).

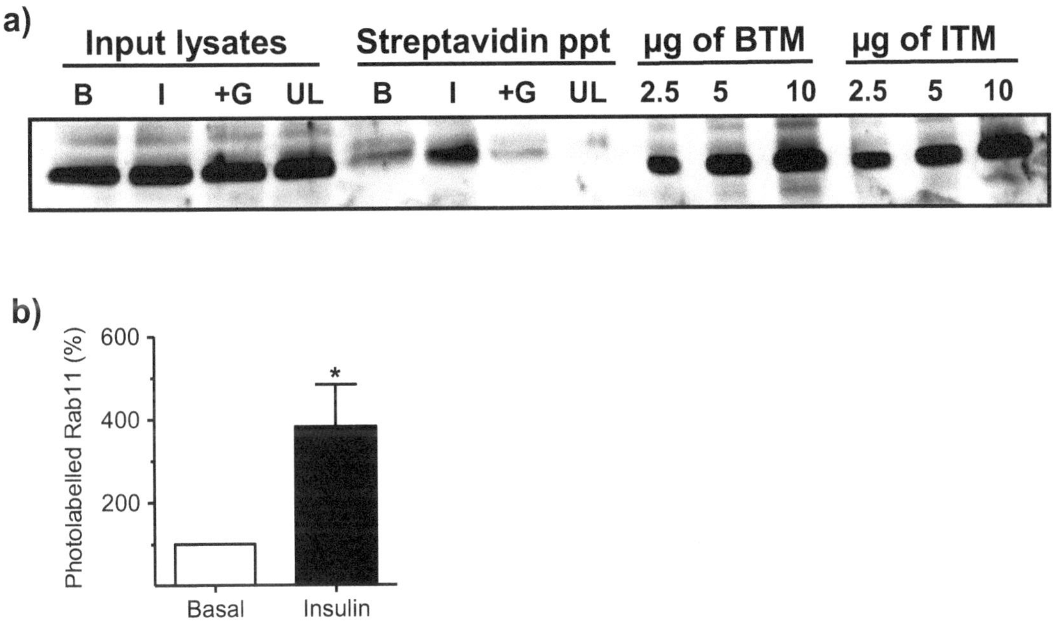

Fig. 3 Example of Bio-ATB-GTP photolabeling of rat adipocyte total membranes and identification of Rab11 GTP loading in response to insulin stimulation. (**a**) Total membrane from rat adipocytes was isolated and labeled with Bio-ATB-GTP photolabel as described in the text. After membrane solubilization in RIPA buffer (input lysate) and streptavidin precipitation (streptavidin ppt) the eluted fractions were resolved by SDS-PAGE and immunoblotted for presence of Rab11 using mouse monoclonal anti-Rab11 antibody (BD Bioscience). To standardize the level of photolabeled Rab11 dilutions of basal and insulin-stimulated total membranes (BTM and ITM, respectively) were also run in the gel. B—basal (unstimulated); I—insulin stimulated; +G—100 molar excess of GTP; UL—unlabeled sample. (**b**) Quantification of the data presented in (**a**). Mean and SEM from at least three independent experiments. *$p < 0.05$ vs. basal

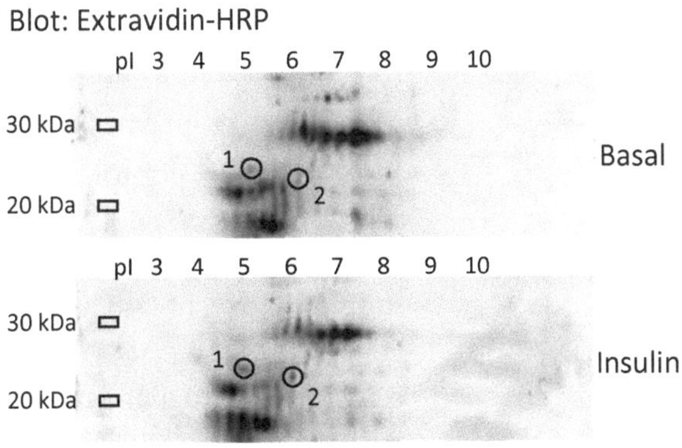

Fig. 4 2D gel analysis of Bio-ATB-GTP-labeled rat adipocyte total membranes. Biotinylated proteins were blotted with extravidin-HRP. Circled spots 1 (Rab3B) and 2 (Rab11) indicate proteins for which the biotin signal was identified as increasing upon insulin stimulation. Reproduced with permission from [12]

2 Materials

All chemical reagents are purchased at the highest purity available. All the aqueous solutions are prepared with double-distilled or MilliQ quality water.

2.1 Preparation of Total Membranes from Rat Adipose Cells

1. Male Wistar rats: 180–220 g.
2. KRH buffer: 140 mM NaCl, 4.7 mM KCl, 2.5 mM $CaCl_2$, 1.25 mM $MgSO_4$, 2.5 mM NaH_2PO_4, 10 mM HEPES, pH 7.4.
3. KRH with 1 % w/v bovine serum albumin (BSA) buffer: KRH buffer supplemented with 1 % w/v BSA.
4. Collagenase solution: Type I collagenase 270 U/mg (Worthington).
5. Digestion buffer: KRH supplemented with 3.5 % w/v BSA and 0.7 mg/mL Type I collagenase.
6. Digestion tubes: 30 mL Universal polystyrene tubes.
7. 20 mL Syringe attached to a stainless steel 13-gauge needle.
8. Melting-point capillary tubes 100 mm × 1.3–1.4 o/d (Bilbate).
9. HES-buffer: 20 mM HEPES pH 7.2, 1 mM EDTA, 255 mM sucrose.
10. Shaking water bath set to 37 °C.
11. Non-shaking water bath set to 37 °C.
12. Potter-Elvehjem homogenizer.
13. Ultracentrifuge.
14. Protein concentration measurement reagent.

2.2 Bio-ATB-GTP Photolabeling of Total Membrane

1. HES buffer: 20 mM HEPES pH 7.2, 1 mM EDTA, 255 mM sucrose.
2. EDTA solution: 25 mM EDTA, pH 7.2.
3. Magnesium solution: 100 mM $MgCl_2$.
4. Bio-ATB-GTP photolabel: The stock is 800 μM in 20 mM sodium bicarbonate.
5. RIPA buffer: 50 mM Tris pH 7.4, 150 mM NaCl, 0.5 % w/v sodium deoxycholate, 0.1 % w/v SDS, 0.1 % v/v NP-40 supplemented with protease inhibitors (without EDTA).
6. Rayonet Photochemical Reactor (Southern New England Ultraviolet Co) with 3000 Å lamps.

2.3 Streptavidin Precipitation and Detection of Bio-ATB-GTP-Labeled GTP-Binding Proteins

1. RIPA buffer: 50 mM Tris pH 7.4, 150 mM NaCl, 0.5 % w/v sodium deoxycholate, 0.1 % w/v SDS, 0.1 % v/v NP-40 supplemented with protease inhibitors (without EDTA).

   RIPA (1:10) buffer: RIPA buffer diluted in PBS, 1 in 10 volumes.

2. Streptavidin-agarose beads (PIERCE).
3. PBS buffer: 12.5 mM Na_2HPO_4, pH 7.2, 154 mM NaCl.
4. SDS sample buffer: 210 mM Tris–HCl, pH 6.8, 0.03 % w/v bromophenol blue, 30 % v/v glycerol, 6 % w/v SDS. Prepare a two times concentrated stock (2× SDS sample buffer) and dilute 1 to 1 with water to get a 1× SDS sample buffer.
5. Dithiothreitol (DTT) solution: 1 M stock of DTT.
6. Dry-heating block.
7. End-to-end tube rotator.
8. Benchtop microfuge.
9. 12 % Tris-glycine gels (*see* **Note 1**) and SDS-PAGE equipment.
10. Semidry electrotransfer apparatus and reagent (*see* **Note 2**).
11. TBS-T buffer: 10 mM Tris–HCl pH 7.4, 0.9 % w/v NaCl, 0.1 % v/v Tween-20.
12. Antibodies against the protein(s) of interest.
13. Enhanced chemiluminescent (ECL) reagent (*see* **Note 3**).
14. Chemiluminescence digital imager or X-ray film developer.

3 Methods

3.1 Preparation of Total Membranes from Primary Rat Adipose Cells (See Note 4)

1. Place epididymal fat pads of male Wistar rats in digestion buffer (2 fat pads are 1 rat equivalent in 2.5 mL of digestion buffer). To get a good membrane preparation we recommend starting with at least 4 fat pads, meaning fat pads from 2 rats.
2. Using fine scissors to mince the fat pads into small pieces, 1–2 mm wide.
3. Place the digestion tubes in a shaking water bath at 37 °C and shake for 45–50 min or until the lumps are digested at a speed of 100–150 rpm.
4. At the end of the digestion time add an equal volume of KRH with 1 % BSA and filter the cell suspension through a 250 μm nylon mesh into new 23 mL polystyrene tubes.
5. Fill the tubes to the top with KRH buffer supplemented with 1 % w/v BSA, place the tubes in the 37 °C water bath without shaking, and leave the cells to float to the top for 1–2 min.
6. Using a 20 mL syringe attached to a 13-gauge needle, remove the buffer below the adipocyte layer. Fill the tube with 1 % w/v BSA in KRH, mix the cells gently by inverting the tube, and leave the cells to float to the top for 2–3 min (in a 37 °C water bath).
7. Repeat **step 6** twice. This will remove any residual collagenase and non-adipose cells from the preparation.

8. Leave the cells to settle for 5 min, and remove some of the buffer to get approximately one volume of buffer for two volumes of cells. Measure the cytocrit of the cell suspension using melting-point capillary tubes. Adjust the cell suspension to 40 % cytocrit (packed cells to buffer).
9. Maintain cells at 37 °C in KRH with 1 % w/v BSA in the non-shaking water bath.
10. Split the cell suspension into the required number of conditions. Make sure that the cells are well mixed before pipetting out the desired volume. Use a cut pipette tip, or a wide-bore pipette tip, to pipette out the cell suspension, avoiding lysing the adipose cells.
11. Treat the cells as desired (e.g., with insulin).
12. At the end of the treatment, wash cells once with KRH buffer without BSA.
13. Resuspend the cells in HES buffer supplemented with protease inhibitors at 18 °C.
14. Leave cells to float in a water bath at 18 °C.
15. Remove buffer and resuspend the cells in fresh HES buffer supplemented with protease inhibitors at 18 °C.
16. Transfer the cells to a precooled Potter-Elvehjem homogenizer at 4 °C.
17. Homogenize the cells using the homogenizer (*see* **Note 5**).
18. The homogenate is spun at 1,000 × *g* for 1 min at 4 °C to remove the fat and the unbroken cells (*see* **Note 6**). This also removes some of the nuclei.
19. The resulting supernatant is spun at 300,000 × *g* for 30 min to give the crude membrane pellet.
20. The pellet is resuspended in HES buffer and homogenized again to make sure that all cells are broken.
21. The resulting suspension is spun again at 300,000 × *g* for 30 min.
22. The pellet is resuspended in HES buffer. The pellet is very dense and usually needs passing through a 27-gauge needle several times to make sure that the membranes are well resuspended.
23. Perform a protein assay and resuspend the pellet to a final concentration of 5–10 mg/mL.
24. Aliquot and snap freeze in liquid nitrogen and store at −80 °C.

3.2 Photolabeling of Total Membranes with Bio-ATB-GTP Photolabel

The Bio-ATB-GTP photolabeling of total membranes is performed in clear 96-well plates, UV grade. All the preparation steps are performed on ice and all the procedures from the moment the photolabel is added to the plate and until the end of the UV irradiation are performed under semidark conditions (*see* **Note 7**).

1. Thaw aliquots of total membranes on ice. Usually for photolabeling of endogenous Rab proteins, we use 300 μg of total membranes per condition. If using membranes from adipocytes that express recombinant proteins, then use 100–150 μg of total membranes.
2. In a pre-chilled 96-well plate, mix the desired amount of total membranes, add the EDTA solution to obtain a final concentration of 2 mM EDTA, and make up the volume to 100 μL with HES buffer. Below is an example of an experimental setup with total membranes at a protein concentration of 10 mg/mL: 30 μL total membranes + 8 μL 25 mM EDTA + 52 μL HES buffer. This accounts for the additional 5 μL volumes in **steps 4** and **5**. Additionally, a control can be added in the experiment when new GTP-binding proteins are studied (*see* **Note 8**).
3. Incubate for 5 min on ice.
4. Under low-light conditions, add 5 μL of 800 μM Bio-ATB-GTP. Mix well with the pipette and incubate for 30 min on ice.
5. Add 5 μL of magnesium solution. Mix well with the pipette and incubate for 10 min on ice.
6. Irradiate, on ice, using 300 nm UV light for 2 min (*see* **Note 9**).
7. Transfer the samples from the wells into 3 mL ultracentrifuge tubes (polycarbonate thick-wall Beckman tubes).
8. Resuspend in 3 mL HES buffer supplemented with 5 mM $MgCl_2$.
9. Spin at 300,000 × *g* for 30 min at 4 °C to pellet the total membrane fraction and remove any unreacted photolabel.
10. Resuspend the total membrane pellet in 1 mL RIPA buffer, transfer in 1.5 mL centrifuge tubes, and lyse for 20 min with end-to-end rotation at room temperature.
11. Pellet the insoluble material at 17,000 × *g* for 20 min at 4 °C. Transfer the supernatant to new tubes. This is now the total membrane lysate, which will be used to precipitate the tagged Rab proteins.

3.3 Streptavidin Precipitation of Bio-ATB-GTP-Labeled Material

1. Pipette 80 μL 50 % w/v streptavidin agarose beads per condition (40 μL of settled beads) into a 1.5 mL centrifuge tube.
2. Wash the beads (spin at 3,500 × *g* for 1 min in benchtop microfuge) by washing twice with 1 mL PBS, and then once with 1 mL RIPA.
3. Add 900 μL total membrane lysate (*see* **Subheading 3.2**) to the beads and rotate overnight at 4 °C on an end-to-end rotator (*see* **Note 10**).

4. On the following day, spin down the beads at 3,500 × *g* for 1 min in benchtop microfuge and collect the supernatants (*see* **Note 11**)
5. Wash the beads twice with 1 mL RIPA buffer, twice with 1 mL dilute RIPA (1:10) buffer, and twice with 1 mL PBS buffer. Do so by spinning down the beads at 3,500 × *g* for 1 min in a benchtop microfuge, remove supernatant, and replenish with new buffer.
6. Elute the proteins from the beads by adding 60 μL 2× SDS sample buffer. Heat for 20 min at 95 °C in a dry-heating block (*see* **Note 12**), vortex, and spin at 3,500 × *g* for 1 min in the benchtop microfuge to pellet the beads. Transfer the supernatant in a clean tube.
7. Repeat **step 6,** but add 40 μL of 1× SDS sample buffer. Collect the supernatant and pool this together with the eluted sample in the previous **step 6**.
8. Add DTT solution to the pooled eluates at a final concentration of 20 mM. Vortex and incubate again for 20 min at room temperature. Spin for 1 min at 16,200 × *g* to discard insoluble material.
9. Load the solubilized material onto a 12 % Tris-glycine SDS polyacrylamide gel (*see* **Notes 1** and **13**).
10. A typical lane sequence is presented in Fig. 3a (also *see* **Notes 14** and **15**).
11. Run the gel in the SDS-PAGE system.
12. Transfer the gels onto a nitrocellulose membrane by semidry electrotransfer for 110 min applying a current of 0.8 mA/cm^2.
13. Immunoblot for the specific Rab proteins of interest (*see* **Note 16**).
14. Block the membrane with nonfat milk.
15. Incubate the membranes overnight at 4 °C with the antibody against the protein of interest (diluted in TBS-T supplemented with 1 % w/v BSA).
16. Wash the blots four times for 5 min with TBS-T.
17. Incubate with secondary antibodies conjugated to peroxidase, diluted in TBS-T supplemented with 5 % w/v nonfat milk, for 1 h at room temperature.
18. Wash the blots six times for 5 min in TBS-T.
19. Use ECL substrate or enhanced substrates if the signal is low (*see* **Note 5**) to generate chemiluminescence and image the emission either on film or in chemiluminescence imager.
20. The generated digital images are analyzed using imaging software (*see* **Note 17**) and the relative intensity of the detected bands is

quantified (e.g., VisionWorksLS from UVP). Examples of the quantification results we obtained for the insulin activation of endogenous Rab11 and heterologously expressed FLAG-tagged Rab3B in primary rat adipocytes are presented in Figs. 3 and 5.

4 Notes

1. We use self-made gels, but precast gels can be used. The percentage of the gel matrix density will depend on the protein of interest; that is, for small GTPases of the Rab family, which all have a molecular weight between 20 and 30 kDa, a 12 % gel is appropriate to obtain a good resolution of the proteins.
2. Our laboratory routinely uses the semidry method of transfer with BioRad semidry blot apparatus. We transfer the protein onto nitrocellulose membranes. However, wet transfer and PVDF membranes should work just as well.
3. Sometimes the protein of interest is of low abundance and is poorly labeled and the classic ECL reagent is not sufficient to detect the signal. In such cases we then use stronger ECL reagents

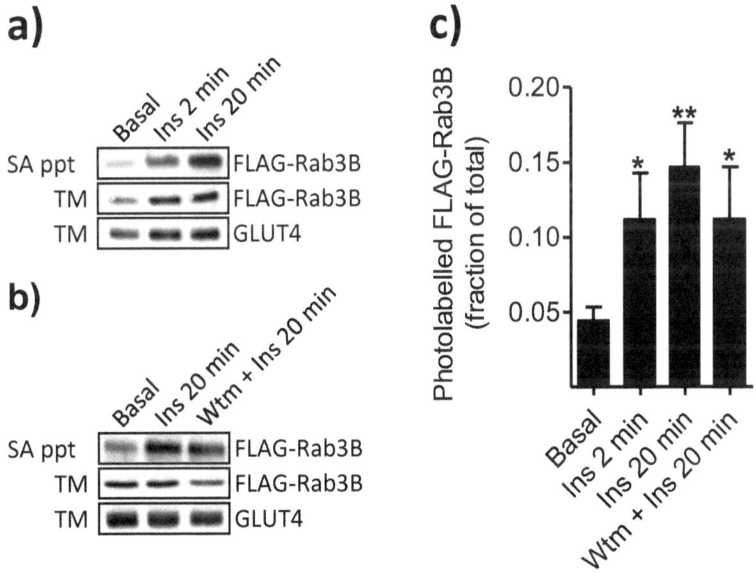

Fig. 5 FLAG-Rab3B is activated upon insulin stimulation. (**a**) Insulin stimulation (Ins) of the Bio-ATB-GTP-loading state of FLAG-Rab3B. Streptavidin-precipitated proteins were detected with anti-FLAG antibody. (**b**) Effect of wortmannin (200 nmol/L for 10 min) (Wtm) on insulin-stimulated FLAG-Rab3B GTP loading. (**c**) Quantification of the Western blotting data presented. Mean and SEM from three independent experiments. *$p < 0.05$ vs. basal. Reproduced with permission from [12]

such as ECL Select from GE Healthcare or SuperSignal West Femto Extended Duration Substrate from Thermo Scientific.

4. We routinely prepare our total membrane from primary rat adipocytes; however total membrane can be prepared in similar way from many different cell types and cultured cell lines.
5. We use a motorized homogenizer at a speed of 500 rpm.
6. If you are performing photolabeling of recombinant proteins expressed in adipocytes, collect aliquots of the supernatants at **steps 11** and **12** in **Subheading 3.1**. This will generate samples that can be analyzed later for the distribution of the recombinant protein between the soluble and membrane fractions (*see* **Note 15**).
7. Stock solutions are kept in amber vials and incubation samples are covered with aluminum foil.
8. When we first study the GTP activation of GTP-binding proteins we usually have two control conditions. The first uses a 100× molar excess of GTP in the incubation mixture to show the specific competitive inhibition of the photolabeling reaction. The second uses an unlabeled control (no Bio-ATB-GTP photolabel added to the reaction mixture) to show the nonspecific detection of extraneous protein carried over during the procedure.
9. We place a precooled metal plate wrapped in aluminum foil under the 96-well plate to reflect the light and maximize the labeling.
10. Make sure that approximately 100 μL of total membrane lysate is collected and later run on the gel to detect the total amount of Rab protein in the membrane lysate.
11. Collect the supernatant and keep until the end of the experiment. If the results are fine then discard these tubes. If the results are unexpected then these samples could be used for troubleshooting of the experiment.
12. Pierce the lid of the tube with a 27-gauge needle to avoid the tube popping open during the incubation.
13. The CBS gel system (VWR) is as wide as a Protean II system from BioRad (16 cm) but with the length of a miniProtean 3 system (6 cm resolving). This allows us to load the entire eluted sample (100 μL) in one well and to run the gel in 1 h. Because the streptavidin precipitation requires two elutions, the final volume of eluted samples is big and cannot be loaded on a normal miniProtean system. If you do not have such a system, then you might need to run the Protean II gel system (BioRad). Alternatively, if the only available option is a minigel system, then smaller elution volumes will have to be used in **steps 6** and 7 in **Subheading 3.3**. Although feasible, this might result in an inefficient elution.

14. *Optional*: We usually run a dilution range of total membranes to standardize how much protein has been labeled.
15. When photolabeling recombinant proteins that have been expressed in adipocytes, all the fractions collected during the total membrane preparation (*see* **Note 6**) are run on a separate gel to make sure that we have equal amounts of recombinant protein expressed in each condition. This will also allow normalization of the photolabeling data for any variation in the expression levels.
16. The protocol given is our laboratory protocol for Western blotting; however any protocol that is optimized for the antibody against the protein of interest would work.
17. For image analyses we use the software provided with our imaging system from UVP, but free quantification software, such as ImageJ National Institute of Health software (http://imagej.nih.gov/ij/), can also be used.

References

1. Stockli J, Fazakerley DJ, James DE (2011) GLUT4 exocytosis. J Cell Sci 124(Pt 24):4147–4159. https://doi.org/10.1242/jcs.097063
2. Koumanov F, Jin B, Yang J, Holman GD (2005) Insulin signaling meets vesicle traffic of GLUT4 at a plasma-membrane-activated fusion step. Cell Metab 2(3):179–189. https://doi.org/10.1016/j.cmet.2005.08.007
3. Huang S, Lifshitz LM, Jones C, Bellve KD, Standley C, Fonseca S, Corvera S, Fogarty KE, Czech MP (2007) Insulin stimulates membrane fusion and GLUT4 accumulation in clathrin coats on adipocyte plasma membranes. MolCell Biol 27(9):3456–3469
4. Lizunov VA, Matsumoto H, Zimmerberg J, Cushman SW, Frolov VA (2005) Insulin stimulates the halting, tethering, and fusion of mobile GLUT4 vesicles in rat adipose cells. J Cell Biol 169(3):481–489
5. Stenkula KG, Lizunov VA, Cushman SW, Zimmerberg J (2010) Insulin controls the spatial distribution of GLUT4 on the cell surface through regulation of its postfusion dispersal. Cell Metab 12(3):250–259. https://doi.org/10.1016/j.cmet.2010.08.005
6. Koumanov F, Holman GD (2007) Thrifty Tbc1d1 and Tbc1d4 proteins link signalling and membrane trafficking pathways. Biochem J 403(2):e9–11. https://doi.org/10.1042/BJ20070271
7. Sano H, Eguez L, Teruel MN, Fukuda M, Chuang TD, Chavez JA, Lienhard GE, McGraw TE (2007) Rab10, a target of the AS160 Rab GAP, is required for insulin-stimulated translocation of GLUT4 to the adipocyte plasma membrane. Cell Metab 5(4):293–303
8. Sano H, Kane S, Sano E, Miinea CP, Asara JM, Lane WS, Garner CW, Lienhard GE (2003) Insulin-stimulated phosphorylation of a Rab GTPase-activating protein regulates GLUT4 translocation. J Biol Chem 278(17):14599–14602
9. Taylor SJ, Resnick RJ, Shalloway D (2001) Nonradioactive determination of Ras-GTP levels using activated ras interaction assay. Methods Enzymol 333:333–342. https://doi.org/10.1016/S0076-6879(01)33067-7
10. Koumanov F, Pereira VJ, Richardson JD, Sargent SL, Fazakerley DJ, Holman GD (2015) Insulin regulates Rab3-Noc2 complex dissociation to promote GLUT4 translocation in rat adipocytes. Diabetologia 58(8):1877–1886. https://doi.org/10.1007/s00125-015-3627-3
11. Hashimoto M, Hatanaka Y, Yang J, Dhesi J, Holman GD (2001) Synthesis of biotinylated bis(D-glucose) derivatives for glucose transporter photoaffinity labelling. Carbohydr Res 331(2):119–127
12. Dumas JJ, Zhu Z, Connolly JL, Lambright DG (1999) Structural basis of activation and GTP hydrolysis in Rab proteins. Structure 7(4):413–423

13. Schwenk RW, Eckel J (2007) A novel method to monitor insulin-stimulated GTP-loading of Rab11a in cardiomyocytes. Cell Signal 19(4):825–830. https://doi.org/10.1016/j.cellsig.2006.10.008
14. Sun Y, Bilan PJ, Liu Z, Klip A (2010) Rab8A and Rab13 are activated by insulin and regulate GLUT4 translocation in muscle cells. Proc Natl Acad Sci U S A 107(46):19909–19914. https://doi.org/10.1073/pnas.1009523107
15. Coppola T, Perret-Menoud V, Gattesco S, Magnin S, Pombo I, Blank U, Regazzi R (2002) The death domain of Rab3 guanine nucleotide exchange protein in GDP/GTP exchange activity in living cells. Biochem J 362(Pt 2):273–279

Chapter 12

Total Internal Reflection Fluorescence Microscopy to Study GLUT4 Trafficking

Sebastian Wasserstrom, Björn Morén, and Karin G. Stenkula

Abstract

Total internal reflection fluorescence (TIRF) microscopy is a powerful method that allows examination of plasma membrane close events in real time. The last decade, the method has successfully been used to explore GLUT4 translocation in adipocytes. Here, we describe the procedure for studying GLUT4 trafficking using TIRF microscopy in isolated primary adipocytes.

Key words Primary adipocytes, Fluorescence, TIRF, Microscopy, GLUT4, GLUT4 storage vesicles, GSV

1 Introduction

In the normal insulin-sensitive state, insulin promotes glucose uptake by translocation of insulin-responsive GLUT4 storage vesicles (GSV) to the cell surface, where GLUT4 facilitates diffusion of glucose into the cell [1]. We and other have applied TIRF microscopy, a powerful method that allows examination of plasma membrane close events in real time, and have made significant progress in understanding the last events involved in insulin-regulated GLUT4 translocation and glucose uptake in both cultured 3T3-L1 adipocytes and primary adipocytes isolated from rodents and humans [2–9]. The first report describing GSV trafficking in primary adipocytes by TIRF microscopy was presented by Lizunov et al. [4], followed by more sophisticated analysis of the actual GSV fusion events [2], single GLUT4 molecule movement in the plasma membrane following fusion [5], and characterization of GLUT4 dynamic and cellular response in human adipocytes [10, 11]. This chapter aims to give a detailed description of TIRF imaging of GLUT4 trafficking using isolated primary adipose cells that express a recombinant GLUT4 tracer. For a deep technical background description of the TIRF imaging field as a whole or the technique in itself, the readers are directed to the excellent description by Poulter et al. [12].

Karin Lindkvist-Petersson and Jesper S. Hansen (eds.), *Glucose Transport: Methods and Protocols*, Methods in Molecular Biology, vol. 1713, https://doi.org/10.1007/978-1-4939-7507-5_12,

2 Materials

2.1 Instrumentation

1. We use a commercial TIRF system based on a Nikon Ti-E eclipse microscope equipped with a 100× Apo TIRF DIC oil immersion objective NA of 1.49 (Nikon Instruments Inc.), an iXon Ultra DU-897 EMCCD camera (Andor Technology Ltd.), and four main laser lines, 405 (Cube, Coherent Inc), 488 (Melles-Griot), 561 (Sapphire, Coherent Inc), and 640 (Cube, Coherent Inc) with corresponding filter sets. The system is equipped with a "perfect focus system," an independent IR laser that detects the position of the coverslip and prevents focus drift during acquisition. For microscope control and image acquisition we use NIS elements (Laboratory Imaging, version 4.50).
2. An EMCCD camera has benefits for both acquisition speed and sensitivity. The settings should be optimized for the biological question addressed. If vesicle traffic over time is being studied the camera settings should be set to an exposure time, low enough to allow fast events to be captured. To minimize cell damage and photo bleaching, optimize camera settings for optimal image quality.
3. For electroporation we use a square-wave electroporator from BTX instruments.

2.2 Buffers for Cell Preparation

Krebs-Ringer's bicarbonate HEPES (KRBH) buffer with bovine serum albumin (BSA) is prepared fresh from three stock solutions (*see* **Note 1**):

1. **Stock 1 (10×)**: 1.2 M NaCl, 40 mM KH_2PO_4, 10 mM $MgSO_4 \cdot 7H_2O$, 10 mM $CaCl_2$. Add 70.08 g NaCl, 5.46 g KH_2PO_4, and 2.46 g $MgSO_4 \cdot 7H_2O$ to 500 mL H_2O and dissolve under mixing with magnetic stirrer. Add 10 mL of 1 M $CaCl_2$ stock (*see* **Note 2**) when the first three salts have dissolved. Add H_2O to a final volume of 1000 mL.
2. **Stock 2 (10×)**: 100 mM $NaHCO_3$. Add 8.4 g $NaHCO_3$ to 1000 mL H_2O and dissolve using a magnetic stirrer.
3. **Stock 3 (10×)**: 300 mM HEPES. Add 71.5 g HEPES to 500 mL H_2O and dissolve with magnetic stirrer. Add H_2O to a final volume of 1000 mL. **Stocks 1–3** are stored at +4 °C.
4. **Adenosine, 200 μM stock (1000×)**: 2.67 mg Adenosine/50 mL H_2O, aliquot and store at −20 °C.
5. To prepare 200 mL KRBH-1 % w/v BSA buffer, dissolve 2 g of BSA in ~100 mL H_2O (*see* **Note 3**). Add 20 mL of each stock solution (**stocks 1–3**). Add 200 μL of 200 μM adenosine stock, and set to pH 7.4 with concentrated NaOH. Adjust volume with H_2O to 200 mL and sterilize using 0.2 μm syringe filters.

6. **Collagenase (type 1) 6 mg/mL (6×)**: Dissolve 60 mg lyophilized collagenase in 10 mL KRBH-1 % w/v BSA buffer. Filter sterilize, aliquot, and store at −20 °C.

2.3 Media for Cell Transfection

1. **N-phenylisopropyladenosine (PIA), 1 mM stock (5000×):** Dissolve 1.93 mg PIA in 5 mL DMSO, aliquot, and store at −20 °C.
2. **DMEM+ buffer:** 50 mL Dulbecco's modified Eagle medium (DMEM, 4.5 g/L glucose), 10 μL of 1 mM PIA, 100 μL gentamicin (50 mg/mL), sterilize using 0.2 μm syringe filters.
3. **DMEM++ buffer:** Dissolve 0.35 g BSA in 10 mL DMEM+ buffer, sterilize using 0.2 μm syringe filters.

3 Methods

3.1 Cell Preparation

Rat adipocytes are isolated according to the protocol described by Rodbell [13]. Usually, we isolate primary adipocytes from epididymal rat fat pads (Sprague-Dawley rats). The procedure, with minor adjustments, is suitable for white adipocytes isolated from mice or human biopsies as well.

1. Dissect the fat pads into a narrow-mouth plastic bottle (30 mL, low-density polyethylene (LDPE)) containing 5 mL of KRBH-1 % w/v BSA buffer (*see* **Note 4**).
2. Mince the tissue, using two scissors that fit into the vial, to a homogenous suspension (*see* **Note 5**).
3. Add 1 mL collagenase buffer and incubate for 45–80 min at 37 °C in a shaking water bath, 120 rpm (*see* **Note 6**).
4. Add pre-warmed (37 °C) KRBH-1 % w/v BSA buffer to the neck of the vial and cover the opening with nylon mesh (250 μm) fastened with rubber band. Swirl the bottle, invert, and gently squeeze twice into a sterile 50 mL tube (*see* **Note 7**).
5. Repeat the swirling and squeeze until the vial is empty.
6. Let the adipocytes float by leaving the tube standing in a water bath at 37 °C for a few minutes, with no shaking.
7. Aspirate the supernatant with a 12-gauge needle attached with a sterile 200 μL pipette tip at the end.
8. Wash by gently adding ~20 mL of KRBH-1 % w/v BSA buffer (*see* **Note 8**).
9. Repeat **steps 6–8** three times.

3.2 Cell Transfection

1. Prepare DNA in DMEM+ buffer, 200 μL/cuvette. Usually, we have 3–6 μg DNA/cuvette depending on the experiment, 4–6 cuvettes/condition (*see* **Note 9**).

2. Aspirate supernatant from adipocytes and wash once in ~10 mL DMEM+ buffer.
3. Remove supernatant, resuspend in DMEM+ buffer, and adjust volume to 40 % packed cell volume/volume buffer (cytocrit).
4. Pour the cells into the pre-sterilized scintillation vial with magnetic stirrer, ~200 rpm.
5. Pipette up-down-up once before pipetting 200 μL of the cell suspension using a wide bore tip to each cuvette.
6. Transfect adipocytes using a square-wave electroporator (200 V, 3 pulses of 12 ms at 1-s intervals for rat adipocytes [4]; 400 V, 1 pulse for human and mouse adipocytes [14]).
7. Carefully pool the cell suspension from the cuvettes into a 15 mL Falcon tube using 200 μL wide bore tips.
8. Aspirate supernatant with a 3-gauge needle attached with a 200 μL pipette tip at the end.
9. Add ~4–5 mL of DMEM++ buffer.
10. Incubate at 37 °C, 5–10 % CO_2, for 18–20 h to allow recombinant protein expression.

3.3 TIRF Imaging of GLUT4

1. Calibrate and align lasers.
2. Temperature of objective and stage should be allowed to equilibrate for at least a couple of hours before running an experiment (*see* **Note 10**). Also make sure that the temperature control unit is accurately calibrated in order to have a correct temperature setting.
3. If the objective lens has a temperature correction ring to compensate for spherical aberration, make sure that it is properly set for the desired temperature with fluorescent beads on a regular basis. Using beads simplifies the process since both focus and optimal laser spread can be calibrated.
4. Make sure that all imaging surfaces, coverslips, and objectives are properly cleaned with lens paper and ethanol and make sure that the surface is dried to ensure best imaging results.
5. Isolated primary adipocytes quickly deteriorate in culture and therefore cells are routinely imaged 18–20 h after electroporation.
6. Mount a glass-bottom dish (designed for microscopy, MatTek Corporation), no. 1.5 thickness, into the dish holder at the scope (*see* **Notes 11** and **12**, also *see* Fig. 1). Add ~10 μL of KRBH-1 % w/v BSA buffer. Place 5 μL packed adipocytes on top of the buffer droplet. Mount the cell chamber by gently placing a Millipore 1 μm membrane and a weight (nut-shaped brass piece) to immobilize the cells to the bottom of the dish. Add KRBH-1 % w/v BSA buffer on top of the membrane and outside the weight (*see* Fig. 1, **right panel**). The membrane

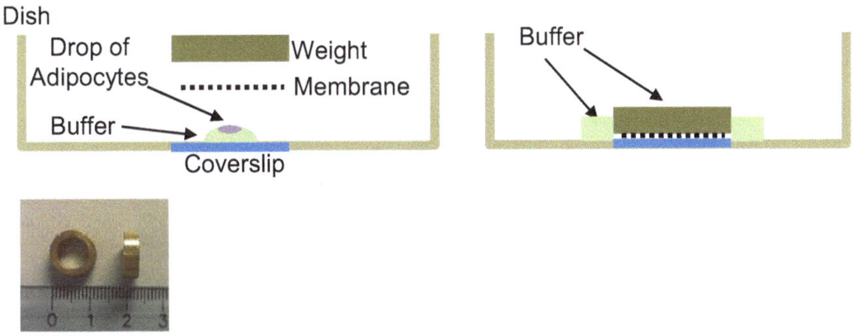

Fig. 1 Visualization chamber. The *left panel* illustrates how to carefully mount the adipocytes in a glass-bottom dish, gently pressing the cells toward the coverslip using a nut-shaped brass weight (size and shape of weight, see below). The weight is shown with scale bar in cm. Add buffer on top of the membrane and outside the weight (shown in *right panel*)

allows addition of stimuli (insulin) without disturbing the cell chamber montage (*see* **Note 13**).

7. Make sure that the TIRF settings are optimized for fast acquisition and at the same time maximize the signal intensity without overly bleaching the sample. When imaging adipocytes expressing HA-GLUT4-GFP, typical settings for our system to achieve a minimum of 10 frames per second are as follow: laser power of 1–2 mW, camera exposure time of 50 ms, maximum readout speed of 17 MHz, EM multiplier at 300, and vertical clock speed of 3.3 μs. Generally, we do not use pixel binning. To avoid the time bottleneck of mechanical filter switching we utilize the camera's triggered acquisition function. Here the camera and software control the sequential firing of selected lasers, and suitable EX/EM filters need to be used to allow the correct wavelengths through.

3.4 Expected Readout of GLUT4 Trafficking

The most prominent effect of insulin on GLUT4 translocation is a decrease of GSV traffic as the GSV tether to the plasma membrane followed by fusion. This can be quantified by measuring the number of trajectories produced during a time-lapse acquisition (*see* Fig. 2), a measurement used in previous studies of GLUT4 trafficking [3, 11]. The actual fusion event can be captured using a pH-sensitive probe (*see* Fig. 3), a tool used by us and others to detect fusion simultaneously with the overall GLUT4 trafficking in the TIRF zone [2, 9, 15]. Insulin-induced GSV fusion is followed by a rapid release of GLUT4 monomers into the plasma membrane. This is detected as an increased plasma membrane intensity, illustrated in Fig. 4, showing snapshot images of an adipocyte expressing GLUT4-GFP, before and after insulin stimulation (*see* Fig. 4).

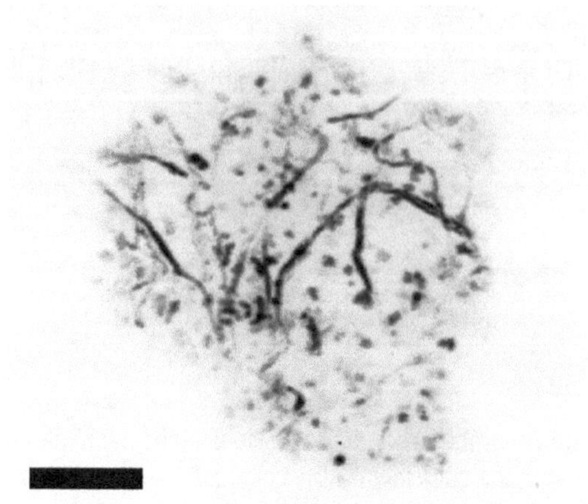

Fig. 2 Maximum-intensity projection. Illustrates GSV trafficking during a 1-min time-lapse acquisition. Scale bar is 5 μm

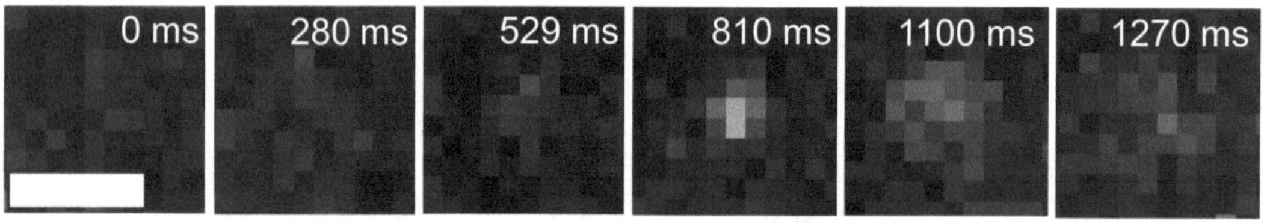

Fig. 3 Example of a fusion event detected using a pH-sensitive probe (IRAP-pHluorin) co-expressed in GLUT4 vesicles. Series of frame-by-frame captions displayed, with a fusion event detected at time 810 ms. Scale bar is 1 μm

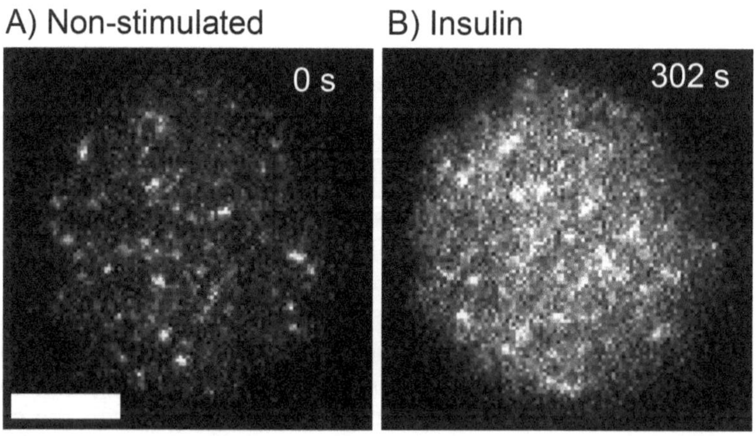

Fig. 4 Images of a primary adipocyte expressing GLUT4-GFP. Shown: (**a**) before (time = 0 s) and (**b**) after insulin stimulation (time = 302 s). Scale bar is 5 μm

4 Notes

1. We use Probumin BSA for working buffers with primary adipocytes. KRBH-BSA buffer should be prepared at the time of experiment, but could be stored at −20 °C and used later.
2. Add the $CaCl_2$ last when all other salts have dissolved to prevent precipitation.
3. Increase BSA concentration to 5 % w/v for mouse or human adipocytes.
4. Always use plasticware (polypropylene or LDPE) when working with adipose tissue/cells.
5. The adipose tissue should be minced thoroughly, for adequate collagenase digestion.
6. Terminate the digestion if too much oil assembles at the top, which indicates cell breakage.
7. For mouse or human adipocytes: Fold a funnel using 400 μm nylon mesh and place on top of a 50 mL Falcon tube, swirl the squeeze bottle, and carefully pour the cells into the mesh funnel until all of the cell suspension has passed through. Mouse and human adipocytes are generally larger than rat adipocytes and tend to be more fragile, and they therefore need to be treated accordingly. The funnel filtration exerts less force on the cells as they pass through by gravity instead of physical pressure.
8. When washing, adding media or buffer to primary adipocytes, pour slowly on the sides of the tube in order to disturb the cells as little as possible.
9. There are several different GLUT4 constructs suitable for live-cell imaging. We have routinely used either HA-GLUT4-GFP or HA-GLUT4-mCherry, where the exofacial HA-tag allows detection of GLUT4 inserted into the membrane following fusion [2]. A photoactivatable probe, EOS, was used for single GLUT4 molecule tracking of monomers in the plasma membrane [5]. Co-expression of fluorescently tagged GLUT4 and IRAP-pHluorin, another GSV-residing molecule carrying a pH-sensitive probe, allows detection of the exact fusion event as the pH increases following GSV fusion with the plasma membrane, and simultaneously monitoring of overall GLUT4 traffic [2, 9]. Also, a double-tagged GLUT4 construct was shown to be useful for characterization of cell heterogeneity [15].
10. To avoid cell stress, we have found that in our system, the adipocytes are best preserved at 35 °C. At this temperature we observe the longest cell durability and maintained vesicle traffic. Having a higher temperature often results in increased cell breakage. Once adipocytes start breaking, their lipid content leaks out into the immediate environment, which rapidly induces further cell breakage.

11. Commercial TIRF systems are usually based on inverted microscope systems, and therefore the buoyancy of the primary adipocytes demands that they are pressed down against the coverslip. This has to be done gently in order to avoid cell breakage.
12. We have not seen any significant differences in image quality when using regular tolerance No. 1.5 thickness coverslips compared to high-tolerance 0.170 mm coverslips. The limited variety of coverslip sizes available with the high-tolerance dishes limits their usefulness in our setup.
13. For shorter experiments (i.e., 10–15 min) that do not require addition of stimuli to the cells during acquisition, a coverslip could replace the membrane and brass weight.

References

1. Bryant NJ, Govers R, James DE (2002) Regulated transport of the glucose transporter GLUT4. Nat Rev Mol Cell Biol 3(4):267–277. https://doi.org/10.1038/nrm782
2. Stenkula KG, Lizunov VA, Cushman SW, Zimmerberg J (2010) Insulin controls the spatial distribution of GLUT4 on the cell surface through regulation of its postfusion dispersal. Cell Metab 12(3):250–259. https://doi.org/10.1016/j.cmet.2010.08.005
3. Lizunov VA, Lisinski I, Stenkula K, Zimmerberg J, Cushman SW (2009) Insulin regulates fusion of GLUT4 vesicles independent of Exo70-mediated tethering. J Biol Chem 284(12):7914–7919. https://doi.org/10.1074/jbc.M806460200
4. Lizunov VA, Matsumoto H, Zimmerberg J, Cushman SW, Frolov VA (2005) Insulin stimulates the halting, tethering, and fusion of mobile GLUT4 vesicles in rat adipose cells. J Cell Biol 169(3):481–489. https://doi.org/10.1083/jcb.200412069
5. Lizunov VA, Stenkula K, Troy A, Cushman SW, Zimmerberg J (2013) Insulin regulates Glut4 confinement in plasma membrane clusters in adipose cells. PLoS One 8(3):e57559. https://doi.org/10.1371/journal.pone.0057559
6. Bai L, Wang Y, Fan J, Chen Y, Ji W, Qu A, Xu P, James DE, Xu T (2007) Dissecting multiple steps of GLUT4 trafficking and identifying the sites of insulin action. Cell Metab 5(1):47–57. https://doi.org/10.1016/j.cmet.2006.11.013
7. Huang S, Lifshitz LM, Jones C, Bellve KD, Standley C, Fonseca S, Corvera S, Fogarty KE, Czech MP (2007) Insulin stimulates membrane fusion and GLUT4 accumulation in clathrin coats on adipocyte plasma membranes. Mol Cell Biol 27(9):3456–3469. https://doi.org/10.1128/MCB.01719-06
8. Xu Y, Rubin BR, Orme CM, Karpikov A, Yu C, Bogan JS, Toomre DK (2011) Dual-mode of insulin action controls GLUT4 vesicle exocytosis. J Cell Biol 193(4):643–653. https://doi.org/10.1083/jcb.201008135
9. Jiang L, Fan J, Bai L, Wang Y, Chen Y, Yang L, Chen L, Xu T (2008) Direct quantification of fusion rate reveals a distal role for AS160 in insulin-stimulated fusion of GLUT4 storage vesicles. J Biol Chem 283(13):8508–8516. https://doi.org/10.1074/jbc.M708688200
10. Lizunov VA, Stenkula KG, Blank PS, Troy A, Lee JP, Skarulis MC, Cushman SW, Zimmerberg J (2015) Human adipose cells in vitro are either refractory or responsive to insulin, reflecting host metabolic state. PLoS One 10(3):e0119291. https://doi.org/10.1371/journal.pone.0119291
11. Lizunov VA, Lee JP, Skarulis MC, Zimmerberg J, Cushman SW, Stenkula KG (2013) Impaired tethering and fusion of GLUT4 vesicles in insulin-resistant human adipose cells. Diabetes 62(9):3114–3119. https://doi.org/10.2337/db12-1741
12. Poulter NS, Pitkeathly WT, Smith PJ, Rappoport JZ (2015) The physical basis of

total internal reflection fluorescence (TIRF) microscopy and its cellular applications. Methods Mol Biol 1251:1–23. https://doi.org/10.1007/978-1-4939-2080-8_1

13. Rodbell M (1964) Metabolism of isolated fat cells. I. Effects of hormones on glucose metabolism and lipolysis. J Biol Chem 239:375–380

14. Stenkula KG, Said L, Karlsson M, Thorn H, Kjolhede P, Gustavsson J, Soderstrom M, Stralfors P, Nystrom FH (2004) Expression of a mutant IRS inhibits metabolic and mitogenic signalling of insulin in human adipocytes. Mol Cell Endocrinol 221(1–2):1–8. https://doi.org/10.1016/j.mce.2004.04.011

15. Burchfield JG, Lu J, Fazakerley DJ, Tan SX, Ng Y, Mele K, Buckley MJ, Han W, Hughes WE, James DE (2013) Novel systems for dynamically assessing insulin action in live cells reveals heterogeneity in the insulin response. Traffic 14(3):259–273. https://doi.org/10.1111/tra.12035

Chapter 13

Translocation and Redistribution of GLUT4 Using a Dual-Labeled Reporter Assay

Robert M. Jackson and Ann Louise Olson

Abstract

It is crucial to determine the regulation of GLUT4 translocation and redistribution to the plasma membrane. The HA-GLUT4-GFP dual-reporter construct has become an important tool in the assessment of GLUT4 recycling in cultured adipocytes and myocytes. Through the use of light microscopy, this reporter construct allows for visualization of GLUT4 specifically at the cell surface or GLUT4 that has recycled from the cell surface while simultaneously marking the total GLUT4 pool. Here, we discuss and outline the general application of this reporter construct and its use in evaluating GLUT4 translocation within cultured adipocytes.

Key words GLUT4, Membrane traffic, Adipocyte, Insulin signaling, Translocation assay

1 Introduction

Obesity is a growing epidemic and known to lead to other complications such as metabolic syndrome, cardiovascular disease, and type 2 diabetes mellitus. Insulin resistance is a hallmark of these diseases, and this is due in part to a failure of insulin to signal glucose uptake. GLUT4, an insulin-dependent facilitative glucose transporter, is an important effector molecule for regulating insulin-dependent glucose uptake in adipose tissue, heart, and skeletal muscle. During obesity and diabetes, both the expression of GLUT4 and the ability of insulin to recruit GLUT4 to the plasma are decreased. The mechanisms of analyzing GLUT4 translocation through cells have evolved greatly over the years. Insulin-stimulated clearance of glucose from the blood was first reported through the use of membrane sub-fractionation [1, 2]. This activity was eventually attributed to a tissue-specific insulin-regulated glucose transporter (IRGT) [3, 4], which would later be cloned and characterized as GLUT4 [5]. This opened the door for more controlled manipulation of GLUT4. Immunogold staining was later used to confirm

Karin Lindkvist-Petersson and Jesper S. Hansen (eds.), *Glucose Transport: Methods and Protocols*, Methods in Molecular Biology, vol. 1713, https://doi.org/10.1007/978-1-4939-7507-5_13,

membrane and adipose localization [6, 7]. A few years later, an HA epitope was inserted between residues 67 and 68 in the first exofacial loop of GLUT4 [8], which allowed this GLUT4-HA construct to be probed for immunofluorescence and viewed at the cell surface via light microscopy. Another group would later generate a GLUT4-GFP chimera to visualize internal GLUT4 traffic in individual cells [9]. These two constructs would eventually be combined to form the HA-GLUT4-GFP reporter construct that has been widely used over the past decade to evaluate GLUT4 translocation to and from the cell surface [10, 11].

The HA-GLUT4-GFP reporter construct is unique in that it allows for visualization of both total GLUT4, via the GFP tag, and GLUT4 that is or has been at the cell surface, via its HA epitope (*see* Fig. 1). The GFP molecule is located at the C-terminus and is positioned in the cytosol. The GFP molecule is able to be excited and emit light from anywhere in the cell and thus serves as a marker of the total GLUT4 within a given cell. The HA epitope, as mentioned above, is inserted into the first exofacial loop. It is positioned on the luminal face of the membrane, and if the cell has not been permeabilized, this epitope is inaccessible to antibodies while GLUT4 is internalized. Insulin stimulation will signal fusion of GLUT4 with the plasma membrane, exposing the HA epitope, making it available for detection via immunostain. Using this dual-reporter construct, our lab and others have been able to detect a five- to tenfold increase in GLUT4 found at the cell surface after acute insulin stimulation [12, 13], which is consistent with earlier reports using other methods [14–17]. In the same study, we were also able to pulse-chase cell surface GLUT4 before it was internalized by probing the HA epitope in live, insulin-stimulated cells and then removing them from insulin stimulation [13, 18].

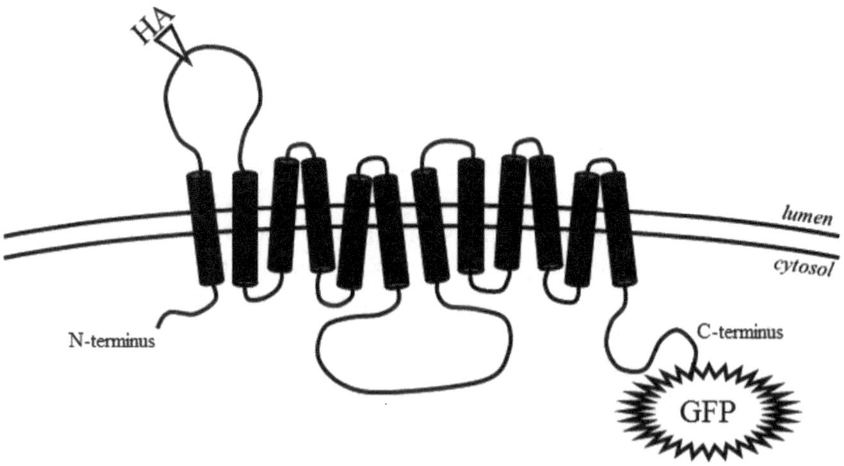

Fig. 1 GLUT4 spans the membrane with 12-transmembrane domains. Both the N- and C-termini are cytosolic. In the dual-reporter construct, a HA epitope has been inserted between amino acids 67 and 68 in the first exofacial loop, while a GFP molecule has been attached to the C-terminus via a linker sequence

Here, we outline the methods that take advantage of this dual-reporter construct HA-GLUT4-GFP for visualization of the exogenous GLUT4 in both the basal and insulin-stimulated states. We describe translocation and recycling methods for use with 3T3-L1 adipocytes, as they are the standard cell line when assessing GLUT4 in a diabetic cell model. We cover cell culture and transfection, as well as fixing and staining for both a translocation assay and a recycling assay. *See* **Note 1** for general comments on confocal microscopy and image quantification.

2 Materials

Prepare all solutions/media just prior to performing each experiment to ensure optimal pH and freshness of the reagents. Prepare and store all reagents and generated solutions at room temperature (unless otherwise indicated). Store cell culture media at 4 °C.

2.1 Cell Culture and Transfection

1. NIH 3T3-L1 mouse pre-adipocyte fibroblasts grown and maintained in 100 mm cell culture dishes (*see* **Note 2**).
2. HA-GLUT4-GFP expression plasmid: Store as an ethanol precipitate.
3. Proliferation medium: Dulbecco's modified Eagle medium (DMEM), 10 % v/v calf serum, 1 % v/v penicillin/streptomycin, 18 mM sodium bicarbonate, 25 mM HEPES, pH to 7.4 with HCl or NaOH, filter sterilized in cell culture hood.
4. Maturation medium: DMEM, 10 % w/v fetal bovine serum (FBS), 1 % v/v penicillin/streptomycin, 18 mM sodium bicarbonate, 25 mM HEPES, pH to 7.4 with HCl or NaOH, filter sterilized in cell culture hood.
5. Starvation medium: F-12 Ham's starvation media, 0.5 % w/v bovine serum albumin (BSA), 14 mM sodium bicarbonate, 25 mM HEPES, pH to 7.4 with HCl or NaOH, filter sterilized in cell culture hood.
6. Electroporation medium: DMEM, 1 % v/v penicillin/streptomycin, 18 mM sodium bicarbonate, 25 mM HEPES, pH to 7.4 with HCl or NaOH, filter sterilized in cell culture hood.
7. Post-electroporation medium: DMEM, 20 % v/v FBS, 1 % penicillin/streptomycin, 18 mM sodium bicarbonate, 25 mM HEPES, pH to 7.4 with HCl or NaOH, filter sterilized in cell culture hood.
8. Mild acid wash: Mix 8 mL of 5 M NaCl, 20 mL of 500 mM MES at pH 5.0, and approximately 150 mL of ddH_2O. Adjust the pH to 5.0 with HCL or NaOH. Bring volume to 200 mL final with ddH_2O. Store at 4 °C.
9. dPBS: Start with 800 mL of distilled water; add 8 g of NaCl, 0.2 g of KCl, 1.44 g of Na_2HPO_4, 0.24 g of KH_2PO_4; adjust

the pH to 7.4 with HCl; and add distilled water to a total volume of 1 L.

10. Trypsin-EDTA solution: 0.05 % w/v Trypsin, 0.53 mM EDTA, without sodium bicarbonate.
11. Collagenase: Type I, resuspended to ~2 mg/mL with 2.5 mL 0.05 % w/v trypsin, 0.53 mM EDTA, without sodium bicarbonate and 2.5 mL dPBS.
12. Insulin solution: Insulin, tissue culture grade.
13. Tabletop centrifuge capable of spinning 15 mL conical tubes at 300 × *g*.
14. Mammalian cell electroporation system and system-matched cuvettes.
15. Chamber culture slides.

2.2 Fixing and Staining

1. Rocking platform shakers: one maintained at room temperature and another maintained at 37 °C.
2. dPBS solution: dPBS, pH 7.4: Stored both at room temperature and at 4 °C.
3. Fixing solution: 1.15 mL of 16 % v/v paraformaldehyde stock diluted in 3.85 mL of room-temperature dPBS for a final volume of 5 mL (3.7 %). Keep at room temperature.
4. Quenching solution: 10 mg of $NaBH_4$ solubilized in 10 mL cold dPBS just prior to use (0.1 % w/v). Keep on ice.
5. Permeabilization solution: 3 mg of saponin dissolved in room-temperature dPBS (0.1 % w/v). Keep at room temperature.
6. Blocking buffer: 10 % w/v BSA in cold dPBS. Keep on ice (*see* **Note 3**).
7. Primary antibody: Anti-HA antibody diluted 1:1000 in cold dPBS containing 1 % w/v BSA. Keep on ice (*see* **Note 3**).
8. Unlabeled antibody: Unconjugated secondary diluted 1:1000 in cold dPBS containing 1 % w/v BSA. Keep on ice (*see* **Note 3**).
9. Secondary antibody: Fluorophore-conjugated secondary diluted 1:400 in cold dPBS containing 1 % w/v BSA. Keep on ice (*see* **Note 3**).
10. Mounting medium.
11. Coverslips.

3 Methods

Perform all procedures handling live cells inside a cell culture hood. Limit cell exposure to light as much as possible once they have been transfected. The following methods will denote quantities and volumes needed for a single 4-well-chamber culture slide.

3.1 Cell Culture and Transfection

1. Grow and maintain 3T3-L1 fibroblasts as per standard protocols in 100 mm cell culture dishes (*see* **Note 4**). Briefly, seed individual 100 mm cell culture dishes with 5 × 10^4 cells. Allow them to expand over 7 days in proliferation medium. Induce differentiation into adipocytes by exchanging the proliferation medium for maturation medium containing 1 μM dexamethasone, 17.5 nM insulin, and 500 μM isobutyl-1-methylxanthine (DII). After 3.5 days, exchange for fresh maturation medium without DII included. Newly formed adipocytes are ready for transfection 2 days after the removal of DII (day-5 adipocytes).
2. Turn on the electroporation system to allow it time to warm up. Set it to pulse at 0.18 kV at 950 μF (*see* **Note 5**).
3. Prepare two 1 μg/μL aliquots of the HA-GLUT4-GFP construct plasmid. Transfer the necessary amount of ethanol precipitate for 50 μg of plasmid into a microcentrifuge tube and pellet it by spinning at max speed (21,000 × *g*) in a tabletop centrifuge. If possible, do this at 4 °C. Resuspend to the desired concentration with 50 μL molecular grade, nuclease-free water (*see* **Note 6**).
4. Measure out 8–12 mg collagenase in a 15 mL conical tube. In the cell culture hood, resuspend the collagenase to ~2 mg/mL.
5. Retrieve one 100 mm cell culture dish of day-5 adipocytes. Aspirate off the media and wash the cells twice with 5 mL dPBS. Aspirate off any remaining dPBS and add the collagenase/trypsin suspension to the cells. Place the plate back in its incubator for 5–10 min or until cells have visibly lifted off the plate. Gently, transfer the cells to the 15 mL conical tube and slowly pellet them via centrifugation at ~300 × *g*. Aspirate the supernatant.
6. Gently resuspend the pelleted cells with 3 mL electroporation medium. Transfer 500 μL of the resuspension into each individual electroporation cuvettes. Add 50 μL of the desired plasmid to each of the cuvettes and gently mix by pipetting several times. Place the cuvettes in the electroporator system and pulse them with 0.18 kV at 950 μF. Gently tap the sides of the cuvettes to drive any generated bubbles to the surface. Leave the cells to recover at room temperature for 10 min.
7. Pool all cuvettes with identical conditions prior to seeding the newly transfected cells onto the chamber culture slides. Note that the cells are still extremely fragile due to the electroporation process and need to be handled with care. Assuming a chamber volume of 1 mL, where 500 μL is the ideal media volume, transfer 200 μL of the pooled cells into a chamber well (*see* **Note 7**). Add 50 μL of fresh electroporation medium to bring the volume to a 250 μL total. Incubate at 37 °C and 5 % CO_2 for 20 min.
8. After the cells have settled on the slide, add 250 μL of post-electroporation medium to bring each chamber to a final

500 μL volume (*see* **Note 8**). Incubate at 37 °C and 5 % CO_2 overnight.

9. Approximately 18 h after transfection starve the cells to induce the basal condition that will serve as the baseline in quantification. To do this, decant the media out of the chamber slides into a waste beaker. Do not aspirate (*see* **Note 9**). Rinse the cells twice with 500 μL mild acid wash, leaving the second application on the cells for 2 min. Rinse the cells twice with 500 μL dPBS. Finally, add 500 μL starvation medium to the cells. Incubate at 37 °C and 5 % CO_2 for 2 h.
10. If the experimental goal is the recycling assay, anti-HA antibody will need to be diluted 1:1000 in these new starvation aliquots.

3.2 Fixing and Staining for Translocation

To reduce photo bleaching over time, work in the dark for the remainder of the protocol as much as possible. At minimum, keep the cells covered from sources of direct light. For the methods outlining the recycling assay, skip ahead to Subheading 3.3.

1. Prepare an aliquot of starvation medium containing 10 nM insulin (*see* **Note 10**). At the end of the 2-h starvation period, replace the media in the chamber slides with either 500 μL fresh starvation medium for the basal condition or 500 μL 10 nM insulin for the stimulated condition. Incubate at 37 °C and 5 % CO_2 for 30 min.
2. During the insulin stimulation period, prepare the fixing solution and blocking buffer. Keep the blocking buffer on ice. Set up three beakers of cold dPBS on ice for sequential washing (*see* **Note 9**).
3. Immediately at the end of the 30-min insulin stimulation, dump the media into a waste beaker and sequentially dip the chamber slide through 3 × ice-cold dPBS rinses. Dab/wick away excess liquid with a chemwipe tissue paper (*see* **Note 11**). Add 500 μL fixing solution to each chamber. Leave on the benchtop at room temperature for 20 min.
4. Halfway through the fixing solution incubation period, prepare the quenching solution (*see* **Note 12**).
5. Decant the fixing solution into the waste beaker and sequentially dip the chamber slide through 3 × ice-cold dPBS rinses. Dab/wick away excess liquid with a chemwipe tissue paper. Add 500 μL cold quenching solution. Place the slide on ice for 30 min, refreshing the quenching solution at 10-min intervals for a total of 3 × 10-min quenches.
6. Decant the quenching solution into the waste beaker and sequentially dip the chamber slide through 3 × ice-cold dPBS rinses. Dab/wick away excess liquid with a chemwipe tissue paper. Add 500 μL blocking buffer. Place on a rocking platform shaker at room temperature for 15 min.

7. Prepare the primary antibody at this time. At the end of the 15 min, dump the blocking buffer into the waste beaker. Dab/wick away excess liquid with a chemwipe tissue paper. Add 500 μL primary antibody. Place on a rocking platform shaker at room temperature for 1 h.
8. As the 1-h incubation ends, prepare the secondary antibody. Dump the primary antibody into the waste beaker and sequentially dip the chamber slide through 3 × ice-cold dPBS rinses. Dab/wick away excess liquid with a chemwipe tissue paper. Add 500 μL secondary antibody. Place on a rocking platform shaker at 37 °C for 30 min.
9. Decant the secondary antibody and sequentially dip the chamber slide through 3 × ice-cold dPBS rinses. Rinse chambers with 500 μL molecular grade water. Dump the water into the waste beaker. Dab/wick away excess liquid with a chemwipe tissue paper.
10. Being careful not to snap the slide, remove the chamber walls from the slide. Again, dab/wick away any remaining liquid with a chemwipe tissue paper (*see* **Note 11**).
11. Add a drop of mounting media to the cells and coverslip them. Cover the slide and place in a drawer to set overnight (*see* **Note 13**).

3.3 Fixing and Staining for Recycling

The recycling assay is designed to probe the recycling pool of HA-GLUT4-GFP during insulin stimulation, allowing the anti-HA antibody to be internalized. The internalized anti-HA antibody functions to label the endosomal compartments to which GLUT4 are recycled once it leaves the cell surface (*see* Fig. 2). Therefore, for the recycling assay to work, anti-HA antibody must be present in the media while insulin stimulation is occurring.

1. Prepare an aliquot of starvation medium containing 10 nM insulin and anti-HA antibody diluted 1:1000 (*see* **Note 10**). At the end of the 2-h starvation period, replace the media in the chamber slides with either 500 μL fresh starvation medium for the basal condition or 500 μL 10 nM insulin for the stimulated condition. Incubate at 37 °C and 5 % CO_2 for 2 h.
2. During the insulin stimulation period, prepare the fixing solution, unlabeled antibody, permeabilization solution, and blocking buffer. Keep the unlabeled antibody and blocking buffer on ice. Set up three beakers of cold dPBS on ice for sequential washing (*see* **Note 9**).
3. Immediately at the end of the 30-min insulin stimulation, dump the media into a waste beaker and sequentially dip the chamber slide through 3 × ice-cold dPBS rinses. Dab/wick away excess liquid with a chemwipe tissue paper (*see* **Note 11**).

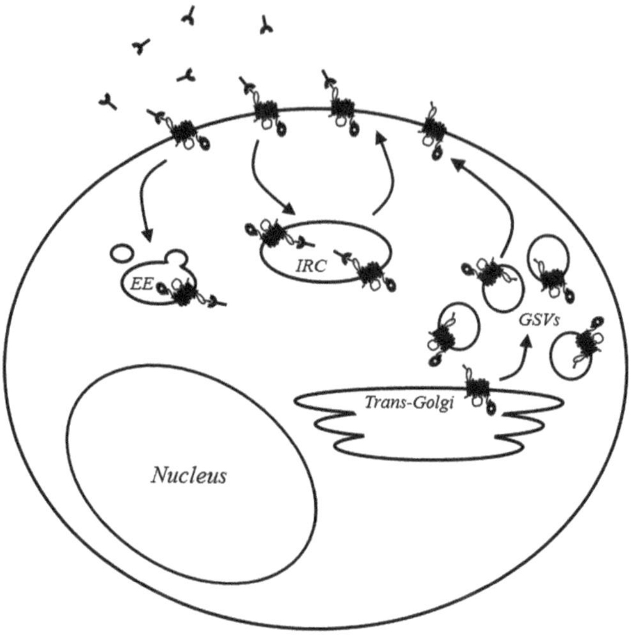

Fig. 2 GLUT4 translocation and HA epitope antibody labeling. As GLUT4 final processing occurs in the trans-Golgi, it is sorted into small GLUT4 storage vesicles (GSVs) where they are primarily stored under basal conditions. Upon insulin stimulation, the GSVs translocate to the cell surface to incorporate their GLUT4 in the plasma membrane. It is at this point that the HA epitope becomes exposed and available for probing. Beyond the initial wave of translocation to the cell surface, GLUT4 will begin to be pulled from the plasma membrane to the insulin-responsive compartment (IRC), from where it can continue to recycle to and from the plasma membrane while insulin stimulation continues. As the insulin signal wanes, GLUT4 will begin to be targeted to the early endosomal compartment (EE) from where it can ultimately be resorted back into GSVs to prepare for the next insulin stimulation. The recycling assay is designed to visualize the early compartments in GLUT4 recycling, such as the IRC and EE

Add 500 μL fixing solution to each chamber. Leave on the benchtop at room temperature for 20 min.

4. Decant the fixing solution into the waste beaker and sequentially dip the chamber slide through 3 × ice-cold dPBS rinses. Dab/wick away excess liquid with a chemwipe tissue paper. Add 500 μL unlabeled antibody to block detection of HA-GLUT4-GFP that has remained at the cell surface. Place on a rocking platform shaker at room temperature for 1 h.
5. Decant the unlabeled antibody into the waste beaker and sequentially dip the chamber slide through 3 × ice-cold dPBS rinses. Dab/wick away excess liquid with a chemwipe tissue paper. Add 500 μL fixing solution to cross-link the unlabeled antibody, locking it in place. Leave on the benchtop at room temperature for 20 min.

6. Halfway through the fixing solution incubation period, prepare the quenching solution (*see* **Note 12**).
7. Decant the fixing solution into the waste beaker and sequentially dip the chamber slide through 3 × ice-cold dPBS rinses. Dab/wick away excess liquid with a chemwipe tissue paper. Add 500 μL cold quenching solution. Place the slide on ice for 30 min, refreshing the quenching solution at 10-min intervals for a total of 3 × 10-min quenches.
8. Decant the quenching solution into the waste beaker and sequentially dip the chamber slide through 3 × ice-cold dPBS rinses. Dab/wick away excess liquid with a chemwipe tissue paper. Add 500 μL permeabilization solution to allow for the internalized anti-HA antibody to be accessible to the labeled secondary antibody. Place on a rocking platform shaker at room temperature for 15 min.
9. Decant the permeabilization solution into the waste beaker and sequentially dip the chamber slide through 3 × ice-cold dPBS rinses. Dab/wick away excess liquid with a chemwipe tissue paper. Add 500 μL blocking solution. Place on a rocking platform shaker at room temperature for 15 min.
10. Decant the blocking buffer into the waste beaker and sequentially dip the chamber slide through 3 × ice-cold dPBS rinses. Dab/wick away excess liquid with a chemwipe paper. Add 500 μL secondary antibody that has a conjugated fluorophore. This will label all anti-HA that had been internalized during the insulin stimulation period. Place on a rocking platform shaker at 37 °C for 30 min.
11. Decant the secondary antibody and sequentially dip the chamber slide through three ice-cold dPBS rinses. Rinse chambers with 500 μL molecular grade water. Dump the water into the waste beaker. Dab/wick away excess liquid with a chemwipe tissue paper.
12. Being careful not to snap the slide, remove the chamber walls from the slide. Again, dab/wick away any remaining liquid with a chemwipe tissue paper (*see* **Note 11**).
13. Add a drop of mounting media to the cells and coverslip them. Cover the slide and place in a drawer to set overnight (*see* **Note 13**).

4 Notes

1. While the HA-GLUT4-GFP can be used in other experimental designs, it enables single-cell assay approaches, lending itself to confocal microscopy imaging. We have access to and image our transfected cells with a Leica SP2 MP confocal microscope, using a 63× objective lens with oil immersion.

It is important to recognize that the cells are undergoing transfection and analysis before they have had a chance to fully mature (*see* **Note 4**). Another pitfall is the low transfection efficiency (*see* **Notes 4–6**). Therefore, stringent morphological criteria should be outlined prior to imaging cells to ensure that a true adipocyte and not a fibroblast is being imaged. In general, verify that it (i) has GFP expression and not just strong background autofluorescence; (ii) is squamous in shape; (iii) exhibits strong perinuclear GFP staining; and (iv) has lipid droplets. Using these criteria, our chosen cells' z-planes (cell thickness) range from 15 to 25 μm. We collect 13 z-stacks (image slices) per cell. More stacks would provide better resolution, but each slice increases the imaging time, which also increases the risk of photo bleaching. Find a balance between resolution and imaging time. For simple quantification, we use the Leica LCS Lite software to generate maximum-intensity projections for each cell's 13 z-stacks. We are then able to outline and isolate each cell to get a SumIntensity value for each color channel in the image. We quantify and represent GLUT4 translocation as a ratio of Alexa 647 to GFP fluorescence (*see* **Note 3**). It is also important to be aware of the geometry of the fluorophore signal. Depending on the experimental conditions, if the GLUT4 is restricted to a particular compartment the signal becomes compacted, while the GLUT4 may not be restricted in another experimental condition, allowing that same number of signaling molecules to spread into nonoverlapping vesicles. Signal from overlapping vesicles can underestimate total protein levels if the pixels become saturated. Assuming equivalent transfection rates, comparing total GFP fluorescence between experimental conditions should reveal whether or not a given condition impacts your signal detection in this manner.

2. NIH 3T3-L1 mouse pre-adipocyte fibroblasts should be grown and maintained in 100 mm cell culture dishes as previously described [19].

3. We use goat serum in our blocking buffer and antibody mixtures. Our secondary is an Alexa 647-conjugated goat anti-mouse antibody. We found matching the species of the blocking buffer with the parent species of the secondary reduced background and nonspecific binding signal. Also, we chose an Alexa 647-conjugated antibody because the excitation and emission spectrums of Alexa 647 and GFP have minimal cross-interference. Lastly, the dilution indicated above is optimized for work with our specific antibodies of choice. Deviations from these antibody dilutions may be required for each antibody.

4. Standard 3T3-L1 differentiation protocols typically continue to supplement their maturation media with 175 nM insulin during days 4–6, following the DII differentiation cocktail during days 1–3. We found that taking day-5 cells for transfection, we were often working with cells that exhibited a muted response to insulin when we would later try to stimulate with 10 nM insulin. Due to this observed temporary insulin resistance, we removed insulin from the maturation medium during days 4 and 5. In addition, 3T3-L1 adipocytes are not completely mature until around days 8 and 9. We found that the transfection rates worsen the older the cells got, while day-5 adipocytes exhibit the highest transfection efficiency. We suspect that this is a result of the presence of more lipid droplets in the more mature adipocytes.
5. The voltage and current settings for optimal electroporation may vary between electroporation systems. It will be important to perform tests to ensure that the cells are being pulsed with the optimal voltage and current. Refer to the electroporation system's manual to evaluate any changes that may need to be made.
6. In our experience, electroporation of 3T3-L1 adipocytes has a low transfection rate. However, it is the most effective transient transfection method available for these cells, as other methods, such as the use of lipofectamine reagents, do not function well in the high-lipid environment of the adipocyte. Therefore, in order to obtain optimal transfection, it will be important to determine the proper cell/plasmid ratio with each electroporation system (*see* **Note 5**).
7. For our experiments, we use 4-well-chamber culture slides from BD Falcon. The volumes and amounts listed are based on our work with those. However, if plating transfected cells in other chambers and/or dishes with different sizes, simply adjust volumes accordingly.
8. Some experimental designs, such as assessing the effect of chronic hyperinsulinemia on GLUT4 redistribution, may require overnight treatment. It is best to prepare the experimental treatment in the post-electroporation media at double the desired final concentration. Once the cells have had the opportunity to recover fully and set down on the slide, addition of the post-electroporation media containing the treatment will dilute it twofold.
9. We have found that using aspiration to remove the various solutions and washes from the chambers dislodged our cells. Over the course of the many steps, aspiration appears to disrupt the integrity of many of the cells near the site of aspiration. We found that simply decanting the liquid from the chambers was sufficient to empty the chambers. In a similar manner, we also

found that dipping the slides into the dPBS washes improved overall cell integrity by removing the strong directed flow that can accompany pipetting.

10. Prepare the 10 nM insulin aliquot as close to the time of use as possible. This is also the point at which anti-HA antibody needs to be incubated with the cells for the recycling assay. This can be expanded to other membrane-trafficking proteins, such as the transferrin receptor [13, 18].
11. When removing liquids from the chambers and the slide with a chemwipe, it is important to not allow the cells to dry out. While decanting and wicking away any excess liquids is essential, such that the integrity and specific dilution of the various solutions are maintained, small amounts of liquid must be left behind to keep the cells from drying out. This is especially important while removing the chambers from the slide, since this is the longest period of time that the cells will not be covered. Have mounting media and coverslips ready at hand to seal the cells quickly after the removal of the chambers.
12. $NaBH_4$ is the quenching reagent that we have found works best for reducing the autofluorescence induced by paraformaldehyde cross-linking, as well as the general autofluorescence the lipid-laden adipocytes exhibit. This is a solid that when dissolved begins to effervesce. For this reason, it should be prepared just before use.
13. We use Prolong Gold with or without DAPI. In our experience with this reagent, the mounting media does not always fully solidify overnight. However, due to its viscous nature as it does set up, any movement of the coverslip after it is laid down risks shearing the cells underneath, making them unfit for imaging. To avoid this, we recommend applying a drop of clear fingernail polish at the corners of the coverslips upon mounting.

References

1. Cushman SW, Wardzala LJ (1980) Potential mechanism of insulin action on glucose transport in the isolated rat adipose cell. Apparent translocation of intracellular transport systems to the plasma membrane. J Biol Chem 255(10):4758–4762
2. Suzuki K, Kono T (1980) Evidence that insulin causes translocation of glucose transport activity to the plasma membrane from an intracellular storage site. Proc Natl Acad Sci U S A 77(5):2542–2545
3. James DE, Brown R, Navarro J, Pilch PF (1988) Insulin-regulatable tissues express a unique insulin-sensitive glucose transport protein. Nature 333(6169):183–185. https://doi.org/10.1038/333183a0
4. Birnbaum MJ (1989) Identification of a novel gene encoding an insulin-responsive glucose transporter protein. Cell 57(2):305–315
5. Fukumoto H, Kayano T, Buse JB, Edwards Y, Pilch PF, Bell GI, Seino S (1989) Cloning and characterization of the major insulin-responsive glucose transporter expressed in human skeletal muscle and other insulin-responsive tissues. J Biol Chem 264(14):7776–7779
6. Slot JW, Geuze HJ, Gigengack S, Lienhard GE, James DE (1991) Immuno-localization of the insulin regulatable glucose transporter in brown adipose tissue of the rat. J Cell Biol 113(1):123–135
7. Smith RM, Charron MJ, Shah N, Lodish HF, Jarett L (1991) Immunoelectron microscopic

demonstration of insulin-stimulated translocation of glucose transporters to the plasma membrane of isolated rat adipocytes and masking of the carboxyl-terminal epitope of intracellular GLUT4. Proc Natl Acad Sci U S A 88(15): 6893–6897

8. Quon MJ, Guerre-Millo M, Zarnowski MJ, Butte AJ, Em M, Cushman SW, Taylor SI (1994) Tyrosine kinase-deficient mutant human insulin receptors (Met1153→Ile) overexpressed in transfected rat adipose cells fail to mediate translocation of epitope-tagged GLUT4. Proc Natl Acad Sci U S A 91(12):5587–5591
9. Oatey PB, Van Weering DH, Dobson SP, Gould GW, Tavare JM (1997) GLUT4 vesicle dynamics in living 3T3 L1 adipocytes visualized with green-fluorescent protein. Biochem J 327(Pt 3):637–642
10. Lampson MA, Racz A, Cushman SW, McGraw TE (2000) Demonstration of insulin-responsive trafficking of GLUT4 and vpTR in fibroblasts. J Cell Sci 113(Pt 22):4065–4076
11. Dawson K, Aviles-Hernandez A, Cushman SW, Malide D (2001) Insulin-regulated trafficking of dual-labeled glucose transporter 4 in primary rat adipose cells. Biochem Biophys Res Commun 287(2):445–454. https://doi.org/10.1006/bbrc.2001.5620
12. Gonzalez E, Flier E, Molle D, Accili D, McGraw TE (2011) Hyperinsulinemia leads to uncoupled insulin regulation of the GLUT4 glucose transporter and the FoxO1 transcription factor. Proc Natl Acad Sci U S A 108(25):10162–10167. https://doi.org/10.1073/pnas.1019268108
13. Tessneer KL, Jackson RM, Griesel BA, Olson AL (2014) Rab5 activity regulates GLUT4 sorting into insulin-responsive and non-insulin-responsive endosomal compartments: a potential mechanism for development of insulin resistance. Endocrinology 155(9):3315–3328. https://doi.org/10.1210/en.2013-2148
14. Satoh S, Nishimura H, Clark AE, Kozka IJ, Vannucci SJ, Simpson IA, Quon MJ, Cushman SW, Holman GD (1993) Use of bismannose photolabel to elucidate insulin-regulated GLUT4 subcellular trafficking kinetics in rat adipose cells. Evidence that exocytosis is a critical site of hormone action. J Biol Chem 268(24):17820–17829
15. Kozka IJ, Clark AE, Reckless JP, Cushman SW, Gould GW, Holman GD (1995) The effects of insulin on the level and activity of the GLUT4 present in human adipose cells. Diabetologia 38(6):661–666
16. Maier VH, Gould GW (2000) Long-term insulin treatment of 3T3-L1 adipocytes results in mis-targeting of GLUT4: implications for insulin-stimulated glucose transport. Diabetologia 43(10):1273–1281. https://doi.org/10.1007/s001250051523
17. Hoehn KL, Hohnen-Behrens C, Cederberg A, LE W, Turner N, Yuasa T, Ebina Y, James DE (2008) IRS1-independent defects define major nodes of insulin resistance. Cell Metab 7(5):421–433. https://doi.org/10.1016/j.cmet.2008.04.005
18. Zeigerer A, Lampson MA, Karylowski O, Sabatini DD, Adesnik M, Ren M, McGraw TE (2002) GLUT4 retention in adipocytes requires two intracellular insulin-regulated transport steps. Mol Biol Cell 13(7): 2421–2435. https://doi.org/10.1091/mbc.E02-02-0071
19. Eyster CA, Duggins QS, Olson AL (2005) Expression of constitutively active Akt/protein kinase B signals GLUT4 translocation in the absence of an intact actin cytoskeleton. J Biol Chem 280(18):17978–17985. https://doi.org/10.1074/jbc.M409806200

Chapter 14

GLUT4 Translocation in Single Muscle Cells in Culture: Epitope Detection by Immunofluorescence

Javier R. Jaldin-Fincati, Philip J. Bilan, and Amira Klip

Abstract

GLUT4 is the major glucose transporter in skeletal muscle. GLUT4 cycles to and from the plasma membrane and its exocytic rate is accelerated by insulin and muscle contraction to achieve a new steady state with more GLUT4 proteins at the muscle cell surface. To gain a better understanding of the molecular and cellular mechanisms that govern GLUT4 protein recycling, we developed an in vitro model in which *myc*-epitope-tagged GLUT4 or GLUT4-GFP is expressed in L6 skeletal muscle cells. The *myc*-epitope is inserted into an exofacial domain that is accessible to anti-*myc* antibodies from the outside of non-permeabilized cells, allowing one to count the number of transporters at the cell surface. This enables one to perform single-cell analysis using confocal fluorescence microscopy to quantify cell surface GLUT4*myc* or GLUT4*myc*-GFP in cells co-transfected with diverse cDNA constructs, treated with siRNAs, or co-stained with antibodies for other proteins of interest. Herein, we describe the methodology to perform these experimental approaches in insulin-stimulated L6 muscle cells.

Key words GLUT4 translocation, GLUT4-GFP, Vesicle traffic, Insulin, L6 muscle cells, Glucose uptake, Skeletal muscle

1 Introduction

Skeletal muscle is the principal consumer of dietary glucose, the bulk of which is stored as glycogen. This homeostatic process is regulated by insulin action. During muscle contraction, skeletal muscles also consume glucose but in this case it is metabolized through glycolysis and the Krebs cycle to generate ATP. The rate-limiting step of glucose utilization by skeletal muscle is glucose transport [1]. In the resting, postabsorptive state, GLUT4 dynamically cycles from internal membrane stores to the cell surface. Insulin primarily accelerates the exocytic rate of GLUT4-containing vesicles. In mature muscle, the number of GLUT4

Javier R. Jaldin-Fincati and Philip J. Bilan contributed equally to this work.

Karin Lindkvist-Petersson and Jesper S. Hansen (eds.), *Glucose Transport: Methods and Protocols*, Methods in Molecular Biology, vol. 1713, https://doi.org/10.1007/978-1-4939-7507-5_14,

glucose transporters at the muscle cell surface is rapidly increased by the mobilization and insertion of intracellular GLUT4-containing vesicles into the two domains of the muscle cell surface: the sarcolemma and the T-tubules [2]. Typically, most muscles examined increase glucose uptake by two- to threefold in response to insulin, and where, measured, this is underscored by a parallel twofold increase in surface GLUT4 [3]. In contrast, during conditions of energy demand, GLUT4 endocytosis from the cell surface is slowed down [4–7], suggesting that exercise may increase GLUT4-containing vesicle exocytosis and delay GLUT4 endocytosis. In the pathological conditions of obesity and type 2 diabetes, the insulin responsiveness of glucose transport (as well as other actions of insulin) is blunted [8]. At the molecular level, insulin signal transduction is reduced [9, 10] resulting in lower than normal GLUT4 translocation to the surface membranes. Therefore, understanding the mechanisms that regulate GLUT4 vesicle traffic to and from the muscle cell surface has major physiological implications on glucose homeostasis in health and disease.

Due to imprecise methods and the architecture of skeletal muscle, it has been complicated to quantify GLUT4 translocation in the intact tissue, although a number of studies have successfully used sophisticated imaging approaches [11, 12]. These strategies detect GLUT4 at the cell periphery, but do not ascertain that the transporter is inserted in the membrane, as required to enact glucose uptake. A membrane-impermeant, glucose photoaffinity label has been used to this end, but this requires laborious membrane solubilization, pull-down of labeled GLUT4 proteins, and gel electrophoresis [13].

To circumvent these impasses, we have generated and improved cellular models of muscle cells in culture suitable to study GLUT4 translocation. L6 muscle cells, originally derived from day-old rat skeletal muscle, propagate as mononucleated myoblasts but upon serum removal differentiate by cellular fusion into multinucleated primary myotubes [14, 15]. L6 myotubes express many proteins typical of skeletal muscle including the GLUT4 glucose transporter [16–18]. Insulin stimulates glucose uptake to a higher degree in differentiated L6 myotubes and GLUT4 expression parallels the acquisition of these characteristics during muscle differentiation [16, 19]. Another transporter, GLUT1, is expressed at low levels in mature skeletal muscle fibers but is abundant in most cells in culture and also contributes to glucose uptake [20]. Over the years we selected an L6 muscle cell clone that expresses relatively lower levels of this transporter.

To more precisely ascertain the regulation of GLUT4 in muscle cells and to better understand the molecular mechanisms of this regulation, we generated an L6 myoblast cell line that stably expresses the GLUT4 protein containing a 14-amino-acid epitope of human *c-myc* within its first exofacial loop [21, 22]. L6-GLUT4*myc* cells differentiate normally from myoblasts to myotubes and they respond to insulin with a twofold stimulation of glucose uptake and twofold translocation of GLUT4*myc* to their cell surface [22]. In L6-GLUT4*myc* myotubes, GLUT4*myc* levels are greater than those of the endogenous GLUT4 or GLUT1 by almost 100-fold and GLUT4*myc* determines glucose influx [23, 24]. Furthermore, GLUT4*myc* recycling is highly similar to that of endogenous GLUT4 as far as the translocation of unlabeled transporter can be ascertained [25]. L6 GLUT4*myc* myotubes respond to insulin by phosphorylation of IRS-1, and activation of phosphatidylinositol 3-kinase and protein kinase B (Akt/PKB) [26], which are necessary events in the stimulation of glucose uptake by insulin [27]. In vivo, excitation-contraction coupling of muscle fibers leads to a rapid rise of intracellular calcium ions, and this response can also be evoked in the muscle cells. Accordingly, we have examined the signal transduction pathways triggered by a rise in cytosolic Ca^{2+} or energy demand that regulate GLUT4 traffic in L6-GLUT4*myc* cells [4, 5].

Interestingly, GLUT4*myc* recycling and its regulation by insulin can be documented in myoblasts, which are already muscle committed and present an insulin-responsive GLUT4-specifc compartment that harbors insulin-regulated aminopeptidase (IRAP), and largely excludes GLUT1 [26, 28]. As such, L6-GLUT4*myc* myoblasts provide a muscle cell system in which GLUT4*myc* traffic can be easily studied. The cells are amenable to diverse molecular analyses due to their unobstructed morphology, and suitability to grow at sub-confluence, which is required for gene transfer and RNAi-mediated knockdown of proteins of interest [26, 29].

Arrival and insertion of the GLUT4*myc* protein at the cell surface are monitored simply by labeling intact cells with anti-*myc* antibodies [30] followed by enzyme-conjugated secondary antibodies in a colorimetric assay [22, 31] or by fluorophore-conjugated secondary antibodies for fluorescence microscopy detection [32]. The latter methodology will be featured in this protocol.

The L6-GLUT4*myc* cell technology has been very instructive for our understanding of insulin- and contraction-regulated GLUT4 traffic and the reader is directed to several reviews on this topic [24, 33–36].

2 Materials

2.1 cDNA Constructs and siRNA Oligomers

1. GLUT4*myc* expression vector: GLUT4*myc* cDNA is created by introducing a 14-codon *myc*-epitope encoding the protein sequence AEEQKLISEEDLLK between Pro^{66} and Gly^{67} of the rat GLUT4 cDNA. GLUT4*myc* cDNA is then cloned into the pEGFP-N1 vector by standard molecular biology protocols to create the GLUT4*myc*-GFP construct that expresses a chimeric protein with GFP fused to the C-terminus of GLUT4*myc*.

Subsequently, GLUT4*myc*-GFP is sub-cloned into pcDNA3 to create the mammalian expression vector pcDNA3-GLUT4*myc*-GFP (*see* **Note 1**).

2. Other constructs carrying genes of interest, such as cDNAs for dominant-negative or constitutively active mutants, fluorescent fusion proteins, or wild-type proteins in mammalian expression vectors.
3. HP Custom or Flexitube siRNA oligomer product lines (Qiagen Sciences) or equivalent, of your choice.

2.2 Cell Lines

1. L6 myoblasts stably expressing GLUT4 and a *myc*-epitope in the first exofacial loop (L6-GLUT4*myc*) and a selection marker that confers blasticidin-HCl resistance (*see* **Note 2**).
2. Immortalized L6 rat skeletal myoblasts, for transiently expressing the pcDNA3-GLUT4*myc*-GFP construct (*see* **Note 2**).

2.3 Cell Culture

1. Alpha modified Eagle's medium with Earle's salts and ribonucleosides, L-glutamine, and deoxyribonucleosides (α-MEM).
2. Cell culture medium: 10 % v/v Fetal bovine serum (FBS), 100 IU/mL penicillin sodium salt, 100 μg/mL streptomycin sulfate, 0.25 μg/mL amphotericin B in α-MEM.
3. Blasticidin-HCl solution: Blasticidin S, hydrochloride, *Streptomyces* sp., sterile-filtered aqueous solution, cell culture tested.
4. Trypsin-EDTA solution: 0.5 % w/v Trypsin, 0.53 mM ethylenediaminetetraacetic acid (EDTA) in sterile deionized water.
5. Insulin solution: Human insulin (Humulin R) 100 IU/mL (Eli Lilly). Prepare working solutions of insulin in α-MEM without FBS to a final concentration of 100 nM or less (1 IU is equivalent to 0.0347 mg of human insulin).

2.4 Transient Transfection of DNA Constructs or siRNA Oligomers

1. Transfection medium: 10 % v/v FBS in Opti-MEM 1. Opti-MEM 1 consists of Eagle's minimum essential media buffered with HEPES and sodium bicarbonate, and supplemented with hypoxanthine, thymidine, sodium pyruvate, L-glutamine, trace elements, and growth factors (Thermo Fisher Scientific).
2. DNA transfection reagent: Fugene HD transfection reagent (Promega) (*see* **Note 3**).
3. siRNA oligomer transfection reagent: jetPRIME® transfection reagent (Polyplus-transfection Inc) (*see* **Note 3**).

2.5 Immunofluore scence

1. Round precision coverslips, No. 1.5 and 18 mm diameter, for high-performance microscopy (Marienfeld-Superior).
2. Cell culture multi-well plates with 12 wells.

3. Cell culture flasks 75 cm^2 (T75).
4. Buffered salt solution: Phosphate-buffered saline (PBS) supplemented with calcium and magnesium (*see* **Note 4**): 137 mM NaCl, 2.7 mM KCl, 1.5 mM KH_2PO_4, 8.1 mM Na_2HPO_4, 0.9 mM $CaCl_2$, 0.5 mM $MgCl_2$, pH 7.4.
5. Paraformaldehyde (PFA) stock solution: 16 % w/v.
6. Fixation solution: 3 % w/v PFA in PBS.
7. Quenching solution: 50 mM NH_4Cl in PBS.
8. Permeabilization solution: 0.1 % w/v Triton X-100 in PBS.
9. Blocking solution: 2 % w/v BSA in PBS.
10. Antibody dilution buffer: 0.2 % w/v BSA in PBS.
11. Primary antibody: Mouse monoclonal anti-*myc* antibody, clone 9E10 (Santa Cruz Biotechnology) and others (e.g., rabbit polyclonal IgG), as required.
12. Secondary antibody: Goat anti-mouse IgG Cy®3-conjugate and other conjugated secondary antibodies (e.g., donkey anti-rabbit IgG Cy®5-conjugate) as required.
13. Fluorescence mounting medium.
14. Parafilm M.
15. Kimwipes EX-L delicate task wipers (Kimberly-Clark Corp.).
16. Spinning disc confocal microscope: Inverted spinning-disc Olympus microscope IX81 equipped with a 60× numerical aperture (NA) 1.35 immersion objective, a CSU-X1-A Yokogawa spinning-disc unit, a Hamamatsu C9100–13 EM-CCD camera and controlled by Volocity 6.1.2 software (PerkinElmer) (*see* **Note 5**).

3 Methods

3.1 Cell Surface GLUT4myc Detection in L6-GLUT4myc Myoblasts

3.1.1 Cell Culture

1. L6-GLUT4*myc* myoblasts are grown in T75 culture flasks with cell culture medium supplemented with 5 μg/mL blasticidin-HCl (to provide selection pressure on the cells to maintain stable GLUT4*myc* expression) under incubation conditions: 37 °C, 5 % CO_2, and 100 % relative humidity. When cells reach ~70 % of confluence they are subcultured for either growing or seeding for experiments.
2. Culture medium in T75 culture flasks containing cells is discarded and the cells are immediately washed once with sterile PBS that is free of calcium and magnesium.
3. PBS is removed and cells are incubated with a minimum volume of a pre-warmed trypsin–EDTA solution for approximately 60 s (until the cells become rounded and start to detach from the

plate, bumping the flask with the heel of the hand will aid cell release).

4. Cells are suspended in 10 mL of cell culture medium supplemented with 5 μg/mL blasticidin-HCl with continuous and gentle up-and-down pipetting.
5. Cells are counted using a Z2 Coulter counter (Beckman Coulter).
6. Cells are seeded at 2.5×10^6 cells/T75 flask for growing or 1.2×10^5 cells/well 12-well plates for transfection experiments.

3.1.2 Transfection of L6-GLUT4myc Myoblasts with siRNA Oligomers

Optional: To be done before DNA transfection (*see* **Note 6**).

1. L6-GLUT4*myc* myoblasts are seeded at 1.2×10^5 cells/well under sterile conditions on 18 mm diameter round coverslips (*see* **Note 7**) in 12-well plates approximately 18–24 h before transfection.
2. On the transfection day, cells are washed twice with warm PBS and 1.0 mL per well of transfection medium is added. The cells are kept under incubation conditions until jetPRIME®–siRNA complex for transfection is prepared following the manufacturer's recommendations (*see* **Note 8**).
3. Add dropwise 100 μL of complexes per well and mix thoroughly with swirling (final siRNA concentration is 200 nM).
4. After 24 h of incubation, the transfection medium containing the complexes is replaced by cell culture medium without blasticidin-HCl.
5. Allow cells to incubate with siRNA for 48 h after transfection (*see* **Note 9**).

3.1.3 Transfection of L6-GLUT4myc Myoblasts with DNA Constructs

Optional procedure, *see* **Note 10**.

1. L6-GLUT4*myc* myoblasts are seeded at 1.2×10^5 cells/well under sterile conditions on 18 mm diameter coverslips (*see* **Note 7**) in 12-well plates approximately 18–24 h before transfection.
2. On the transfection day, cells are washed twice with warm PBS and 950 μL per well of transfection medium is added. The cells are kept under incubation conditions until Fugene HD/DNA complexes for transfection are prepared following the manufacturer's recommendations (*see* **Note 11**). If cells are also to be co-transfected with another DNA construct, *see* **Note 12.** If cells are also to be transfected with siRNA, follow the specific indications for that procedure and *see* **Notes 8** and **9**.
3. Add dropwise 50 μL of the transfection DNA complexes per well to the cells, and mix thoroughly with swirling.

4. After 4 h of incubation, the transfection medium containing the complexes is replaced by cell culture medium without blasticidin-HCl.
5. Perform DNA transfection 24 h before the detection and quantification of GLUT4*myc* levels at the plasma membrane.

3.1.4 Immunofluorescence for Detection of GLUT4myc at the Plasma Membrane

1. L6-GLUT4*myc* myoblasts are incubated in serum-free α-MEM for at least 3 h and no more than 6 h before the start of the experiment (*see* **Note 13**). Carry an extra coverslip (basal, non-transfected) that will receive only the secondary conjugated antibody and will be used for background estimation.
2. Small-molecule inhibitors or activators are incubated with the cells for the specified period of time during the serum-deprivation phase and together with the stimulus if required (*see* **Note 14**).
3. Then, serum-deprived cells are washed twice with serum-free α-MEM and stimulated or not (basal condition) with working solutions of insulin (up to 100 nM) for 10 min at 37 °C.
4. After stimulation with insulin, the cells are rapidly washed three times with cold PBS and then fixed with ice-cold fixation solution for 30 min on ice.
5. After fixation, cells are washed once with PBS at room temperature and then incubated with quenching solution for 15 min at room temperature.
6. Then the cells are washed three times with PBS and incubated with blocking solution for 30 min at room temperature.
7. After the blocking step, the cells are incubated for 1 h at 37 °C with a dilution (1:100) of primary anti-*myc* antibody prepared in antibody dilution buffer. Put 100 μL of the antibody dilution on clean Parafilm and place each coverslip facedown upon its antibody droplet (*see* **Note 15**).
8. Following incubation with the primary antibody dilution, the coverslips are returned to the plate and washed three times with antibody dilution buffer for 5 min each with continuous shaking.
9. Afterwards, the cells are incubated with a dilution (1:1000) of a secondary goat-anti-mouse Cy®3-conjugate for 45 min at 37 °C prepared in antibody dilution buffer. Also, prepare the control coverslip (non-transfected, basal) that will only receive the Cy®3-fluorophore secondary antibody conjugate to be used to calculate background fluorescence. Place 100 μL of the antibody dilution on clean Parafilm and place each coverslip facedown on it.
10. After the incubation with the secondary antibody the coverslips are returned to the plate and washed five times with PBS for 5 min each with continuous shaking.

11. If other proteins, either endogenous or transfected proteins, need to be detected, follow **steps 12–14 of** Subheading 3.1.4, since cell permeabilization is required to immunostain these proteins with primary antibodies; otherwise, jump to **step 16 of** Subheading 3.1.4.
12. Incubate coverslips with permeabilization solution for 15 min at room temperature (*see* **Note 16**).
13. Wash the cells three times with PBS and incubate with blocking solution for 30 min at room temperature.
14. After the blocking step, the cells are incubated for 1 h at 37 °C with a rabbit primary antibody against the intracellular protein, diluted in antibody dilution buffer. Put 100 μL of the antibody dilution on clean Parafilm and place each coverslip facedown upon its antibody droplet. Then the coverslips are returned to the plate and washed three times with PBS for 5 each min at room temperature.
15. Afterwards, the cells are incubated with a dilution (1:1000) of a secondary donkey-anti-rabbit Cy®5-conjugate (if necessary, *see* **Note 17**) for 45 min at 37 °C prepared in antibody dilution buffer. Also, prepare a control coverslip (non-transfected, basal) that will receive the Cy®5-fluorophore secondary antibody conjugate. Put 100 μL of the antibody dilution on clean Parafilm and place each coverslip facedown on it. Then the coverslips are returned to the plate and washed three times with PBS for 5 min each at room temperature.
16. Lift each coverslip with forceps and remove PBS excess by dabbing the edge of the coverslip on a static-free Kimwipe tissue paper. Mount the samples by inverting each coverslip into a drop (25 μL) of fluorescence mounting medium on glass microscope slides.
17. Samples are allowed to air-dry overnight before visualizing in the microscope. The samples can be stored at −20 °C and protected from light for up to 1 month.

3.2 Cell Surface GLUT4myc-GFP Detection in L6 Cells

3.2.1 Cell Culture

1. L6 myoblasts are grown in T75 culture flasks with cell culture medium under incubation conditions: 37 °C, 5 % CO_2, and 100 % relative humidity. When cells reach ~70 % of confluence they are subcultured for either growing or seeding for experiments.
2. Culture medium in T75 culture flasks containing cells is discarded and the cells are immediately washed once with sterile PBS that is free of calcium and magnesium.
3. PBS is removed and cells are incubated with a minimum volume of a pre-warmed trypsin–EDTA solution for approximately 60 s (until the cells become rounded and start to detach

from the plate, bumping the flask with the heel of the hand will aid cell release).

4. Cells are suspended in 10 mL of cell culture medium with continuous and gentle up-and-down pipetting.
5. Cells are counted using a Z2 Coulter counter (Beckman Coulter).
6. Cells are seeded at 2.5×10^6 cells per T75 flask for growing or 1.2×10^5 cells/well (M12) for transfection experiments.

3.2.2 Transfection of L6 Myoblasts with siRNA Oligomers (Optional, to Be Done Before DNA Transfection, See **Note 6***)*

1. L6 myoblasts are seeded at 1.2×10^5 cells/well under sterile conditions on the round precision coverslips for microscopy (*see* **Note** 7) in 12-well plates approximately 18–24 h before transfection.
2. On the transfection day, cells are washed twice with warm PBS and 1.0 mL per well of complete transfection medium is added. The cells are kept under incubation conditions until jetPRIME®–siRNA complex for transfection is prepared following the manufacturer's recommendations (*see* **Note 8**).
3. Add dropwise 100 μL of complexes per well and mix thoroughly with swirling (final siRNA concentration is 200 nM).
4. After 24 h of incubation, the transfection medium containing the complexes is replaced by cell culture medium.
5. Allow cells to incubate with siRNA for 48 h after transfection (*see* **Note 9**).

3.2.3 Transfection of L6 Myoblasts with the GLUT4myc-GFP DNA Construct

Optional procedure, *see* **Note 10**.

1. L6 myoblasts are seeded at 1.2×10^5 cells/well under sterile conditions on 18 mm diameter coverslips (*see* **Note** 7) in 12-well plates approximately 18–24 h before transfection.
2. On the transfection day, cells are washed twice with warm PBS and 950 μL per well of transfection medium is added. The cells are kept under incubation conditions until Fugene HD/GLUT4*myc*-GFP DNA complex for transfection is prepared following the manufacturer's recommendations (*see* **Note 11**). If cells are also to be co-transfected with another DNA construct, *see* **Note 12**. If cells are also to be transfected with siRNA, follow the specific indications for that procedure and *see* **Notes 8** and **9**.
3. Add dropwise 50 μL of complex per well to the cells, and mix thoroughly with swirling.
4. After 6 h of incubation, the transfection medium containing the complex is replaced by cell culture medium.
5. The experiment for detection and quantification of GLUT4*myc*-GFP levels at the plasma membrane is performed 24 h after transfection.

3.2.4 Immunofluorescence for Detection of GLUT4myc-GFP at the Plasma Membrane

1. L6-GLUT4*myc*-GFP is incubated in serum-free α-MEM for at least 3 h and no more than 6 h before the start of the experiment (*see* **Note 13**). Carry an extra coverslip (basal, non-transfected) that will receive only the secondary conjugated antibody and will be used for background estimation.
2. Small-molecule inhibitors or activators are incubated with the cells for the specified period of time during the serum-deprivation phase and together with the stimulus if required (*see* **Note 14**).
3. Then, serum-deprived cells are washed twice with serum-free α-MEM and stimulated or not (basal condition) with working solutions of insulin (up to 100 nM) for 10 min at 37 °C.
4. After stimulation with insulin, the cells are rapidly washed three times with cold PBS and then fixed with cold fixation solution for 30 min on ice.
5. After fixation, cells are washed once with PBS at room temperature and then incubated with quenching solution for 15 min at room temperature.
6. Then the cells are washed three times with PBS and incubated with blocking solution for 30 min at room temperature.
7. After the blocking step, the cells are incubated for 1 h at 37 °C with a dilution (1:100) of primary anti-*myc* antibody prepared in antibody dilution buffer. Put 100 μL of the antibody dilution on clean Parafilm and place each coverslip facedown upon its antibody droplet (*see* **Note 15**).
8. Following incubation with the primary antibody dilution, the coverslips are returned to the plate and washed three times with antibody dilution buffer for 5 min each with continuous shaking.
9. Afterwards, the cells are incubated with a dilution (1:1000) of a secondary goat-anti-mouse Cy®3-conjugate for 45 min at 37 °C prepared in antibody dilution buffer. Also, prepare the control coverslip (non-transfected, basal) that will only receive the Cy®3-fluorophore secondary antibody conjugate to be used to calculate background fluorescence. Place 100 μL of the antibody dilution on clean Parafilm and place each coverslip facedown on it.
10. After the incubation with the secondary antibody the coverslips are returned to the plate and washed five times with PBS for 5 min each with continuous shaking.
11. If other proteins, either endogenous or transfected proteins, need to be detected, follow **steps 12–14 of** Subheading 3.2.4, since cell permeabilization is required to immunostain these proteins with primary antibodies; otherwise, jump to **step 16 of** Subheading 3.2.4.

12. Incubate coverslips with permeabilization solution for 15 min at room temperature (*see* **Note 16**).
13. Wash the cells three times with PBS and incubate with blocking solution for 30 min at room temperature.
14. After the blocking step, the cells are incubated for 1 h at 37 °C with a rabbit primary antibody against the intracellular protein, diluted in antibody dilution buffer. Put 100 μL of the antibody dilution on clean Parafilm and place each coverslip facedown upon its antibody droplet. Then the coverslips are returned to the plate and washed three times with PBS for 5 each min at room temperature.
15. Afterwards, the cells are incubated with a dilution (1:1000) of a secondary donkey-anti-rabbit Cy®5-conjugate (if necessary, *see* **Note 17**) for 45 min at 37 °C prepared in antibody dilution buffer. Also, prepare a control coverslip (non-transfected, basal) that will receive the Cy®5-fluorophore secondary antibody conjugate. Put 100 μL of the antibody dilution on clean Parafilm and place each coverslip facedown on it. Then the coverslips are returned to the plate and washed three times with PBS for 5 each min at room temperature.
16. Lift each coverslip with forceps and remove PBS excess by dabbing the edge of the coverslip on a static-free Kimwipe tissue paper. Mount the samples by inverting each coverslip into a drop (30 μL) of fluorescence mounting medium on glass microscope slides.
17. Samples are allowed to air-dry overnight before visualizing in the microscope. The samples can be stored at −20 °C and protected from light for up to 1 month.

3.3 Acquisition and Quantification of Confocal Images

1. Acquisition settings (*see* **Note 18**): Before starting the image acquisition, sampling parameters need to be defined: The excitation wavelength for GFP-, Cy®3-, and Cy®5-tagged proteins are 489 nm, 548 nm, and 649 nm, respectively. For this protocol, any set of laser lines that excite the fluorophores mentioned above can be used. Define the sampling density applying the Nyquist-Shannon sampling theorem (*see* **Note 19**). For the microscope used in this protocol, the minimal voxel size for reaching the sampling criteria is 45 × 45 × 147 nm (x, y, z) for the green channel, 50 × 50 × 160 nm (x, y, z) for the red channel, and 60 × 60 × 196 nm (x, y, z) for the far-red channel. Take into account that every user should adjust all global acquisition geometry parameters based on the channel parameters of the confocal microscope available to pursue the protocol (*see* **Note 19**). Adjust exposure time and camera gain sensitivity in order to avoid acquiring saturated pixels. Set those parameters per channel using the sample with the highest intensity of fluorescence.

2. Image acquisition: Acquire the images keeping the preestablished acquisition settings constant between the different experimental conditions to be compared.
3. Immunofluorescence of L6-GLUT4*myc*: First acquire xy-plane images along the entire z-plane (red channel and if necessary far-red channel) of a non-transfected, basal condition coverslip incubated only with conjugated secondary antibodies. These images are used to estimate the level of background fluorescence generated by the nonspecific binding of the secondary antibody to the cells (measured in the red channel and/or the far-red channel in non-permeabilized or permeabilized cells). Acquire xy-plane images along the entire z-plane (red channel and if necessary far-red channel) in all experimental conditions. For quantitative purposes at least 30 cells per condition need to be imaged. Perform at least three independent experiments for assessing significant differences.
4. Immunofluorescence of L6 cells transfected with GLUT4*myc*-GFP: First acquire xy-plane images along the entire z-plane (red channel and if necessary far-red channel) of a non-transfected, basal condition coverslip incubated only with conjugated secondary antibodies, but not primary antibodies. These images are used for estimating the level of background fluorescence generated by the nonspecific binding of the secondary antibodies to the cells (measured in the red channel and/or the far-red channel in non-permeabilized or permeabilized cells) (*see* **Note 20**). With the experimental samples, acquire xy-plane images along the entire z-plane (green channel, red channel, and, if necessary, far-red channel) of transient transfected cells under basal and stimulated conditions, avoiding seeking out transfected cells displaying the strongest expression levels. For quantitative purposes at least 30 cells per conditions need to be imaged (*see* Fig. 1a). Perform at least three independent experiments for assessing significant differences.
5. Image quantification: Generate z-projections of each z-stack image using Volocity 6.1.2 software (PerkinElmer, Waltham, MA, USA).
6. Export the z-projections as TIFF files. Using ImageJ software (NIH) convert images to grey scale (one-channel images) or separate channels using split-channel function. Save each channel image as a separate TIFF file. Perform background subtraction using the rolling-ball background subtraction function (use the same rolling ball radius for all the samples that are to be compared).
7. Using the threshold function select each cell (outlining each cell's contour) that will be quantified as a region of interest (ROI) and add it to the ROI manger.

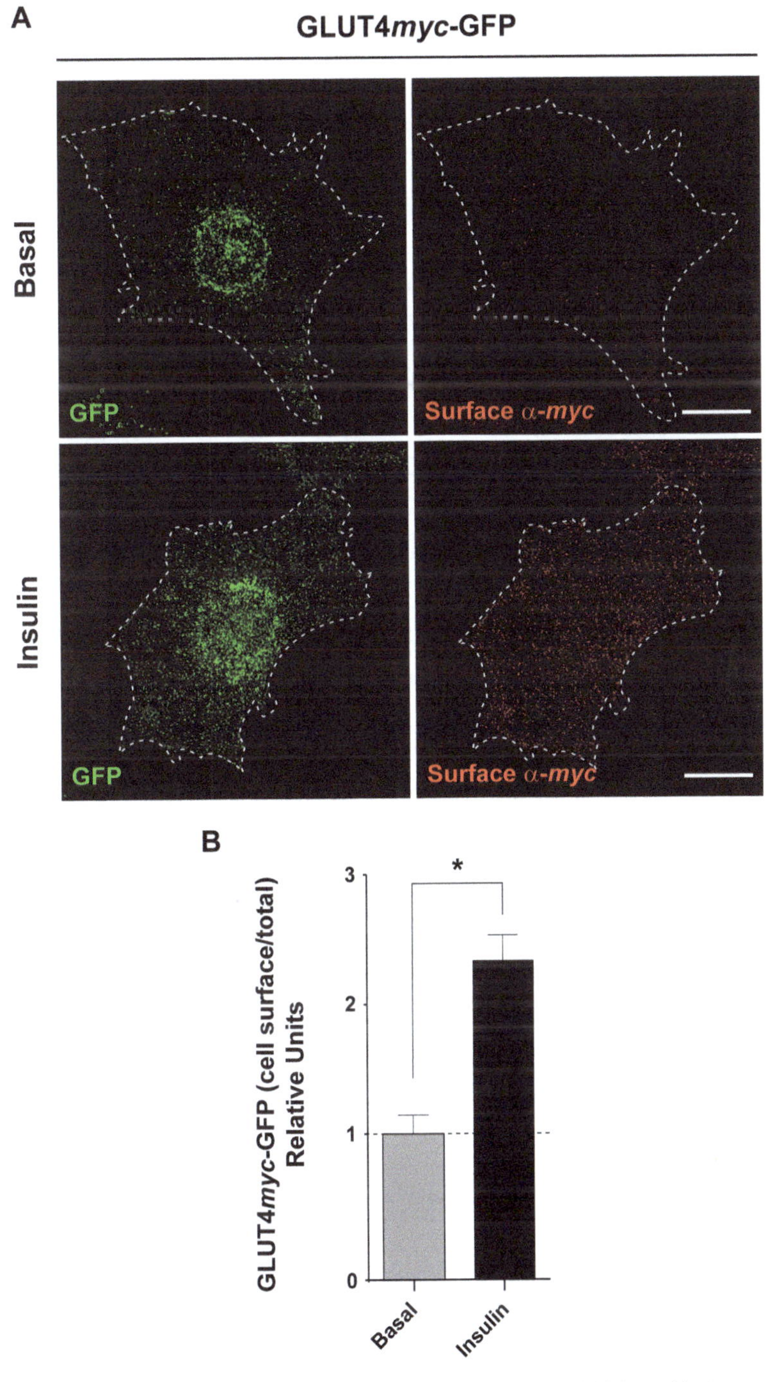

Fig. 1 Detection of cell surface GLUT4*myc*-GFP in intact cells after insulin stimulation. (**a**) L6 myoblasts grown on cover-slips were transiently transfected with GLUT4*myc*-GFP (24 h), then serum-deprived for 3 h, and incubated with insulin (100 nM, 10 min) or without and processed as described (current protocol). The figure shows a representative single composite of stacked xy-projections of confocal images in transfected cells. The total expression level of GLUT4*myc*-GFP is shown in the *left panels* (*green* channel, GFP) and the cell surface GLUT4*myc*-GFP (antibody labeling of exofacial –*myc* epitope) is shown in the *right panels* (*red* channel, surface α-*myc*). *Dashed lines* define regions of interest (ROIs). Scale bar represents 10 μm. (**b**) The graph shows the fold increase in the ratio of mean pixel intensities in the ROI for the *red* channel (cell surface GLUT4*myc*-GFP) normalized to the mean pixel intensities of the *green* channel (total GLUT4*myc*-GFP) for cells treated with insulin relative to non-treated (basal) cells (mean ± SD, $n = 3$, $^*p < 0.05$)

8. Determine the mean grey value within each ROI. Calculate the averages and SEM of the mean grey values per condition including backgrounds. Final expression levels of GLUT4*myc* at the cell surface are calculated after subtracting background estimations, as appropriate. For L6 cells transfected with GLUT4*myc*-GFP, it is necessary to normalize the intensity of fluorescence of the red channel (surface α-*myc*) with the intensity of fluorescence of the green channel (GLUT4*myc*-GFP). Finally, the results are expressed like fold changes with respect to the basal condition (*see* Fig. 1b).

4 Notes

1. The constructs are available upon request. Experimental details on generation of GLUT4*myc* cDNA can be found in reference [37] and the mammalian expression vector pcDNA3-GLUT4*myc*-GFP in reference [38]. If some experiments require co-transfection of DNA constructs that express a GFP-fusion protein, GLUT4*myc*-mCherry replaces GLUT4*myc*-GFP [39].
2. Both the L6-GLUT4*myc* Rat Myoblast Cell Line and the L6 Rat Myoblast Cell Line can be obtained commercially (Kerafast). Establishment of these cell lines is described in details elsewhere [22, 26, 32].
3. The L6 cells perform best in the media stated. Fugene HD transfection reagent for DNA and jetPRIME® transfection reagent for siRNA oligomers perform well with L6 and L6-GLUT4*myc* myoblasts. Other reagents can be substituted with alternatives and proper evaluation.
4. Calcium and magnesium are important for cell adhesion as these ions allow optimal activity of integrins and cadherins.
5. While we use a spinning-disc confocal microscope, any type of confocal microscope can be used with this protocol.
6. This procedure is performed in cases where it is desirable to determine the participation of different proteins on GLUT4 translocation.
7. Coverslip preparation (cleaning and sterilization): Make 1 M HCl in a glass container and put coverslips (#1.5) in it. Heat the acid to 50–60 °C for 4–16 h with occasional agitation. Cool to room temperature. Rinse out 1 M HCl with filtered (0.45 μm) MilliQ H_2O. Wash the coverslips extensively in MilliQ H_2O (two times 30 min each) with continuous agitation. Be sure to wash out the acid between stuck coverslips. Wash the coverslips with 70 % EtOH 2 h with continuous agitation. Wash the cov-

erslips with 95 % EtOH 1 h with continuous agitation. Rinse out 95 % EtOH with MilliQ H_2O. Dry coverslips between sheets of Whatman filter paper or light-duty tissue wipers. Bake in a glass container with foil for 2–4 h at 180 °C.

8. Complex reagent/siRNA oligomer preparation: Prepare 100 μL of siRNA (220 pmol) in jetPRIME® buffer (from 20 μM stock) in Opti-MEM I per coverslip. For controls, prepare coverslips that are not transfected and transfected with a non-targeting siRNA at the same final concentration. Add 3 μL volume of transfection reagent (jetPRIME®) Vortex for 10 sec and spin down briefly. Incubate 10–15 min at room temperature.
9. It is recommended to prepare an extra coverslip or an extra well to assess efficiency of siRNA-mediated knockdown of the endogenous protein by immunofluorescence or lysis and immunoblotting, respectively. The siRNA and cDNA transfections are rarely performed together in L6-GLUT4*myc* cells.
10. This procedure is performed in cases where it is desirable to determine the participation of different kind of proteins in GLUT4 translocation (using dominant-negative and -positive versions) as well as when it is necessary to identify intracellular compartments avoiding immunostaining techniques (using fluorescently tagged proteins).
11. Complex reagent/DNA construct preparation: Prepare 50 μL of plasmid solution (1 μg of DNA) in Opti-MEM I per coverslip. If a DNA construct is co-transfected with GLUT4*myc*-GFP, use GLUT4*myc*-GFP at 0.5 μg and the other construct between 0.5 and 1 μg, as empirically determined. Add the necessary volume of Fugene HD to get a reagent/DNA ratio of 2:1. Mix carefully by vortexing briefly and spin down. Incubate for 10 min at room temperature.
12. If co-transfections are performed, prepare a separate coverslip to assess efficiency of the co-transfected cDNA with the GLUT4*myc*-GFP cDNA.
13. Pre-incubation of cells in serum-free medium prior to insulin addition ensures that insulin or other growth factors in the serum do not artificially modify the basal steady-state cell surface levels of GLUT4*myc* (or GLUT4*myc*-GFP).
14. Incubation time with cell-permeant small-molecule inhibitors/activators prior to insulin addition is typically 30 min, but shorter incubation times can be empirically determined. DMSO vehicle should be no greater than 0.1 % w/v. The compatibility of higher amounts of DMSO or other solvents would have to be empirically determined. Inhibitors/activators that are cell-permeant peptides may require longer times

of incubation, such as 1–2 h, as recommended or tested. The serum-free condition of this incubation eliminates unwanted binding of small molecules to serum albumin. All inhibitors/activators should be included during the insulin incubations. Basal cells should be incubated with vehicle alone for the same times as the inhibitor/activators.

15. Incubation of coverslips face down on Parafilm allows one to minimize the volumes of the precious primary and secondary antibody solutions.

16. DNA constructs that are transfected into L6-GLUT4*myc* cells or co-transfected with GLUT4*myc*-GFP in L6 cells are detected by antibodies that need access to intracellularly localized epitopes on the wild-type protein or epitope tags (e.g., HA, FLAG). In these cases, permeabilization of the cells with Triton X-100-containing permeabilization solution for 15 min at room temperature could be required. Epitope tags are a special case, because the best primary antibodies for HA or FLAG are mouse monoclonal. In this case, good anti-*myc* epitope rabbit polyclonal antibodies are available from Sigma-Aldrich.

17. If primary antibodies were used to detect intracellular epitopes, secondary fluorophore conjugates in addition to Cy®3-goat-anti-mouse will be necessary. We have written our protocol with the Cy®5 secondary antibody conjugate as the first option. We have used Alexa Fluor® 488 and Alexa Fluor® 647 secondary antibody conjugates with success. Alexa Fluor® 488 cannot be used with GLUT4*myc*-GFP because of emission spectra coincidence.

18. This section contains acquisition settings and quantification methods for all the possibilities previously described in this protocol. For different fluorophores, fluorophore combinations, and/or acquisition systems, further optimization is required.

19. There are some deconvolution software companies that offer online tools for calculating Nyquist rate that can be used to define the minimal voxel size for image sampling (https://svi.nl/NyquistCalculator).

20. For estimating the level of background fluorescence generated by transfection process, acquire xy-plane images along the entire z-plane (red channel and if necessary far-red channel) of specimens (basal condition) with transient transfected cells not incubated with neither primary nor secondary antibodies. The level of background fluorescence is measured in the red channel and/or the far-red channel in non-permeabilized or permeabilized cells.

Acknowledgements

The methodology described in this chapter was developed over the years with support from grants from the Canadian Institutes of Health Research and the Canadian Diabetes Association.

References

1. Kubo K, Foley JE (1986) Rate-limiting steps for insulin-mediated glucose uptake into perfused rat hindlimb. Am J Phys 250(1 Pt 1):E100–E102
2. Marette A, Burdett E, Douen A, Vranic M, Klip A (1992) Insulin induces the translocation of GLUT4 from a unique intracellular organelle to transverse tubules in rat skeletal muscle. Diabetes 41(12):1562–1569
3. Marette A, Richardson JM, Ramlal T, Balon TW, Vranic M, Pessin JE, Klip A (1992) Abundance, localization, and insulin-induced translocation of glucose transporters in red and white muscle. Am J Phys 263:C443–C452
4. Antonescu CN, Diaz M, Femia G, Planas JV, Klip A (2008) Clathrin-dependent and independent endocytosis of glucose transporter 4 (GLUT4) in myoblasts: regulation by mitochondrial uncoupling. Traffic 9(7):1173–1190. https://doi.org/10.1111/j.1600-0854.2008.00755.x
5. Li Q, Zhu X, Ishikura S, Zhang D, Gao J, Sun Y, Contreras-Ferrat A, Foley KP, Lavandero S, Yao Z, Bilan PJ, Klip A, Niu W (2014) Ca(2)(+) signals promote GLUT4 exocytosis and reduce its endocytosis in muscle cells. Am J Physiol Endocrinol Metab 307(2):E209–E224. https://doi.org/10.1152/ajpendo.00045.2014
6. Wijesekara N, Tung A, Thong F, Klip A (2006) Muscle cell depolarization induces a gain in surface GLUT4 via reduced endocytosis independently of AMPK. Am J Physiol Endocrinol Metab 290(6):E1276–E1286
7. Yang J, Holman GD (2005) Insulin and contraction stimulate exocytosis, but increased AMP-activated protein kinase activity resulting from oxidative metabolism stress slows endocytosis of GLUT4 in cardiomyocytes. J Biol Chem 280(6):4070–4078
8. Klip A, Marette A, Dimitrakoudis D, Ramlal T, Giacca A, Shi ZQ, Vranic M (1992) Effect of diabetes on glucoregulation. From glucose transporters to glucose metabolism in vivo. Diabetes Care 15(11):1747–1766
9. Chiu TT, Jensen TE, Sylow L, Richter EA, Klip A (2011) Rac1 signalling towards GLUT4/glucose uptake in skeletal muscle. Cell Signal 23(10):1546–1554. https://doi.org/10.1016/j.cellsig.2011.05.022
10. Patel N, Huang C, Klip A (2006) Cellular location of insulin-triggered signals and implications for glucose uptake. Pflugers Arch 451(4):499–510
11. Lauritzen HP (2013) Insulin- and contraction-induced glucose transporter 4 traffic in muscle: insights from a novel imaging approach. Exerc Sport Sci Rev 41(2):77–86. https://doi.org/10.1097/JES.0b013e318275574c
12. Sylow L, Nielsen IL, Kleinert M, Moller LL, Ploug T, Schjerling P, Bilan PJ, Klip A, Jensen TE, Richter EA (2016) Rac1 governs exercise-stimulated glucose uptake in skeletal muscle through regulation of GLUT4 translocation in mice. J Physiol 594(17):4997–5008. https://doi.org/10.1113/JP272039
13. Karlsson HK, Chibalin AV, Koistinen HA, Yang J, Koumanov F, Wallberg-Henriksson H, Zierath JR, Holman GD (2009) Kinetics of GLUT4 trafficking in rat and human skeletal muscle. Diabetes 58(4):847–854. https://doi.org/10.2337/db08-1539
14. Klip A, Logan WJ, Li G (1982) Hexose transport in L6 muscle cells. Kinetic properties and the number of [3H]cytochalasin B binding sites. Biochim Biophys Acta 687:265–280
15. Yaffe D (1968) Retention of differentiation potentialities during prolonged cultivation of myogenic cells. Proc Natl Acad Sci U S A 61:477–483
16. Mitsumoto Y, Burdett E, Grant A, Klip A (1991) Differential expression of the GLUT1 and GLUT4 glucose transporters during differentiation of L6 muscle cells. Biochem Biophys Res Commun 175:652–659
17. Mitsumoto Y, Liu Z, Klip A (1993) Regulation of glucose and ion transporter expression during controlled differentiation and fusion of rat skeletal muscle cells. Endocr J 1:307–315
18. Shainberg A, Yagil G, Yaffe D (1969) Control of myogenesis in vitro by Ca 2 + concentration in nutritional medium. Exp Cell Res 58:163–167
19. Mitsumoto Y, Klip A (1992) Development regulation of the subcellular distribution and

glycosylation of GLUT1 and GLUT4 glucose transporters during myogenesis of L6 muscle cells. J Biol Chem 267:4957–4962

20. Murata H, Hruz PW, Mueckler M (2000) The mechanism of insulin resistance caused by HIV protease inhibitor therapy. J Biol Chem 275:20251–20254
21. Kishi K, Muromoto N, Nakaya Y, Miyata I, Hagi A, Hayashi H, Ebina Y (1998) Bradykinin directly triggers GLUT4 translocation via an insulin- independent pathway. Diabetes 47:550–558
22. Wang Q, Khayat Z, Kishi K, Ebina Y, Klip A (1998) GLUT4 translocation by insulin in intact muscle cells: detection by a fast and quantitative assay. FEBS Lett 427(2):193–197
23. Huang C, Somwar R, Patel N, Niu W, Torok D, Klip A (2002) Sustained exposure of L6 myotubes to high glucose and insulin decreases insulin-stimulated GLUT4 translocation but upregulates GLUT4 activity. Diabetes 51:2090–2098
24. Rudich A, Klip A (2003) Push/pull mechanisms of GLUT4 traffic in muscle cells. Acta Physiol Scand 178:297–308
25. Li D, Randhawa VK, Patel N, Hayashi M, Klip A (2001) Hyperosmolarity reduces GLUT4 endocytosis and increases its exocytosis from a VAMP2-independent pool in L6 muscle cells. J Biol Chem 276:22883–22891
26. Ueyama A, Yaworsky KL, Wang Q, Ebina Y, Klip A (1999) GLUT4myc ectopic expression in L6 myoblasts generates a GLUT4-specific pool conferring insulin sensitivity. Am J Phys 277(3 Pt 1):E572–E578
27. Wang Q, Somwar R, Bilan PJ, Liu Z, Jin J, Woodgett JR, Klip A (1999) Protein kinase B/Akt participates in GLUT4 translocation by insulin in L6 myoblasts. Mol Cell Biol 19:4008–4018
28. Sumitani S, Ramlal T, Liu Z, Klip A (1995) Expression of syntaxin 4 in rat skeletal muscle and rat skeletal muscle cells in culture. Biochem Biophys Res Commun 213(2):462–468. https://doi.org/10.1006/bbrc.1995.2154
29. Sun Y, Bilan PJ, Liu Z, Klip A (2010) Rab8A and Rab13 are activated by insulin and regulate GLUT4 translocation in muscle cells. Proc Natl Acad Sci U S A 107(46):19909–19914. https://doi.org/10.1073/pnas.1009523107
30. Evan GI, Lewis GK, Ramsay G, Bishop JM (1985) Isolation of monoclonal antibodies specific for human c-myc proto- oncogene product. Mol Cell Biol 5:3610–3616
31. Ishikura S, Antonescu CN, Klip A (2010) Documenting GLUT4 exocytosis and endocytosis in muscle cell monolayers. Curr Protoc Cell Biol Chapter 15:Unit 15.15. https://doi.org/10.1002/0471143030.cb1515s46
32. Foster LJ, Li D, Randhawa VK, Klip A (2001) Insulin accelerates inter-endosomal GLUT4 traffic via phosphatidylinositol 3-kinase and protein kinase B. J Biol Chem 276:44212–44221
33. Dugani CB, Klip A (2005) Glucose transporter 4: cycling, compartments and controversies. EMBO Rep 6(12):1137–1142
34. Foley K, Boguslavsky S, Klip A (2011) Endocytosis, recycling, and regulated exocytosis of glucose transporter 4. Biochemistry 50(15):3048–3061. https://doi.org/10.1021/bi2000356
35. Klip A, Sun Y, Chiu TT, Foley KP (2014) Signal transduction meets vesicle traffic: the software and hardware of GLUT4 translocation. Am J Physiol Cell Physiol 306(10):C879–C886. https://doi.org/10.1152/ajpcell.00069.2014
36. Zaid H, Antonescu CN, Randhawa VK, Klip A (2008) Insulin action on glucose transporters through molecular switches, tracks and tethers. Biochem J 413(2):201–215. https://doi.org/10.1042/BJ20080723
37. Kanai F, Nishioka Y, Hayashi H, Kamohara S, Todaka M, Ebina Y (1993) Direct demonstration of insulin-induced GLUT4 translocation to the surface of intact cells by insertion of a c-myc epitope into an exofacial GLUT4 domain. J Biol Chem 268:14523–14526
38. Boguslavsky S, Chiu T, Foley KP, Osorio-Fuentealba C, Antonescu CN, Bayer KU, Bilan PJ, Klip A (2012) Myo1c binding to submembrane actin mediates insulin-induced tethering of GLUT4 vesicles. Mol Biol Cell 23(20):4065–4078. https://doi.org/10.1091/mbc.E12-04-0263
39. Sun Y, Jaldin-Fincati J, Liu Z, Bilan PJ, Klip A (2016) A complex of Rab13 with MICAL-L2 and alpha-actinin-4 is essential for insulin-dependent GLUT4 exocytosis. Mol Biol Cell 27(1):75–89. https://doi.org/10.1091/mbc.E15-05-0319

Chapter 15

Glucose Transport: Methods for Interrogating GLUT4 Trafficking in Adipocytes

Dougall M. Norris, Tom A. Geddes, David E. James, Daniel J. Fazakerley, and James G. Burchfield

Abstract

In this chapter we detail methods for the systematic dissection of GLUT4 trafficking. The methods described have been optimized for cultured 3T3-L1 adipocytes, but can be readily adapted to other cell types.

Key words GLUT4, Glucose transport, GSV, Immunoprecipitation, Kinetics, Trafficking, Imaging, Fluorescence, Calibration, Fractionation

1 Introduction

Postprandial transport of glucose into peripheral tissues such as muscle and adipose is an essential part of glucose homeostasis. This process is largely controlled by the actions of the peptide hormone insulin, which augments glucose transport into target tissues by stimulating the redistribution of the glucose transporter GLUT4 from intracellular storage vesicles to the cell surface. The trafficking of GLUT4 to and from the plasma membrane (PM) is a form of regulated exocytosis [1, 2], and requires the movement of GLUT4 through numerous sorting and trafficking steps along with the integration of complex metabolic and molecular signals. Defective GLUT4 trafficking in adipose and muscle tissue is a primary defect in insulin resistance, a major risk factor for the development of metabolic diseases, particularly type 2 diabetes. Therefore, there is great interest in understanding the mechanisms governing GLUT4 traffic and how this process is altered in insulin resistance. In this chapter we provide protocols for the systematic dissection of GLUT4 trafficking, and cover methods for subcellular fractionation, GLUT4 recycling, as well as fixed-cell and live-cell imaging. Here we give a brief introduction to each of the methods presented in this book chapter, followed by step-by-step protocols.

Karin Lindkvist-Petersson and Jesper S. Hansen (eds.), *Glucose Transport: Methods and Protocols*, Methods in Molecular Biology, vol. 1713, https://doi.org/10.1007/978-1-4939-7507-5_15,

1.1 Studying Endogenous GLUT4 by Subcellular Fractionation

Immuno-electron microscopy techniques have revealed that the majority of GLUT4 resides in tubulo-vesicular elements called GLUT4 storage vesicles (GSVs) in unstimulated cells. These 50–70 nm GSVs are found throughout the cytoplasm [3]. Insulin stimulates the release and translocation of GSVs to the cell surface where they fuse with the PM. Interestingly, GSVs do not appear to be regenerated until the system returns to the basal state [4], implying that increased cell surface GLUT4 in insulin-stimulated cells is not maintained by continual delivery of GSVs. Rather, after the initial GSV exocytosis burst, GLUT4 continuously cycles between the PM and recycling endosomes [5, 6].

The trafficking of endogenous GLUT4 has classically been studied using differential centrifugation techniques coupled to Western blotting to monitor GLUT4 abundance in specific subcellular fractions. Here, cell homogenates are separated by density to yield fractions enriched for specific subcellular compartments of differing densities. The low-density microsomal (LDM) fraction contains small vesicles, including GSVs. Accordingly, GLUT4 is highly enriched in LDM fractions from unstimulated cells. Insulin-stimulated translocation of GSVs to the cell surface can be monitored by sub-fractionation as a loss of GLUT4 within the LDM fraction and concomitant increase in GLUT4 abundance in the PM fraction (*see* Fig. 1).

1.2 Studying Endogenous GLUT4 by Immuno-Isolation of GLUT4 Storage Vesicles

As the GSV compartment plays a significant role in how GLUT4 responds to insulin, it is useful to study this compartment in greater detail. Specifically, subcellular fractionation techniques are used to generate a GSV-enriched fraction and an immunoadsorption step is added to precipitate GLUT4-containing membranes (namely GSVs) from this fraction using a GLUT4-specific antibody (e.g., monoclonal 1F8 clone [7]). An appropriate immunoglobulin G isotype should be used as a negative control. Antibodies are immobilized on suitable inert beads or equivalent. Magnetic beads or other nonporous matrix is recommended to minimize nonspecific binding. Magnetic beads also eliminate the need for centrifugation.

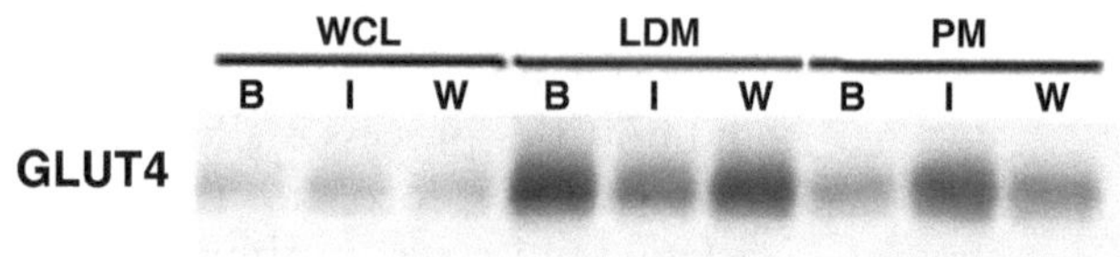

Fig. 1 Example of subcellular fractionation. 3T3-L1 adipocytes where either left untreated (B), stimulated with 1 nM insulin (I), or stimulated with 1 nM insulin in the presence of the PI3K inhibitor Wortmannin (W). Subcellular fractionation was performed as described and the abundance of GLUT4 within the whole-cell lysate (WCL), low-density microsome (LDM), and plasma membrane fractions were determined by Western blot using an anti-GLUT4 antibody

Immuno-isolation of GSVs can be undertaken from either the purified LDM fraction or the supernatant following the pelleting of the HDM fraction (referred to as the post-HDM supernatant). While both of these starting materials are enriched for GSVs, it is likely that the additional centrifugation step required to pellet LDMs will increase loss of proteins peripherally bound to GSVs as well as potentially promote nonspecific membrane aggregation. As such we recommend using the post-HDM supernatant as the starting material for immuno-isolation of GSVs.

1.3 Studying Kinetic Parameters of GLUT4 Trafficking

Endpoint measures of surface GLUT4 abundance are informative, but measuring trafficking kinetics can be more revealing when trying to understand how GLUT4 trafficking is altered in response to stimuli and under pathogenic conditions. GLUT4 has a complex trafficking itinerary including biogenesis of the specialized GSVs, translocation to the cell surface, GSV tethering and docking via the interactions between vesicles and the cell surface, fusion of lipid bilayers, internalization of GLUT4 from the cell surface within endocytic vesicles, and sequestration of GLUT4 into the GSV compartment. Despite this complexity, the steady-state level of cell surface GLUT4 under any condition is effectively determined by the size of the pool of GLUT4 free to recycle to and from the plasma membrane, the rate constant for GLUT4 exocytosis, and the rate constant for GLUT4 endocytosis. In unstimulated cells, these parameters favor the retention of GLUT4 at intracellular sites. These parameters are altered to maintain higher levels of PM GLUT4 under certain conditions, such as in the presence of insulin. For example, insulin stimulation augments PM GLUT4 via dose-dependently enhancing the magnitude of the pool of GLUT4 recycling with the PM [6], increasing the rate constant for GLUT4 exocytosis (k_{ex}), and, in some cases, lowering the rate constant for int ernalization/endocytosis (k_{in}, k_{endo}) under steady-state conditions in 3T3-L1 adipocytes [6, 8]. Similar studies have been performed in cultured myotubes to investigate GLUT4 trafficking in response to distinct stimuli [9, 10]. The Mastick and Coster laboratories have taken these analyses further to interrogate additional steps in GLUT4 trafficking in adipocytes [11].

These kinetics are referred to as steady-state kinetics, as they refer to the size of the GLUT4 pool and rate constants for GLUT4 trafficking that maintain a constant GLUT4 PM abundance. The time taken to transit between two steady states, for example during initial phases of insulin stimulation, is also of interest (half-time, $t_{1/2}$), but determination of rate constants for exo- and endocytosis cannot be readily inferred under these conditions.

Determination of pool size and kinetic rate constants has been aided by the development of GLUT4 reporter constructs containing epitope tags within the first exofacial loop (e.g., hemagglutinin (HA) or Myc; HA/Myc-GLUT4). The exofacial tag enables antibody labeling of GLUT4 molecules only when GLUT4 is inserted into

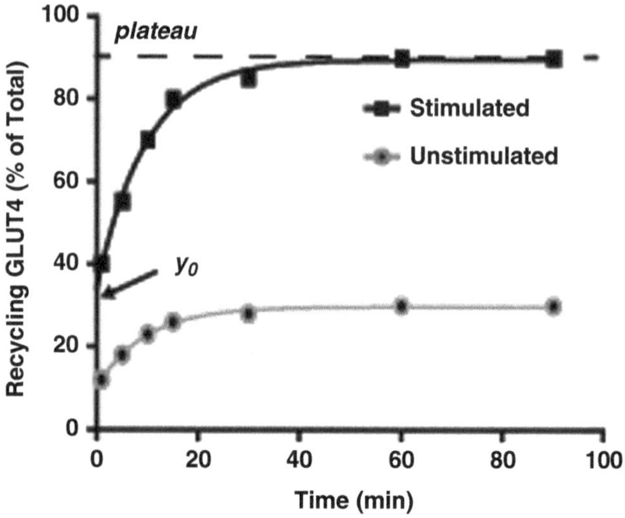

Fig. 2 Example data from anti-HA antibody uptake assays in unstimulated and stimulated 3T3-L1 adipocytes expressing HA-GLUT4. Three kinetic parameters can be derived from these data: the steady-state plasma membrane level of HA-GLUT4 is determined by the Y-intercept; the total amount of GLUT4 in the recycling pathway and free to recycling with the plasma membrane is defined by the plateau; the rate constant for exocytosis (k_{ex}) is derived fitting the curve as described in the text

the plasma membrane and the epitope is exposed to the extracellular milieu. This is useful for quantitation of GLUT4 at the plasma membrane of living or fixed intact cells and for use in antibody uptake assays from which steady-state kinetic parameters can be calculated [6, 12]. Here k_{ex} and k_{endo} values are determined as "bulk" rate constants. These do not refer to a specific process, but are rather the rate-limiting steps in this aspect of GLUT4 trafficking. It is possible that the limiting step for these constants may be different between basal and insulin-stimulated cells.

The principle of an antibody uptake assay to determine the kinetic parameters for GLUT4 trafficking is to bring cells to a steady state (e.g., serum-starved, insulin stimulation) and then incubate cells in the continuous presence of a saturating concentration of anti-HA antibody. In this assay, the high concentration of anti-HA antibody in the extracellular media allows the assumption that all HA-GLUT4 that appears at the cell surface is labeled with antibody (*see* Fig. 2). Antibody uptake is therefore a direct measurement of the appearance of the epitope-tagged GLUT4 at the membrane, and so GLUT4 exocytosis. The amount of antibody-bound GLUT4 plateaus when all GLUT4 available to recycle to the plasma membrane has recycled via the plasma membrane. These data are fitted to the equation [13]:

$$y = y_0 + (\text{plateau} - y_0) \times \left(1 - e^{(-k_{ex}.x)}\right).$$

where y_0 is the steady-state PM levels of GLUT4, the *plateau* is the total amount of GLUT4 recycling with the PM under these conditions, and k_{ex} is the rate constant for exocytosis. These parameters are directly determined from this fit. The fact that the system is at steady state means the delivery of GLUT4 to the surface must equal the removal of GLUT4 from the PM. The rate constant for GLUT4 internalization (k_{endo}) can therefore be derived from

$$k_{endo} = \frac{k_{ex}(\text{plateau} - y_0)}{y_0}$$

Kinetic studies were initially carried out with cell-impermeable photo-reactive glucose analogues to covalently label endogenous GLUTs prior to the development of HA/Myc-GLUT4 [14–16]. This approach requires additional Western blotting steps to specifically probe GLUT4.

1.4 Imaging GLUT4 in Live Cells

Fluorescence-based imaging of GLUT4 in both live and fixed cells has provided detailed insight into the complexities of GLUT4 trafficking. Fixed-cell imaging is typically used for basic localization and co-localization studies, and to correlate GLUT4 trafficking with the phosphorylation states of key signaling nodes. Imaging of GLUT4 in live cells provides the ability to interrogate GLUT4 kinetics, dissect individual trafficking steps, and analyze single-cell responses. The development of these methodologies has revealed that substantial single-cell heterogeneity underpins the population response of GLUT4 to insulin stimulation [2]. This means that all studies that have utilized single-cell analysis of GLUT4 trafficking to infer mechanistic details should be treated with caution.

We recently engineered a dual-color GLUT4 probe that contains a pH-sensitive fluorophore (ecliptic-pHluorin) in the first exofacial loop (luminal and/or extracellular) of GLUT4 and a second fluorophore (tdTomato) at the C-terminus [2]. This probe can be imaged using a variety of imaging modalities, each delivering different benefits. Preliminary studies and high-throughput studies can be performed using wide-field epifluorescence microscopy (epiM). Multi-position low-temporal-frequency TIRF (LF-TIRFm) can be used to dissect trafficking to the PM from fusion and high-temporal-frequency single-cell TIRF (HF-TIRFm) or spinning disk confocal can be utilized to interrogate the behavior of individual vesicles.

When imaged by wide-field epifluorescence the pHluorin provides a measure of GLUT4 in the PM and the tdTomato provides a measure of total expression levels and accounts for changes in cell size and imaging artifacts. Thus, the pHluorin:tdTomato ratio provides a reliable measure of PM-localized GLUT4 normalized to total expression. The system can be calibrated using a series of pH-based buffer exchanges to accurately assess the percentage of GLUT4 on the PM with respect to the total. The methodology for this is detailed below.

TIRF or evanescent wave microscopy is a technique that can be used to excite only those molecules that are on or in close proximity to the plasma membrane. In comparison to EpiM, TIRF delivers improved resolution of PM and sub-PM structures and as such is ideal for imaging exocytosis. We routinely perform two styles of TIRF experiment. Multi-position experiments with a relatively low temporal frequency (~1 frame/min, LF-TIRF) are ideal for distinguishing between GSVs moving to the membrane from those undergoing fusion with the PM. In this paradigm, the tdTomato signal is sensitive to its cellular localization and is a measure of GLUT4 recruitment to the PM while the pHluorin signal is a marker of fusion. HF-TIRFm experiments are performed on single cells imaged at high speed (~10 frames/s). These experiments are used to look at and characterize individual vesicle trafficking events such as fusion with the PM [2]. This is time consuming with a low throughput but enables the visualization and quantification of individual trafficking and fusion events. For a detailed experimental protocol of TIRF microscopy experiments *see* [17].

The key to these experiments is healthy cells expressing the exogenous GLUT4 reporter. The choice of transduction method will depend on the type of cells you are imaging. For mitotic cells (such as myoblasts, fibroblasts), lipofection is an effective transient, nonviral method, and for the production of cell lines with prolonged expression, lentivirus, adenovirus, and retrovirus can be utilized. Nondividing cells (such as 3T3-L1 adipocytes or primary cells) can be electroporated for transient expression or infected with retrovirus or for prolonged expression. For further information regarding viral transduction *see* [18, 19].

Given the potential effects of long-term expression of exogenous genes on pre-adipocyte growth and differentiation, our lab typically uses electroporation to deliver purified DNA into differentiated 3T3-L1 adipocytes. Electroporation (also known as electropermeabilization) is a method for efficiently introducing nucleic acids and other molecules into a variety of cell types including mammalian cells. The molecular principles governing this process are still not completely understood and are covered extensively elsewhere [20]. In brief, the exposure of cells to a short, high electric field results in the stochastic formation of pores in the plasma membrane and an increase in cell permeability. If the exposure is not too high or too long, the process is reversible. In the case of nucleic acids, the electric field also facilitates entry via electrophoresis [21].

1.5 Perspectives and Future Development

In reality no one method for assessing GLUT4 trafficking is perfect with each displaying its own set of limitations (*see* Table 1). Subcellular fractionation is limited by cross contamination of organelles among the various fractions. Vesicle immunoadsorption is limited by the fact that GLUT4 is targeted to multiple compartments including the TGN, many different kinds of endosomes, as well as GSVs and it is difficult to differentiate them using this approach.

Table 1
Methodologies for interrogating GLUT4 trafficking

Method	Major uses	Strengths	Weaknesses
Subcellular fractionation and GSV Immunoprecipitation	• Enrichment of GSVs • Co-localization	• Endogenous GLUT4	• Limited kinetics/ time course • Population only • Requires very high numbers of cells
GLUT4 recycling assay	• Revealing how PM GLUT4 levels are maintained	• Measurement of kinetics parameters of GLUT4 trafficking	• Overexpressed GLUT4
Fixed-cell imaging	• Localization • Co-localization	• Endogenous or overexpressed GLUT4 • Flexible	• Limited kinetics/ time course • Reliant on good antibodies to endogenous GLUT4 • Fixation-associated artifacts
Live-cell imaging	• Single-cell kinetics • Heterogeneity • Temporal kinetics • Screening	• Detailed kinetic data on single cells • Power of paired statistics • Flexible experimental design • High temporal resolution	• Lower throughput • Time consuming • Overexpressed GLUT4

Antibody labeling studies are useful for analyzing steady-state turnover but ultimately it would be ideal to know what happens during transition phases and these studies are very limited here. Microscopy is one of the most advanced approaches but again it has a number of limitations. First, the behavior of tagged molecules compared to that of the endogenous protein is often poorly understood. Moreover, the level of overexpression of these reporters and how this affects trafficking is often underappreciated. Ideally one would create cell lines where these reporters are knocked into the endogenous locus. Secondly, these studies often (but not always) involve selection of cells in the population and we have observed that this could be problematic because of the variability in response between different cells. Thirdly is the laborious nature of analysis. This calls for the development of automated data analysis methodologies. Finally, such experiments would ideally be performed in tissue of animals. With recent developments in multiphoton technology, as well as production of recombinant animals, these sorts of approaches should be feasible.

2 Materials

2.1 Preparation for Subcellular Fractionation of Endogenous GLUT4

2.1.1 Equipment

1. Ultra-speed centrifuge, rotor, and tubes capable of 235,200 × *g*.
2. "Rubber policeman" scrapers, 1 per condition (or wash and reuse).
3. Pestle homogenizer or equivalent.

2.1.2 Buffers, Cell Medium, and Reagents

1. Cell culture medium: Dulbecco's modified Eagle medium (DMEM), 0.2 % w/v BSA.
2. Phosphate-buffered saline (PBS): 137 mM NaCl, 2.7 mM KCl, 10 mM Na_2HPO_4, 1.8 mM KH_2PO_4, pH 7.4.
3. HEPES-EDTA-sucrose buffer (HES): 10 mM HEPES pH 7.4, 1 mM EDTA, 250 mM sucrose.
4. Sucrose solution: 10 mM HEPES, 0.05 mM EDTA, 1.12 M sucrose.
5. Protease inhibitors (recommended).
6. Phosphatase inhibitors (recommended).

2.2 Preparation for Immuno-Isolation of GLUT4 Storage Vesicles

2.2.1 Equipment

1. Magnetic separation beads.
2. Magnet or magnetic separation rack.
3. Tube rotator stored at 4 °C.
4. Cut pipette tips or wide-bore pipette tips for viscous fluids.
5. 30-gauge needle, vacuum trap, and syringe or similar.

2.2.2 Buffers and Reagents

1. PBS buffer: 137 mM NaCl, 2.7 mM KCl, 10 mM Na_2HPO_4, 1.8 mM KH_2PO_4, pH 7.4.
2. HES buffer: 10 mM HEPES pH 7.4, 1 mM EDTA, 250 mM sucrose.
3. HES+ buffer: HES, 150 mM NaCl, 0.1 % w/v BSA. Prepare stock solution of 1.5 M NaCl and 1 % w/v BSA in ddH_2O. Dilute 10× into HES.
4. Laemmli sample buffer (2×): 0.001 % w/v Bromophenol blue, 20 % v/v glycerol, 4 % w/v SDS, Tris–HCl, 126 mM, pH 6.8, containing reducing agent (e.g., DTT, TCEP).

5. BSA solution: 10 % w/v BSA.
6. Primary antibody: Anti-GLUT4 antibody.
7. Secondary antibody: Mouse immunoglobulin G (mouse IgG).

2.3 Preparation for the Study of GLUT4 Trafficking

2.3.1 Equipment

1. Water bath: Set the water bath temperature at 37 °C.
2. Multi-well plates: 96-well clear-bottom black-walled plates for culturing cells.
3. Multichannel pipette.
4. Fluorescent plate reader with appropriate excitation and emission filters.

2.3.2 Buffers and Reagents

1. Medium: DMEM (minus bicarbonate), 20 mM HEPES, 0.2 % w/v BSA.
2. PBS buffer: 137 mM NaCl, 2.7 mM KCl, 10 mM Na_2HPO_4, 1.8 mM KH_2PO_4, pH 7.4.
3. Fixative: Paraformaldehyde (PFA).
4. Serum: Normal swine-serum.
5. Primary antibody: Anti-HA antibody.
6. Secondary antibody: Fluorescent-conjugated anti-mouse IgG antibody. This will depend on your plate reader filter sets. We recommend the use of longer wavelength fluorophores to reduce both plate and cellular autofluorescence (e.g., CF™ 555; CF™ 647).

2.4 Preparation for Imaging GLUT4 in Live Cells

2.4.1 Biological Material

1. Healthy differentiated 3T3-L1 adipocytes ½ –1× 10 cm dishes per electroporation (*see* **Note 1**).
2. High-quality (260:280 > 1.8), high-concentration (>4 mg/mL) stocks of plasmid DNA (e.g., GLUT4-pHluorin; GLUT4-pHluorin:tdTomato) (*see* **Note 2**).

2.4.2 Equipment

1. Centrifuge.
2. Electroporator (ECM 830 Square Wave Electroporation System, BTX Molecular Delivery Systems).
3. Electroporation cuvettes, 0.4 cm (BioRad).
4. Coverslips: 42 × 0.17 mm.
5. Imaging chamber, e.g., POC (PeCon), or glass-bottomed petri dishes (35 mm).
6. Inverted microscope with cell incubator set at 37 °C. Turn on at least 1 h prior to experiment to allow temperature to equilibrate throughout the system. TIRF capability (only required for TIRF). Suitable filter sets.
7. Fluidic system (to achieve buffer exchange).

2.4.3 Buffers and Reagents

1. PBS buffer: 137 mM NaCl, 2.7 mM KCl, 10 mM Na_2HPO_4, 1.8 mM KH_2PO_4, pH 7.4.
2. Electroporation solution: 20 mM HEPES, 135 mM KCl, 2 mM $MgCl_2$, 0.5 % v/v Ficol 400, 1 % v/v DMSO, pH 7.6. The solution is based on cytomix described by [22], and can be made up in bulk and stored at 4 °C for >6 months.
3. ATP-glutathione stock solution: 200 mM ATP, 500 mM glutathione, pH 7.4. Stored in single-use aliquots at −20 °C or below.
4. Matrigel™ (BD Biosciences) or suitable extracellular matrix protein.
5. Trypsin-EDTA solution (10×): 0.25 % Trypsin, 0.05 % EDTA.
6. Culture medium: DMEM.
7. Serum: Fetal bovine serum (FBS).
8. Albumin: BSA > 98 %. It is important to use a highly purified grade of BSA. We routinely batch test to ensure undesirable basal activation of insulin signaling due to trace amounts of growth factors.
9. Insulin stock solution: 70 μM Insulin in HCl.
10. Modified Krebs Ringer phosphate buffer (KRP) buffer: 120 mM NaCl, 0.6 mM Na_2HPO_4, 0.4 mM NaH_2PO_4, 6 mM NaCl, 1.2 mM $MgSO_4$, 12.5 mM HEPES, 1 mM $CaCl_2$, 10 mM glucose, 0.2 % w/v BSA, pH 7.4 or another suitable imaging buffer.
11. Calibration buffer 1: 120 mM NaCl, 0.6 mM Na_2HPO_4, 0.4 mM NaH_2PO_4, 6 mM KCl, 1.2 mM $MgSO_4$, 12.5 mM MES, 1 mM $CaCl_2$, 10 mM glucose, 0.2 % w/v BSA, pH 5.5.
12. Calibration buffer 2: 70 mM NaCl, 50 mM NH_4Cl, 0.6 mM Na_2HPO_4, 0.4 mM NaH_2PO_4, 6 mM KCl, 1.2 mM $MgSO_4$, 12.5 mM bicine, 0.2 % w/v BSA, pH 9.0.
13. Calibration buffer 3: 70 mM NaCl, 50 mM NH_4Cl, 0.6 mM Na_2HPO_4, 0.4 mM NaH2PO4, 6 mM KCl, 1.2 mM $MgSO_4$, 12.5 mM MES, 0.2 % w/v BSA, pH 5.0.

3 Methods

3.1 Subcellular Fractionation of GLUT4

This protocol for subcellular fractionation (sub-fractionation) in adipocytes is adapted from the protocol described by McKeel and Jarett [23] and adapted by Simpson et al. [24]. The fractions produced by this method include whole-cell homogenate; nuclear-mitochondrial (NM) fraction; plasma membrane (PM); high-density microsomal (HDM) membranes; low-density microsomal (LDM) membranes; and cytosol.

The number of cells required will depend on the intended downstream application. As a general starting point we recommend at least 3 × 15 cm dishes (approx. 25,000,000 cells) per condition to ensure ample PM and HDM fractions per preparation. All steps from 8 onwards, unless otherwise noted, are carried out on ice at 4 °C.

1. Seed 3T3-L1 fibroblasts into 15 cm dishes.
2. Differentiate 3T3-L1 fibroblasts into adipocytes using established protocols (*see* **Note 3**).
3. Experiments should be performed between days 8 and 12 post-differentiation (*see* **Note 4**).
4. Pretreat cells as required depending on experimental design, e.g., chronic treatments to alter insulin sensitivity.
5. Wash cells in step-down cell culture medium to remove serum.
6. Add step-down DMEM to cells and leave in incubator for 2 h.
7. Add insulin at concentrations between 0 and 100 nM 20 min. Half-maximal effects on glucose transport are usually observed at concentrations of 0.1–1 nM with maximum effects being observed at 100 nM.
8. Place on dishes on ice and wash three times in ice-cold PBS. Aspirate carefully after the final wash.
9. Add 1.5 mL of ice-cold HES buffer onto cells.
10. Scrape cells and buffer using a rubber policeman and transfer homogenate to a 15 mL tube.
11. Homogenize by applying 15–20 strokes with a pestle homogenizer and collect in a 15 mL tube. Wash homogenizer with a further 1.5 mL of ice-cold HES and collect in the same tube.
12. Take a 200 μL aliquot of homogenate at this point for determination of whole-cell GLUT4 expression levels. Snap-freeze in liquid nitrogen and store in −20 °C.
13. Centrifuge homogenate at 500 × *g* for 10 min to remove unhomogenized cells. Transfer the supernatant to a fresh tube, discarding the pellet.
14. Centrifuge supernatant at 13,550 × *g* for 12 min, yielding a pellet containing the nucleus, mitochondria, and plasma membrane and a supernatant containing microsomal membranes and cytosol. Retain the supernatant for further fractionation (**step 20 below**).
15. Resuspend pellet in 1 mL HES and centrifuge at 13,550 × *g* for 12 min. Discard the supernatant.
16. Resuspend pellet in 1 mL HES.
17. Carefully layer samples into 11 mL of sucrose solution (*see* **Note 5**).

18. Centrifuge at 111,160 × *g* for 60 min in a swing-out rotor. This produces a nuclear-mitochondrial (NM) pellet as well as a cloudy layer at the interface of the sucrose cushion and loaded sample. The interface contains plasma membrane (PM).
19. Carefully remove the PM interface by slow pipetting, taking a maximum of 1 mL and taking care to avoid taking up the cushion. Thoroughly mix in >6 mL HES and centrifuge at 235,200 × *g* for 15 min. Resuspend the pellet in 80 μL HES (**PM fraction**) (*see* **Note 6**).
20. Remove the sucrose solution and discard. Resuspend NM pellet in HES (**NM fraction**), snap freeze in liquid nitrogen, and store in −20 °C.
21. Centrifuge the retained initial supernatant (**step 13 above**) at 21,170 × *g* for 17 min, yielding an HDM pellet and supernatant containing LDM and cytosol.
22. Resuspend HDM pellet in HES and centrifuge as in the previous step. Resuspend this HDM pellet in 100 μL HES (**HDM fraction**).
23. Centrifuge the LDM/cytosol supernatant at 235,200 × *g* for 75 min, producing an LDM pellet and cytosol. Retain 1 mL cytosol if desired (**CYTOSOL fraction**).
24. Resuspend LDM pellet in HES and centrifuge as in the previous step. Resuspend in 100 μL HES (**LDM fraction**).
25. Apply Western blotting to assess GLUT4 abundance in each fraction (*see* **Note 7**).

3.2 Immuno-Isolation of GLUT4 Storage Vesicles

We recommend at least 2 × 15 cm dishes per condition. All steps unless otherwise noted are carried out on ice at 4 °C. If using nonmagnetic beads, undertake all separation steps by centrifuging at 2,000 × *g* for 2 min at 4 °C.

1. Transfer 40 μL of magnetic bead slurry per immunoprecipitation (IP) per condition +1 to a 1.5 mL tube (beads for mouse IgG controls may be included at this step).
2. Wash beads twice by adding cold PBS up to a total volume of 1.5 mL, applying magnetic field to separate beads, and aspirating PBS.
3. Resuspend beads in 100 μL PBS per IP condition +1. Transfer 100 μL beads per IP to separate tubes for each antibody.
4. Spike in BSA to a final concentration of 2 % to minimize nonspecific binding.
5. Add 2 μg antibody per IP and rotate for 1 h at 4 °C (*see* **Note 8**).
6. Apply magnetic field to separate beads.
7. Split beads into separate tubes for each IP condition (100 μL per condition/tube). Store in cold PBS to prevent drying out.

Apply magnetic field to separate beads and aspirate PBS immediately before use.

8. Add SM in equal amounts to each IP (*see* **Note 9**). Add HES+ buffer to a final volume of 1 mL. Rotate at 4 °C for 1–2 h, or overnight.
9. Apply magnetic field to separate beads. Retain 400 μL supernatant per IP. Aspirate remaining supernatant.
10. Wash three times by resuspending in HES+ buffer and applying magnetic field to separate beads.
11. Dry beads by aspirating using a 30-gauge needle.
12. Add 2 × Laemmli sample buffer containing reducing agent (e.g., TCEP or DTT) and place samples on a heating block at 37 °C for 45 min (*see* **Note 7**).
13. Apply magnetic field to remove beads and retain supernatant.
14. Proceed as normal for Western blot preparation. Load approximately 10 μL eluent.

3.3 Method for Studying Kinetic Parameters of GLUT4 Trafficking

This protocol describes the methodology for antibody uptake assay into cultured cells using fluorescence antibodies and a fluorescence plate reader and the analysis of GLUT4 trafficking parameters (total pool size, k_{ex}, k_{endo}) in 3T3-L1 adipocytes stably expressing HA-GLUT4 via retrovirus as described in Govers et al., 2004 [6], and Coster et al., 2004 [13]. Alternative methods for quantifying antibody uptake such as flow cytometry and fluorescence microscopy have been described [5, 8]. The 96-well format permits measurement of antibody uptake at five time points in unstimulated (e.g., columns 1–5) and stimulated cells (e.g., columns 6–10), and determination of total HA-GLUT4 expression (e.g., columns 11 and 12). Each measurement is performed in quadruplicate. To control for nonspecific antibody uptake, the same experiment is performed in adipocytes not expressing HA-GLUT4 (i.e., expressing a control vector) or antibody uptake is assessed with an isotype control antibody in HA-GLUT4-expressing adipocytes. This control should be performed at each time point, and for anti-HA antibody binding to fixed cells.

The experiment should be designed to permit measurement of the total amount of HA-GLUT4 expression.

1. Express epitope-tagged GLUT4 in 3T3-L1 cells (e.g., HA-GLUT4). We use 3T3-L1 cells retrovirally expressing HA-GLUT4, albeit HA-GLUT4 can be delivered via alternate means (*see* **Note 10**).
2. Seed 3T3-L1 fibroblasts into black-walled clear-bottom 96-well plates.
3. Differentiate 3T3-L1 fibroblasts into adipocytes using standard protocols (*see* **Note 1**). Experiments should be performed between days 8 and 12 post-differentiation (*see* **Note 11**).

4. Pretreat cells as required depending on experimental design (e.g., chronic treatments to alter insulin sensitivity).
5. Wash cells in step-down DMEM (DMEM (minus bicarbonate), 20 mM HEPES, 0.2 % BSA) to remove serum (*see* **Note 12**).
6. Add 100 μL step-down DMEM to cells and leave in 37 °C water bath for 2 h.
7. Add insulin to final concentration of 100 nM 20 min prior to addition of anti-HA antibody, to achieve the new steady state.
8. Add anti-HA antibody to cells. **Steps** 7 and **8** should be repeated for each time point since one 96-well plate can generate a complete time course of antibody uptake. All treatments should be staggered so that they end at the same time (*see* **Note 13**).
9. Move plate to ice and wash cells three times with 200 μL of ice-cold PBS.
10. Fix cells with 100 μL 3 % v/v PFA for 20 min at room temperature. All subsequent steps are performed at room temperature.
11. Cells were blocked and permeabilized with 50 μL 5 % normal swine serum in PBS with 0.1 % w/v saponin.
12. Incubate wells for determination of total HA-GLUT4 expression with anti-HA antibody (1:1000) in 2 % normal swine serum in PBS (30 μL) for 1 h (*see* **Note 14**).
13. Wash wells for determination of total HA-GLUT4 expression three times with 200 μL PBS to remove unbound anti-HA antibody.
14. Incubate all wells with fluorescently labeled secondary antibody (1:100) in 2 % serum, diluted into PBS, for 1 h.
15. Wash all wells three times with 200 μL PBS to remove unbound antibody.
16. Add 100 μL PBS to each well and read in plate reader with appropriate excitation and emission filters (*see* **Note 15**).

3.4 Live-Cell Imaging GLUT4

The protocol described below uses a single square wave pulse of 200 mV, but many variables affect the efficiency of electroporation and conditions should be optimized for each cell line and instrument. ***Before starting: Ensure that all buffers are made up and at room temperature.***

Depending on the application, one 6-well plate will provide enough cells for 1–2 electroporations. For optimal results, 3T3-L1 cells should be 95–100 % differentiated (*see* **Note 11**). 3T3-L1 cells should be electroporated 6–8 days post-differentiation in order to prevent loss of cells, as they become more buoyant at later times due to lipid accumulation.

3.4.1 Preparation of Matrigel-Coated Coverslips or Imaging Chambers

1. Thaw Matrigel on ice at 4 °C. This will take about 2 h. Matrigel is best stored in small single-use aliquots at −80 °C.
2. The next day, dilute Matrigel 1:50 into ice-cold PBS using pre-chilled pipettes (chilled by pipetting ice-cold PBS several times).
3. Aliquot a small amount (~200 μL) into the center of the coverslip or dish (2 mL per 42 mm coverslip).
4. Leave at room temperature for 2 h.
5. Wash twice with sterile, room-temperature PBS to remove unbound material. Ideally, coverslips/dishes should be prepared fresh and used within a few days of preparation.
6. Matrigel-coated coverslips/dishes should be washed twice in culture media prior to use.

3.4.2 Electroporation

1. Wash cells 2–3 times with room-temperature PBS.
2. Trypsinize cells from each dish using trypsin–EDTA solution (diluted in PBS from 10× stock) for 5–10 min in a 37 °C incubator (check trypsinization by gently tapping the cells and assessing the effectiveness of dislodgment using a microscope) (*see* **Note 16**).
3. Resuspend cells in 3 mL DMEM containing 10 % FCS to every 1 mL 5× trypsin–EDTA and transfer to a 50 mL Falcon tube and make up the volume to 30 mL with DMEM culture medium (dishes can be pooled here) (*see* **Note 17**).
4. Centrifuge cells at 150 × *g*, room temperature, for 5 min. Higher speed spins will decrease the number of viable adipocytes in the pellet, reducing cell number and electroporation efficiency.
5. Aspirate medium and resuspend cells in 30 mL of PBS.
6. Centrifuge cells at 150 × *g*, for 5 min.
7. Aspirate medium and resuspend cells in 30 mL of PBS.
8. Centrifuge cells at 150 × *g*, for 5 min. During this spin, thaw single-use 100× ATP–glutathione stock solution and add to electroporation solution. Filter sterilize using a 0.2 μm filter.
9. Add 10–40 μg of plasmid to each electroporation cuvette.
10. Aspirate medium and resuspend cells in the 400 μL electroporation buffer per electroporation.
11. Transfer 0.4 mL volumes of resuspended cells to each cuvette containing the DNA. Flick gently to mix (*see* **Note 18**). Continue with next step immediately.
12. Electroporate each cuvette with a single 20 ms, 200 V square wave pulse (*see* **Note 19**).

13. Remove the white viscous material (representing a mixture of lipid and denatured protein from dead adipocytes) floating on the surface with an aspirator pipette, being very careful not to aspirate the rest.
14. Immediately add 2 mL of DMEM culture media containing 10 % FCS and GlutaMAX and mix gently.
15. Seed ~100–200 μL into the center Matrigel-coated dishes.
16. Refresh with culture media approximately 1 h after seeding, longer if density is low. Bring media up to normal culture volume for the given coverslip/dish.

3.4.3 Preparation for Microscopy

Cells can be used for microscopy from 12–72 h post-electroporation, although generally we leave them for a minimum of 24 h to ensure good recovery, attachment, and stable expression levels.

1. Prepare a working stock of insulin (3×) diluted from 70 μM stock on the day of the experiment.
2. Wash the cells three times in serum-free DMEM.
3. Add a suitable amount of serum-free DMEM and incubate for approximately 2–3 h at 37 °C and 10 % CO_2 before imaging or stimulation.
4. Add 2 mL of KRP/0.2 % w/v BSA/10 mM glucose pH 7.4 buffer or suitable imaging buffer (*see* **Note 1**).
5. Wash the bottom of the coverslip carefully with dilute detergent on a kimwipe tissue paper.
6. Repeat this with RO water and dry carefully.
7. Cells are now ready for imaging.

3.4.4 Performing Calibration Experiments

This involves a number of buffer exchanges and as such is achieved best using perfusion (*see* **Note 20**).

1. Select healthy, suitably transfected cells using both brightfield and fluorescence modalities. Ideally use a low magnification high NA objective (Nikon CFI Plan Apo VC 20×/NA 0.75) (*see* **Note 21**).
2. Set up imaging parameters (*see* Table 2 and **Note 22**).
3. Begin image acquisition.
4. Change media to pH 5.5 MES-buffered KRP (*see* **Note 23**).
5. Change media back to imaging buffer.
6. Perform experiment.
7. Change media to NH_4Cl pH 9.0 bicine buffer. This gives the maximum (***Max***) fluorescence signal of the pHluorin. The NH_4Cl will equilibrate the pH across all cellular compartments.

8. Change media to NH_4Cl pH 5.0 MES buffer KRP. This quenches all fluorescence of the pHluorin and gives minimum (***Min***) fluorescence signal of the pHluorin.
9. Analyze data.
10. The percentage of total GLUT4 protein on the surface ($P_{surface}$) at time n (relative to the total) is given by

$$P_{\text{surface}} = \frac{Xn - \text{Int}}{(\max - \min) \times F_{7.4}} \times 100$$

where $F_{7.4}$ is the fraction of pHluorin molecules fluorescing at pH 7.4 (the pH of the external media). This can be determined using the Henderson–Hasselbalch equation in the form

$$F_{(pH)} = \frac{1}{1 - 10^{\left(pKa_{\text{pHlourin}} - pH\right)}}.$$

Table 2
Experimental parameters for live GLUT4 imaging

Parameter	High resolution—single cells	Low resolution—multiple cells
Imaging modality	TIRF or spinning disk confocal microscopy offer improved signal-to-noise ratios for investigating single-vesicle dynamics at high resolution	Epifluorescence for whole-cell high-throughput imaging, TIRFM allows for sensitive plasma membrane quantification
Magnification	100× NA 1.45	Less magnification, more cells per field of view. Objective-based TIRFM requires an NA > 1.45
Spatial resolution (binning)	Resolution is crucial, binning will reduce ability to resolve fusion events and small structures	Binning highly effective at reducing laser power and file size. This decreases phototoxicity and analysis time, respectively
Temporal resolution (frame rate and exposure time)	For fusion and trafficking events we recommend at least 10 frames/s	Compromise between the number of cells or positions and the minimum requirement for temporal information (typically 1 frame/min)
Excitation power	This should be set to the minimum level required to achieve a clean signal above background	
Camera	High-quantum-efficiency camera required. High-end sCMOS (preferred) or emCCD	High-end sCMOS preferred

4 Notes

1. Here follow critical considerations for all experiments: **Guidelines for the assessment of cellular health/viability:** Familiarize yourself with the typical morphology of your cell line of interest. Regular monitoring of key morphological characteristics such as cell size, shape, attachment, blebbing, and growth rate can help to identify and prevent cellular stress. Given that between 5 and 30 % of cell lines worldwide are contaminated with mycoplasma, test regularly for mycoplasma (see [25, 26]). The biggest source of contamination is human skin. Good aseptic technique is essential. Avoid the use of antibiotics in culture media as this promotes sloppy technique and can mask underlying infections. Cross contamination of cell culture lines is still a common problem in labs around the world [27, 28]. Authenticate cell lines. **Buffers:** Standard TC media is bicarbonate buffered, designed to be pH 7.4 when incubated at 37 °C in 5 % CO_2. This media is not suitable for experiments performed in a non-CO_2 environment. It is worth to note that the original 3T3 line was established in 10 % CO_2. Bicarbonate buffering can be exchanged for alternative suitable buffers, such as HEPES; however, the absence of bicarbonate will alter cell physiology. A standard imaging buffer we have used routinely is a modified Krebs Ringer phosphate buffer KRP (120 mM NaCl, 0.6 mM Na_2HPO_4, 0.4 mM NaH_2PO_4, 6 mM NaCl, 1.2 mM $MgSO_4$, 12.5 mM HEPES, 1 mM $CaCl_2$, 10 mM glucose, 0.2 % w/v BSA, pH 7.4). Consider the use of substrates (glucose; amino acids; vitamins). **Incubation conditions:** Irrespective of the type of experiment, environmental control (temperature, gas levels, and humidity) and choice of media are fundamental variables that require careful consideration.
2. Dilute or poor-quality DNA will reduce efficiency.
3. Please refer to reference [29] for extended details.
4. GLUT4 expression is induced during adipocyte differentiation, as is GSV biogenesis. A high efficiency of differentiation is vital for consistent data.
5. Resuspend pellet extremely thoroughly, pipetting up/down 25–30 times using a P1000 pipette. When layering onto the sucrose solution (cushion), pipette slowly and evenly down the side of the tube to ensure that a sharp interface between the sample and cushion is maintained.
6. Dilute and mix PM sample very thoroughly in HES to ensure that any high sucrose cushion transferred is diluted so that it does not interfere with pelleting of the PM.
7. Do not boil samples prior to Western blot. Boiling can result in aggregation of GLUT4. We recommend heating at 37 °C for 45 min.

8. Optimization of antibody and SM protein amounts for GSV depletion should be undertaken for antibodies and cell lines used.
9. It can be a good idea to pre-clear SM with separate blocked beads (no antibody) to reduce nonspecific interaction.
10. More information can be found in reference [30]. Moreover, expression of HA-GLUT4 should be assessed as very high levels of ectopic expression can lead to incorrect localization of HA-GLUT4 and high basal PM levels. If using a retroviral delivery, different viral titers can be assessed for optimal expression levels. It is important to assess expression in single cells by immunofluorescence, since assessment of expression at the population level by Western blot can mask highly variable expression between individual cells.
11. GLUT4 trafficking kinetics are markedly different between fibroblasts and adipocytes [6]. A high efficiency of differentiation is vital for accurate and consistent data. Here follow tips on the culturing of 3T3-L1s to achieve high levels of differentiation:

    Cell line tips for 3T3-L1 cells: Having observed spontaneous adipogenesis in small numbers of cells in confluent cultures of the murine 3T3 cell line, Green and Kehinde isolated a number of 3T3 subclones that were highly predisposed to accumulating lipid [31, 32]. They characterized two of these lines named 3T3-L1 and 3T3-L2. 3T3-L1s have since become the go-to cell line of choice for the study of adipogenesis and adipocyte function in vitro.

Improving differentiation of 3T3-L1 cells: All of the protocols described in this chapter require highly efficient and reproducible differentiation of cells (>95 %; see Shewan et al., 2000, for the differentiation protocol [29]). To achieve this 3T3-L1 cells must be carefully looked after. Cells are cultured in high-glucose DMEM supplemented with glutaMAX (glutamate degrades to ammonia and can alter cell physiology [33]) and 10 % FCS at 37 °C in 10 % CO_2. Fetal calf serum batches should be tested to ensure optimal differentiation efficiency. We routinely screen 4–5 different FCS sources/batches. Cell passage—like all cultured cells, mutations that alter phenotype will accrue over time. We recommend that passaging is kept to a reasonable number. We limit our cells to 15–20 passages post-thawing. After this we observe reduced insulin sensitivity, insulin-stimulated glucose uptake, and differentiation potential. Splitting cells too harshly or allowing them to grow too densely, while passaging, will reduce differentiation efficiency. Split cells no more than once every 2 days, ideally at ratios of 1:2–1:6, allowing a maximum confluence of that displayed (*see* Fig. 3a). It is best if cells are seeded into 6- or 12-well plates, rather than 10 cm dishes. Feed confluent cells a day prior to

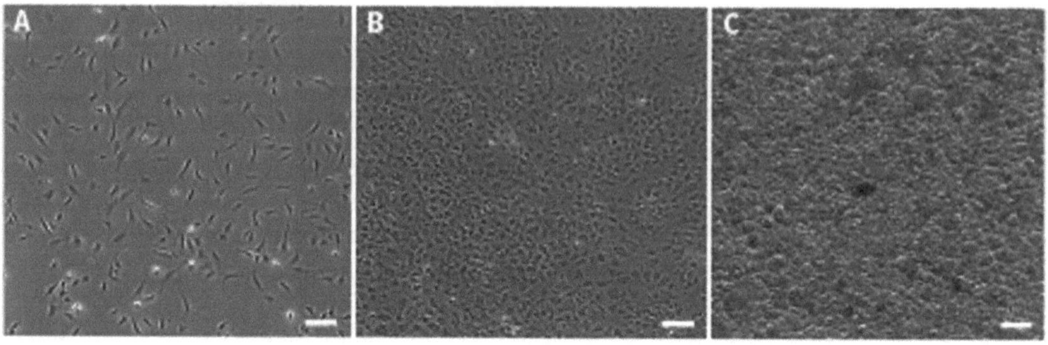

Fig. 3 Examples of 3T3-L1 cells in culture. (**a**) Depicts the maximum density to which 3T3-L1 cells should be grown while passaging. (**b**) Demonstrates the "angry" appearance cells develop around 5 days post-seeding when they are ready to differentiate. (**c**) 3T3-L1 cells 6 days post-differentiation, where 90–95 % of cells display an accumulation of lipid. Scale bars represent 200 μm

differentiation. Begin differentiation 5–7 days post-seeding once cells develop an "angry" appearance (*see* Fig. 3b).

12. Since the assay requires multiple additions, this assay should not be performed in a CO_2 incubator. Perform assay in HEPES-buffered DMEM in a water bath.
13. Test the concentration of anti-HA antibody required to capture all HA-GLUT4 that traffics via the PM, as a rough guide 25–50 μg/mL. For data analysis it is assumed that all unlabeled HA-GLUT4 is immediately labeled with antibody upon exposure of the HA epitope at the plasma membrane, and that all labeling of GLUT4 occurs at the plasma membrane and that any labeling that may occur inside the cell following fluid-phase antibody uptake is negligible.
14. Concentration of anti-HA antibody required to fully capture total HA-GLUT4 levels should be tested. If pretreating cells then determination of total HA-GLUT4 expression should be performed in both control and treated cells to control for changes in HA-GLUT4 expression.
15. Data from this assay should be analyzed as described in [9, 13]. The size of the recycling pool and k_{ex} can be directly calculated by fitting the data to a single exponential. Methods to directly measure GLUT4 internalization have been described [9].
16. Excessive trypsinization will result in a loss of cell viability and a poor electroporation efficiency.
17. FCS is included here to inhibit remaining trypsin. The alternative is to use 2 % BSA in either PBS or DMEM.
18. It is essential that the DNA and cells are well mixed.
19. An increased duration can improve efficiency of larger constructs if needed; however this will be at the cost of cell viability.

20. For more details see Burchfield et al., 2013 [2].

21. Cells should be chosen first and foremost on a healthy bright-field appearance. Second to this is an appropriate expression level and localization. Expression levels should be selected that are the lowest that are detectable above background.

22. Cellular health and viability is critical for the success of live-cell imaging. Experimental design should be tailored to the biological question and desired data output, in order to maximize data yield while maintaining cell viability. The requirement for increased spatial and temporal resolution comes at the cost of cell number and throughput, in addition to analysis time, and thus determining the required output is essential. Therefore, when designing a live-cell imaging experiment there are a number of key parameters that require consideration (*see* Table 2). The combination of these settings will have a dramatic impact on the cell viability, photobleaching, and ultimately quality of your data. Excitation intensity, exposure time, and image frequency and experiment duration determine the amount of electromagnetic radiation each cell is exposed to. Consequently this impacts photobleaching and cell viability. Excitation intensity must be kept to the absolute minimum required to generate a reasonable image. Optimization should be performed. Cells can be exposed to the desired intensity for up to an hour. Cells should be monitored for signs of phototoxicity and photobleaching. It is important to pay attention to how cells look pre- and post-imaging to assess any potential damage, and modify your protocol accordingly.

23. This should be transient and take no more than 30 s. Lower pH buffers will significantly decrease both cytosolic and endosomal pH. This process quenches the surface-derived pHluorin fluorescence. The remaining signal (***Int***) is a combination of fluorescence from the pHluorin in internal compartments, autofluorescence, and background fluorescence.

References

1. Burchfield JG, Lopez JA, Mele K, Vallotton P, Hughes WE (2010) Exocytotic vesicle behaviour assessed by total internal reflection fluorescence microscopy. Traffic 11(4):429–439. https://doi.org/10.1111/j.1600-0854.2010.01039.x
2. Burchfield JG, Lu J, Fazakerley DJ, Tan SX, Ng Y, Mele K, Buckley MJ, Han W, Hughes WE, James DE (2013) Novel systems for dynamically assessing insulin action in live cells reveals heterogeneity in the insulin response. Traffic 14(3):259–273. https://doi.org/10.1111/tra.12035
3. Martin S, Millar CA, Lyttle CT, Meerloo T, Marsh BJ, Gould GW, James DE (2000) Effects of insulin on intracellular GLUT4 vesicles in adipocytes: evidence for a secretory mode of regulation. J Cell Sci 113(Pt 19): 3427–3438
4. Stenkula KG, Lizunov VA, Cushman SW, Zimmerberg J (2010) Insulin controls the spatial distribution of GLUT4 on the cell surface through regulation of its postfusion dispersal. Cell Metab 12(3):250–259. https://doi.org/10.1016/j.cmet.2010.08.005

5. Muretta JM, Romenskaia I, Mastick CC (2008) Insulin releases Glut4 from static storage compartments into cycling endosomes and increases the rate constant for Glut4 exocytosis. J Biol Chem 283(1):311–323. https://doi.org/10.1074/jbc.M705756200
6. Govers R, Coster AC, James DE (2004) Insulin increases cell surface GLUT4 levels by dose dependently discharging GLUT4 into a cell surface recycling pathway. Mol Cell Biol 24(14):6456–6466. https://doi.org/10.1128/MCB.24.14.6456-6466.2004
7. James DE, Brown R, Navarro J, Pilch PF (1988) Insulin-regulatable tissues express a unique insulin-sensitive glucose transport protein. Nature 333(6169):183–185. https://doi.org/10.1038/333183a0
8. Martin OJ, Lee A, McGraw TE (2006) GLUT4 distribution between the plasma membrane and the intracellular compartments is maintained by an insulin-modulated bipartite dynamic mechanism. J Biol Chem 281(1):484–490. https://doi.org/10.1074/jbc.M505944200
9. Fazakerley DJ, Holman GD, Marley A, James DE, Stockli J, Coster AC (2010) Kinetic evidence for unique regulation of GLUT4 trafficking by insulin and AMP-activated protein kinase activators in L6 myotubes. J Biol Chem 285(3):1653–1660. https://doi.org/10.1074/jbc.M109.051185
10. Wijesekara N, Tung A, Thong F, Klip A (2006) Muscle cell depolarization induces a gain in surface GLUT4 via reduced endocytosis independently of AMPK. Am J Physiol Endocrinol Metab 290(6):E1276–E1286. https://doi.org/10.1152/ajpendo.00573.2005
11. Brewer PD, Habtemichael EN, Romenskaia I, Mastick CC, Coster AC (2014) Insulin-regulated Glut4 translocation: membrane protein trafficking with six distinctive steps. J Biol Chem 289(25):17280–17298. https://doi.org/10.1074/jbc.M114.555714
12. Davey JR, Humphrey SJ, Junutula JR, Mishra AK, Lambright DG, James DE, Stockli J (2012) TBC1D13 is a RAB35 specific GAP that plays an important role in GLUT4 trafficking in adipocytes. Traffic 13(10):1429–1441. https://doi.org/10.1111/j.1600-0854.2012.01397.x
13. Coster AC, Govers R, James DE (2004) Insulin stimulates the entry of GLUT4 into the endosomal recycling pathway by a quantal mechanism. Traffic 5(10):763–771. https://doi.org/10.1111/j.1600-0854.2004.00218.x
14. Jhun BH, Rampal AL, Liu H, Lachaal M, Jung CY (1992) Effects of insulin on steady state kinetics of GLUT4 subcellular distribution in rat adipocytes. Evidence of constitutive GLUT4 recycling. J Biol Chem 267(25):17710–17715
15. Satoh S, Nishimura H, Clark AE, Kozka IJ, Vannucci SJ, Simpson IA, Quon MJ, Cushman SW, Holman GD (1993) Use of bismannose photolabel to elucidate insulin-regulated GLUT4 subcellular trafficking kinetics in rat adipose cells. Evidence that exocytosis is a critical site of hormone action. J Biol Chem 268(24):17820–17829
16. Yang J, Holman GD (1993) Comparison of GLUT4 and GLUT1 subcellular trafficking in basal and insulin-stimulated 3T3-L1 cells. J Biol Chem 268(7):4600–4603
17. Burchfield JG, Lopez JA, Hughes WE (2012) Using total internal reflection fluorescence microscopy (TIRFM) to visualise insulin action. In: Badoer E (ed) T visualization techniques, Neuromethods, vol 70. Humana Press, New York, pp 97–109
18. Schaffer DV, Koerber JT, Lim KI (2008) Molecular engineering of viral gene delivery vehicles. Annu Rev Biomed Eng 10:169–194. https://doi.org/10.1146/annurev.bioeng.10.061807.160514
19. Sinn PL, Sauter SL, PB MC Jr (2005) Gene therapy progress and prospects: development of improved lentiviral and retroviral vectors--design, biosafety, and production. Gene Ther 12(14):1089–1098. https://doi.org/10.1038/sj.gt.3302570
20. Chen C, Smye SW, Robinson MP, Evans JA (2006) Membrane electroporation theories: a review. Med Biol Eng Comput 44(1–2):5–14. https://doi.org/10.1007/s11517-005-0020-2
21. Sukharev SI, Klenchin VA, Serov SM, Chernomordik LV, Chizmadzhev Yu A (1992) Electroporation and electrophoretic DNA transfer into cells. The effect of DNA interaction with electropores. Biophys J 63(5):1320–1327. https://doi.org/10.1016/S0006-3495(92)81709-5
22. van den Hoff MJ, Moorman AF, Lamers WH (1992) Electroporation in 'intracellular' buffer increases cell survival. Nucleic Acids Res 20(11):2902
23. McKeel DW, Jarett L (1970) Preparation and characterization of a plasma membrane fraction from isolated fat cells. J Cell Biol 44(2):417–432
24. Simpson IA, Yver DR, Hissin PJ, Wardzala LJ, Karnieli E, Salans LB, Cushman SW (1983) Insulin-stimulated translocation of glucose transporters in the isolated rat adipose cells: characterization of subcellular fractions. Biochim Biophys Acta 763(4):393–407

25. Nikfarjam L, Farzaneh P (2012) Prevention and detection of mycoplasma contamination in cell culture. Cell J 13(4):203–212
26. Olarerin-George AO, Hogenesch JB (2015) Assessing the prevalence of mycoplasma contamination in cell culture via a survey of NCBI's RNA-seq archive. Nucleic Acids Res 43(5):2535–2542. https://doi.org/10.1093/nar/gkv136
27. MacLeod RA, Dirks WG, Matsuo Y, Kaufmann M, Milch H, Drexler HG (1999) Widespread intraspecies cross-contamination of human tumor cell lines arising at source. Int J Cancer 83(4):555–563
28. Masters JR (2002) HeLa cells 50 years on: the good, the bad and the ugly. Nat Rev Cancer 2(4):315–319. https://doi.org/10.1038/nrc775
29. Shewan AM, Marsh BJ, Melvin DR, Martin S, Gould GW, James DE (2000) The cytosolic C-terminus of the glucose transporter GLUT4 contains an acidic cluster endosomal targeting motif distal to the dileucine signal. Biochem J 350(Pt 1):99–107
30. Shewan AM, van Dam EM, Martin S, Luen TB, Hong W, Bryant NJ, James DE (2003) GLUT4 recycles via a trans-Golgi network (TGN) subdomain enriched in Syntaxins 6 and 16 but not TGN38: involvement of an acidic targeting motif. Mol Biol Cell 14(3):973–986. https://doi.org/10.1091/mbc.E02-06-0315
31. Green H, Kehinde O (1975) An established preadipose cell line and its differentiation in culture. II. Factors affecting the adipose conversion. Cell 5(1):19–27
32. Green H, Meuth M (1974) An established preadipose cell line and its differentiation in culture. Cell 3(2):127–133
33. Ozturk SS, Palsson BO (1990) Chemical decomposition of glutamine in cell culture media: effect of media type, pH, and serum concentration. Biotechnol Prog 6(2):121–128. https://doi.org/10.1021/bp00002a005

Chapter 16

Proximity Ligation Assay to Study the GLUT4 Membrane Trafficking Machinery

Dimitrios Kioumourtzoglou, Gwyn W. Gould, and Nia J. Bryant

Abstract

In this chapter a detailed protocol of proximity ligation assay (PLA) is described thoroughly. PLA is a technique that allows detection of protein associations in situ, providing a sensitive and selective approach for protein-protein interaction studies. We demonstrate the technique by applying it for trafficking studies of the facilitative glucose transporter GLUT4. Trafficking of GLUT4 from perinuclear depots to the plasma membrane is regulated by insulin in adipocytes and muscle cells, and mediated by formation of functional SNARE complexes containing Syntaxin4, SNAP23, and VAMP2. The Sec1/Munc18 (SM) protein Munc18c also plays a key role in insulin-stimulated GLUT4 translocation via a series of different interactions with the SNARE complex and/or with the SNARE proteins individually. Studying the interactions that occur between SNARE proteins themselves and also with Munc18c in insulin-responsive cells is critical to further understand SNARE protein function and GLUT4 trafficking mechanism in general.

Key words Proximity ligation assay, PLA, Protein-protein interactions, GLUT4 trafficking, GSVs, SNARE proteins, SNAP23, Munc18c

1 Introduction

A major action of insulin is to reduce elevated plasma glucose levels by increasing the rate of glucose transport into fat and muscle. This is mediated through the facilitative glucose transporter GLUT4. In the absence of insulin ~95 % of GLUT4 shows steady-state localization to intracellular compartments. Insulin stimulation results in GLUT4 redistribution to the cell surface via alterations in membrane trafficking [1, 2]. This process is perturbed in the insulin-resistance underlying type 2 diabetes and thus understanding the trafficking itinerary of GLUT4 is an important research goal. GLUT4 continually cycles through the plasma membrane and numerous compartments of the endosomal system both in the presence and absence of insulin [1, 3]. In the absence of insulin, GLUT4 is efficiently internalized by a fast trafficking loop between the plasma membrane and early/recycling

Karin Lindkvist-Petersson and Jesper S. Hansen (eds.), *Glucose Transport: Methods and Protocols*, Methods in Molecular Biology, vol. 1713, https://doi.org/10.1007/978-1-4939-7507-5_16,

endosomes from where it is sorted into a pool of insulin-responsive vesicles referred to as GLUT4 storage vesicles (GSVs) [3]. These vesicles are the direct source of GLUT4 mobilized to the cell surface in response to insulin. Fusion of GSVs with the plasma membrane is a key terminal step in insulin-regulated glucose transport. This fusion event is mediated by SNARE proteins, Syntaxin4, SNAP23 (t-SNAREs located in plasma membrane), and VAMP2 (v-SNARE anchored to the GLUT4-carrying vesicles) [4], through the formation of a very stable, SDS-resistant [5] ternary complex of their SNARE domains which provides the mechanical force that overcomes the energy barrier for fusion [6]. The whole process is regulated by a series of accessory proteins in which Munc18c (member of Sec1p/Munc18 family) has a predominant role [7, 8]. Munc18c interacts both with SNARE proteins Syntaxin4 and VAMP2 and with SNARE complex through a series of different binding modes which seem to tightly regulate SNARE proteins function both spatially and temporally [9–11]. Understanding the dynamics of these protein interactions and how they change upon insulin stimulation is necessary to fully comprehend the trafficking of GLUT4 transporter in insulin-responsive cells.

Various techniques have been developed to investigate protein associations. The vast majority of these depend on either probe-based targeting or direct labeling of the interacting proteins. In the case of probe-based targeting assays a high-affinity reagent is used (usually an antibody against the protein of interest) which might be labeled with a reporting agent (typically a fluorophore or an enzyme). The most common probe-based technique for protein interaction investigation is co-immunoprecipitation which uses an antibody to pull down the protein of interest along with interacting partners from cell lysates. The major disadvantage of this technique is its extreme invasiveness, as it requires cell destruction. This method can give deceptive results since protein localization in the cell, which often regulates and protects specificity of protein interaction, is disrupted. Immunohistochemistry and immunofluorescence are technical processes, which use a reporter molecule on the antibody specific to target protein, can overcome these problems, and allow detection of proteins in situ. Nonetheless these techniques have other limitations that are related to resolution of the obtained signal. For example in the case of immunofluorescence, the best microscopic resolution that can be achieved is 200 nm (maximum resolution of light microscopy), which subsequently indicates whether the two proteins of interest are within this distance of each other. These resolution limitations can be overcome by using direct targeting methods such as Förster resonance energy transfer (FRET), bioluminescent resonance energy transfer (BRET), or split yellow fluorescent protein (YFP) that use fusion constructs linked to reporter molecules. These methods are more quantitative and the proximity of the two proteins can be defined more precisely since the energy transfer from donor reporter to acceptor reporter is only possible in a range of 5–10 nm [12].

Nonetheless, the accuracy of these assays is compromised by their requirement for tagging proteins, which can potentially modify protein's native structure, function, and interactome.

PLA is a technique that can be used for protein interaction studies since it allows detection of proteins in close proximity [13]. This relatively new assay combines all the advantages of the two previously mentioned approaches, which are the use of endogenous native proteins (probe-tagged method) and the sensitivity and accuracy (direct labeling, split reporter method). This technique can be applied directly on fixed cells that have been grown in chamber slides and have been treated (insulin stimulation) according to the requirements of the experiment. After fixation cells are permeabilized and blocked using the appropriate reagent and then are incubated with primary antibodies that detect the interacting proteins and have been raised in different host species. Subsequently after the incubation and washing of the unbound amounts of primary antibodies, cells are treated with proximity probes. These probes are secondary antibodies specific to recognize the primaries and each of them is covalently attached to a single-stranded oligonucleotide. Following the incubation the excess proximity probes are washed and the connector oligonucleotides are added. In the case of interacting proteins the binding of the oligo-tagged secondary antibodies to adjacent primary antibodies allows hybridization of two connector oligonucleotides. Succeeding hybridization, enzymatic ligation of the stably hybridized connector oligonucleotides produces a single-stranded circular DNA molecule. This can be used as template for rolling circle amplification (RCA) after the addition of φ29 DNA polymerase. The polymerase reaction can be primed from one of the oligonucleotides attached to the secondary antibody (the other oligonucleotide has its 5′ end modified and it is not permissive for nucleotide addition and extension) [13]. During the incubation, as the amplification continues, the polymerase constantly replaces the newly synthesized strand producing an elongated single-stranded polynucleotide whose sequence consists of complementary repeats of the single-stranded circular template. The polymerization product can be visualized by the addition of single-stranded fluorescently labeled detection oligonucleotides, which are complementary to a sequence that is repeated within the elongated single-stranded polynucleotide. This allows hybridization of a large number of detection oligonucleotides, which provides signal amplification. Finally after the last washes and mounting of the slides, samples are ready for microscopy using a conventional fluorescent microscope. Considering that the maximum distance for two proteins to give a positive signal is determined exclusively by the size of both primary and secondary antibodies and the length of the oligonucleotides, typically 10–15 nm [14, 15], every single pair of interacting proteins will appear as a single fluorescent dot under the microscope. This means that obtained data can be easily enumerated and statistically analyzed (*see* Fig. 1).

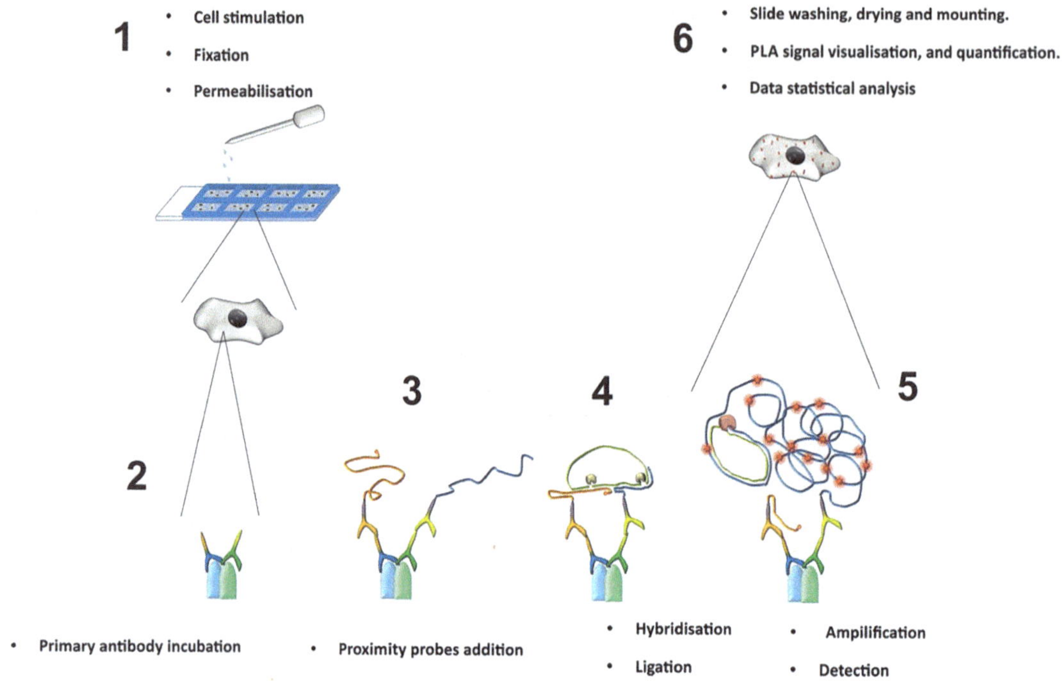

Fig. 1 PLA major steps and overall principle. Cells, 3T3-L1 fibroblasts or adipocytes, are grown on eight-chamber slides and stimulated with insulin or not according to the experiment. Cells are washed, fixed, and permeabilized according to the protocol (*1*). Primary antibodies detecting the proteins of interest are added (*2*). If the proteins are in close proximity then the secondary antibodies (proximity probes) covalently attached to single-stranded oligonucleotides (*3*) are able to stabilize the connector oligonucleotides via hybridization and form a circular single-stranded DNA molecule as a ligation product (*4*). This can function as template for RCA and the final amplification product can be detected with the addition of detector oligonucleotides that can be hybridized to the elongated DNA single strand (*5*). After the final washes, drying, and mounting every single pair of interacting proteins can be visualized as single dot under fluorescent microscope. The obtained numerical data of the signal can be further statistically analyzed (*6*)

2 Materials

2.1 Cell Growth Media

All cell culture media are filter sterilized through a 0.22 μm pore size filter.

1. **3T3-L1 fibroblast growth medium**: 10 % v/v Newborn calf serum (NCS) in Dulbecco's modified Eagle medium (DMEM).
2. **3T3-L1 adipocyte growth medium**: 10 % v/v Fetal bovine serum (FCS) in DMEM.
3. **3T3-L1 differentiation medium 1**: 0.5 mM 3-Isobutyl-1-methyxanthine (IBMX), 0.25 mM dexamethasone, 1 μM insulin in 3T3-L1 adipocyte growth media.
4. **3T3-L1 differentiation medium 2**: 1 μM Insulin in 10 % v/v fetal bovine serum (FCS) containing DMEM.

2.2 Solutions and Buffers

In all the solutions mentioned in this protocol high-purity sterile water is used as a dissolving agent. All the solutions are filtered through 0.22 μm filter and brought to room temperature before use.

1. **PBS buffer**: 85 mM NaCl, 1.7 mM KCl, 5 mM Na_2HPO_4, 0.9 mM KH_2PO_4, pH 7.4. Stored at 4 °C up to 6 months.
2. **PFA fixative solution**: 3 % w/v Paraformaldehyde. Stored at −20 °C up to 6 months.
3. **GLY**: 20 mM Glycine in PBS. Stored at 4 °C up to 4 weeks.
4. **BSA/GLY solution**: 2 % (w/v) Bovine serum albumin (BSA), 20 mM glycine in PBS. Store at 4 °C up to 4 weeks.
5. **BSA/GLY/SAP solution**: 0.1 % w/v Saponin in BSA/GLY solution. Can be stored at 4 °C up to 4 weeks.
6. **Wash buffer A**: 100 mM Tris–HCl, 150 mM NaCl, and 0.05 % w/v Tween-20, pH 7.4. Can be stored at 4 °C up to 6 months.
7. **Wash buffer B**: 200 mM Tris–HCl and 100 mM NaCl, pH 7.5. Can be stored at 4 °C up to 6 months.
8. **TBS-T**: 20 mM Tris–HCl, 137 mM NaCl, 0.1 % v/v Tween-20 pH 7.5. Can be stored at 4 °C up to 6 months.

2.3 Reagents

All the kit reagents are stored at −20 °C and thawed on ice before their use.

1. **Primary antibodies**: Selected primary antibodies raised against the proteins of interest in different host species (e.g., mouse and rabbit).
2. **PLA probe MINUS**: Secondary antibody raised against one of the primaries (e.g., either anti-mouse or anti-rabbit), covalently attached to a single-stranded oligonucleotide whose 5′ end is chemically modified (nonpermissive for elongation) (part of Duolink in situ fluorescence kit).
3. **PLA probe PLUS**: Secondary antibody raised against one of the primaries (e.g., either anti-rabbit or anti-mouse), covalently attached to a single-stranded oligonucleotide (permissive for elongation) (part of Duolink in situ fluorescence kit).
4. **Ligation (5×)**: Contains the connector oligonucleotides that hybridize to the PLA probes and all components needed for ligation except the ligase (part of Duolink in situ fluorescence kit).
5. **Ligase**: Contains the ligase (1 U/μL) (part of Duolink in situ fluorescence kit).
6. **Amplification (5×)**: Contains all components needed for rolling circle amplification except the polymerase. Oligonucleotide probes labeled with a fluorophore that hybridize to the RCA product are also included (part of Duolink in situ fluorescence kit).

7. **Polymerase**: Contains the polymerase (10 U/μL) (part of Duolink in situ fluorescence kit).
8. **Mounting medium with DAPI**: Duolink in situ mounting media with DAPI is aqueous and does not solidify (part of Duolink in situ fluorescence kit).

2.4 Equipment

1. 0.22 μm Pore size filter.
2. 8-Well chamber slide (Thermo Scientific Nunc™ Lab-Tek™ II Chamber Slide™ System 8-wells).
3. 10 cm plastic Petri dishes.
4. Vacuum aspirator.
5. Coplin staining jars.
6. Shaker.
7. Humidity chamber (moist chamber).
8. Freeze block for enzymes.
9. Incubator, 37 °C.
10. Pipettes (covering the range from 1 to 1000 μL).
11. Coverslips compatible with fluorescence microscopy.
12. High-purity water (sterile filtered, Milli-Q® or similar).
13. Fluorescence microscope equipped as follows: Excitation/emission filters compatible with fluorophore and nuclear stain excitation/emission, camera, and software for image acquisition (e.g., Zeiss LSM Pascal Exciter fluorescence system with a 63× oil immersion objective).

2.5 Software

1. Blobfinder version 3.2 (*see* **Note 1**).
2. Imaging analysis LSM software (Zeiss).

3 Methods

Cells, 3T3-L1 fibroblasts or differentiated adipocytes (see **Note 2**), are grown on Labtech eight-chamber slides using either fibroblast or adipocyte growth medium accordingly (500 μL per well). Cells that are to be stimulated with insulin (or not, in the case of basal controls) are "serum starved" for at least 2 h by replacing their serum-containing medium with serum-free DMEM. Cells are then stimulated by adding 100 nM insulin into the chamber wells and incubated at 37 °C for various time periods (based on the experiment—typical stimulation time 5 min). Basal control cells are left untreated or treated with vehicle.

3.1 Fixing and Blocking

1. Transfer chamber slides containing insulin-stimulated and unstimulated fibroblast or adipocyte 3T3-L1 cells from the cell incubator and place them on the bench.
2. Aspirate carefully the medium from the chamber wells (*see* **Note 3**).
3. Wash the cells carefully twice with PBS 400 μL (*see* **Note 3**).
4. Add 200 μL of PFA fixative solution to fix the cells. Incubate at room temperature for 20–30 min.
5. Wash twice with 400 μL GLY solution to quench the PFA.
6. Add 200 μL of BSA/GLY/SAP solution to block and permeabilize the cells. Incubate for 30 min at 37 °C in humidity chamber (*see* **Note 4**).

3.2 Primary Antibodies (See Note 5)

1. Dilute the primary antibodies to a suitable concentration (typical range 1:50–1:200) in BSA/GLY/SAP solution.
2. Aspirate the blocking solution from the wells. Try to obtain an equal residual volume on each well on slide in order to ensure reproducibility. Do not allow the samples to dry before adding the primary antibodies, as this will cause background.
3. Add 100 μL of the primary antibody solution to each sample.
4. Incubate in a humidity chamber. Use optimal time for your primary antibodies, usually ON at 37 °C (*see* **Note 6**).

3.3 PLA Probes

1. Mix and dilute the two PLA probes in a 1:5 ratio in BSA/GLY/SAP solution. For example, for a 40 μL reaction take 8 μL of PLA probe PLUS stock, 8 μL of PLA probe MINUS stock, and 24 μL of BSA/GLY/SAP solution. Leave the antibody mixture for 20 min at room temperature.
2. Aspirate the primary antibody solution from the wells.
3. Remove the well walls (*see* **Note 7**) from each slide and wash the slides using SAP/BSA/GLY solution (*see* **Note 8**). Washing should be performed in a Coplin staining jar, with a minimum volume of 70 mL (enough to cover the full length of the slides), on a shaker with gentle agitation for at least 10 min. The wash buffers should be at room temperature before use.
4. Add the diluted PLA probe solution (40 μL per well). Place the slides into humidity chamber.
5. Incubate the slides for 1 h at 37 °C.

3.4 Ligation (See Note 9)

1. Thaw the ligation (5×) reagent, vortex it, make 1:5 dilution in high-purity water, and mix. It is advised not to add the ligase until immediately before the addition of the mixture to the samples. Take the volume of ligase into consideration when calculating the amount of water added. For example for a 40 μL reaction volume, take 8 μL of the ligation (5×) reagent and 31 μL of high-purity water.

2. Aspirate gently the PLA probe solution from the slides.
3. Wash the slides in wash buffer A for 2 × 10 min with gentle agitation.
4. Remove the ligase from the freezer and keep it on ice or in a freeze box. Add ligase to the ligation solution from step 1 at a 1:40 dilution and vortex. For example for a 40 μL reaction volume, add 1 μL of ligase to 39 μL of ligation solution. Transfer the ligase back to the freezer.
5. Add the ligation-ligase solution to each sample (40 μL per well).
6. Place the slides in a preheated humidity chamber and incubate for at least 30 min at 37 °C.

3.5 Amplification (See Notes 9 and 10)

1. Thaw the amplification (5×) reagent, vortex it, make 1:5 dilution in high-purity water, and mix. You should add the polymerase immediately before the addition of the mixture to the sample. Take the volume of polymerase into account when calculating the amount of water added. For a 40 μL reaction volume, take 8 μL of the amplification (5×) reagent and 31.5 μL of high-purity water.
2. Aspirate carefully the ligation-ligase solution from the slides.
3. Wash the slides in wash buffer A twice for at least 5 min with gentle agitation. Aspirate all the remaining wash solution after the last washing.
4. Remove the polymerase from the freezer and keep it on ice or alternatively place it into a freezing block (−20 °C). Add polymerase to the amplification solution from **step 1** at a 1:80 dilution and vortex. For a 40 μL reaction volume, add 0.5 μL of polymerase to 39.5 μL of amplification solution.
5. Add the amplification-polymerase solution to each sample (40 μL per well).
6. Place the slides in a preheated humidity chamber and incubate for 2 h at 37 °C (*see* **Note 11**).

3.6 Final Wash Step (See Note 10)

1. Prepare 0.01× wash buffer B.
2. Aspirate the amplification-polymerase solution from the slides.
3. Wash the slides in wash buffer B for 2 × 10 min with gentle agitation.
4. Wash the slides in 0.01× wash buffer B for 1 min with gentle agitation.
5. Aspirate the remaining 0.01× wash buffer B from the slides.
6. Let the slides dry at room temperature in the dark (*see* **Note 12**).

3.7 Preparation for Imaging (See Note 10)

1. Mount your slides with a coverslip using a minimal volume of Duolink in situ mounting medium with DAPI. Make sure that no air bubbles get caught under the coverslip (*see* **Note 13**).

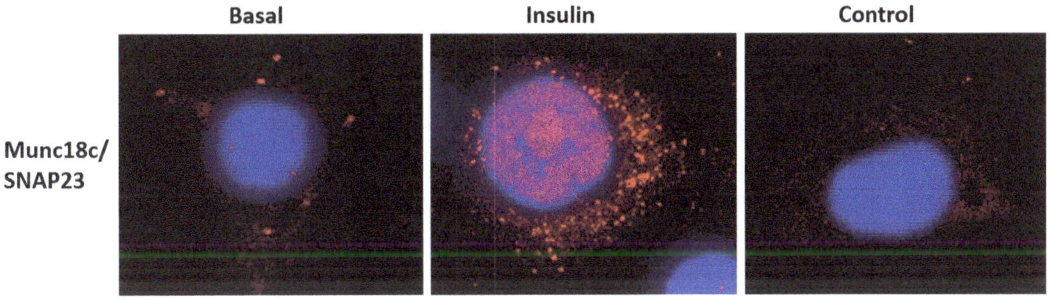

Fig. 2 SNAP23/Munc18c PLA. PLA using antibodies against SNAP23 (mouse) and Munc18c (rabbit) was performed in 3T3-L1 fibroblasts treated with insulin (100 nM for 5 min) or not (basal). *Red spots* correspond to protein-protein interaction couples. *Blue*: DAPI-stained nuclei. The control shown represents the omission of the primary antibody against Munc18c

2. Incubate for approximately 15 min in the dark.
3. Visualize the cells using a 63× oil immersion objective lens fitted to a Zeiss LSM Pascal Exciter confocal fluorescence microscope (*see* **Note 14**). Images can be analyzed using LSM software (Zeiss).

3.8 Image Analysis

1. After imaging, store the slides at −20 °C in the dark. Signal is stable for up to 2–3 weeks.
2. Pictures obtained (*see* Fig. 2) can be further analyzed using Blobfinder software (signal estimation). The following parameters should be used and kept constant throughout all the figure analysis: Blob threshold: 120 (arbitrary units). Minimum nucleus size: 100 pixels. Cytoplasm size: 200 pixels. Blob size: 3 × 3 pixels. The signal can be normalized to the signal obtained from the technical negative controls if necessary (*see* **Note 15**).

4 Notes

1. See reference [16] for further details.
2. Fibroblasts are grown to confluence and subsequently starved for 3 days (no medium replacement). On the day of differentiation cells are washed with serum-free DMEM and differentiation medium 1 is added to the flask (500 μL per well). After 3 days the medium is carefully replaced by differentiation medium 2 (500 μL per well) and after 2 days cells start being fed normally with adipocyte growth media (replacing the media every other day until the adipocytes were used for experiments (typically on the 8th–12th day after differentiation).
3. Differentiated adipocytes are loosely attached to the bottom of the chambers. Aspiration of the medium and addition of the

washing buffers or other reagents should be done carefully and gently in order to avoid any cell loss.

4. For the humidity chamber, a 10 cm plastic Petri dish can be used with a small piece of cotton soaked in distilled sterile water.
5. The success of PLA is largely based on the specificity and binding affinity of the primary antibodies. It is essential that the primary antibodies are carefully chosen, thoroughly checked, and optimized (dilution factor, diluting agent, incubation period, blocking, etc.) by both immunodetection (Western blot) using cell lysates and most importantly by immunofluorescence microscopy.
6. One-hour incubation can also be performed. If that is the case the antibody/solution volume can then be reduced to 40 μL.
7. Extra care should be paid during the removal of the walls. Slides might crack because of the uneven pressure. Use of a razor to remove any residual adhesive material after the removal of the walls is strongly suggested in order to achieve a flat surface, which is crucial for the proper placement of the coverslip during the last step of the protocol.
8. Alternatively, TBS-T can be used when there is strong background signal due to unspecific binding of the primary antibodies.
9. Use open droplet reactions without a coverslip and perform all incubations in a humidity chamber. The volume examples are based on 40 μL reaction volume, suitable for 1 cm^2 reaction area. Adjust the volumes corresponding to your specific enclosed reaction area.
10. During the amplification step light-sensitive reagents are used. Use of a foil-wrapped shallow box to cover the slides during the washes is strongly suggested.
11. Amplification time is a determining factor for the average size of the individual signal. In case of a strong signal due to high protein abundance and/or primary antibody efficiency, reduced incubation time is suggested.
12. In order to speed up the drying procedure the slides can be immerged in ethanol momentarily after the last wash and then can be let dry in the dark.
13. It is recommended the mounting medium to be placed on the coverslip (8 drops of ~10 μL—each of them corresponding to one of the 8 wells) and then the coverslip can be flipped and carefully placed on the top of the slide with the help of a pair of fine forceps. Since the mounting medium does not solidify, nail polish can be used to seal the edges.
14. Conventional fluorescent microscope with the appropriate filters depending on the fluorophore of the detection nucleotide can also be used.

15. As an immunofluorescence-based method, PLA faces the same common problems with other immunodetection approaches such as incomplete immobilization, reorganization, cytoskeletal changes, epitope accessibility, and antibody "patching." These difficulties can be overcome by the use of appropriate technical controls (omission of one of the primary antibodies) and the fact that PLA is used to compare the difference, if any, of interactions in two different conditions (insulin stimulation-basal conditions).

References

1. Bryant NJ, Govers R, James DE (2002) Regulated transport of the glucose transporter GLUT4. Nat Rev Mol Cell Biol 3:267–277. https://doi.org/10.1038/nrm782
2. Bryant NJ, Gould GW (2011) SNARE proteins underpin insulin-regulated GLUT4 traffic. Traffic 12:657–664. https://doi.org/10.1111/j.1600-0854.2011.01163.x
3. Kandror KV, Pilch PF (2011) The sugar is sIRVed: sorting Glut4 and its fellow travelers. Traffic 12:665–671. https://doi.org/10.1111/j.1600-0854.2011.01175.x
4. Thurmond DC, Ceresa BP, Okada S et al (1998) CELL BIOLOGY AND METABOLISM: regulation of insulin-stimulated GLUT4 translocation by Munc18c in 3T3L1 adipocytes regulation of insulin-stimulated GLUT4 translocation by Munc18c in 3T3L1 adipocytes. J Biol Chem 273:33876–33883. https://doi.org/10.1074/jbc.273.50.33876
5. Hayashi T, McMahon H, Yamasaki S et al (1994) Synaptic vesicle membrane fusion complex: action of clostridial neurotoxins on assembly. EMBO J 13:5051–5061
6. Pobbati AV, Stein A, Fasshauer D (2006) N- to C-terminal SNARE complex assembly promotes rapid membrane fusion. Science 313:673–676. https://doi.org/10.1126/science.1129486
7. Toonen RFG (2003) Role of Munc18-1 in synaptic vesicle and large dense-core vesicle secretion. Biochem Soc Trans 31:848–850. https://doi.org/10.1042/BST0310848
8. Tamori Y, Kawanishi M, Niki T et al (1998) Inhibition of insulin-induced GLUT4 translocation by Munc18c through interaction with syntaxin4 in 3T3-L1 adipocytes. J Biol Chem 273:19740–19746. https://doi.org/10.1074/jbc.273.31.19740
9. Brandie FM, Aran V, Verma A et al (2008) Negative regulation of syntaxin4/SNAP-23/VAMP2-mediated membrane fusion by Munc18c in vitro. PLoS One 3:1–7. https://doi.org/10.1371/journal.pone.0004074
10. Aran V, Brandie FM, Boyd AR et al (2009) Characterization of two distinct binding modes between syntaxin 4 and Munc18c. Biochem J 419:655–660. https://doi.org/10.1042/BJ20082293
11. Kioumourtzoglou D, Gould GW, Bryant NJ (2014) Insulin stimulates Syntaxin4 SNARE complex assembly via a novel regulatory mechanism. Mol Cell Biol 34:1271–1279. https://doi.org/10.1128/MCB.01203-13
12. Vogel S, Thaler C, Koushik S (2006) Fanciful FRET. Sci STKE re2:1–8. https://doi.org/10.1126/stke.3312006re2
13. Söderberg O, Gullberg M, Jarvius M et al (2006) Direct observation of individual endogenous protein complexes in situ by proximity ligation. Nat Methods 3:995–1000. https://doi.org/10.1038/nmeth947
14. Söderberg O, Leuchowius KJ, Gullberg M et al (2008) Characterizing proteins and their interactions in cells and tissues using the in situ proximity ligation assay. Methods 45:227–232. https://doi.org/10.1016/j.ymeth.2008.06.014
15. Weibrecht I, Leuchowius K-J, Clausson C-M et al (2010) Proximity ligation assays: a recent addition to the proteomics toolbox. Expert Rev Proteomics 7:401–409. https://doi.org/10.1586/epr.10.10
16. Allalou A, Wählby C (2009) BlobFinder, a tool for fluorescence microscopy image cytometry. Comput Methods Prog Biomed 94:58–65. https://doi.org/10.1016/j.cmpb.2008.08.006

Chapter 17

Quantification of Cell-Surface Glucose Transporters in the Heart Using a Biotinylated Photolabeling Assay

Zahra Maria and Véronique A. Lacombe

Abstract

The biotinylated photolabeling assay enables quantification of cell-surface glucose transporters (GLUTs). This technique has been successfully applied to quantify the cell-surface GLUT protein content in striated muscles and adipose tissue, as a means to evaluate GLUT trafficking. Here, we describe the detailed method of quantifying the cell-surface content of several GLUT isoforms (1, 4, 8, and 12) in isolated cardiac myocytes, as well as in the intact perfused atria and ventricle.

Key words GLUTs, Biotinylation assay, Bis-mannose photolabel, Immunoblot, Myocyte, Heart, Atria, Ventricle, Muscle

1 Introduction

Glucose uptake from the bloodstream into the cell is the rate-limiting step of cardiac glucose utilization. A family of membrane transporter proteins known as GLUTs tightly controls glucose transport across the cell surface. So far, 14 GLUT isoforms have been identified. Based on the similarity in their sequences, they have been classified into three classes: Class 1 (GLUTs 1–4 and GLUT 14), Class 2 (GLUT-5, -7, -9, and -11), and the novel Class 3 (GLUT-6, -8, -10, -12 and HMIT) [1, 2]. The Class 1 glucose transporters consist of the most prominently expressed GLUTs, while the Class 3 GLUTs are the most novel isoforms that remain to be fully investigated. Importantly, GLUT1, a ubiquitously expressed isoform, is a membrane-bound GLUT that allows the influx of glucose into the cell at a basal rate. Other GLUTs, such as GLUT4, require activation by insulin to translocate from their intracellular vesicles (inactive site) to the cell surface (active site) in order to facilitate glucose diffusion in insulin-sensitive tissues (i.e., striated muscle and adipose tissue). Importantly, GLUT-1, -4, -8, and -12 are reportedly expressed in the cardiac tissue [3–5]. The biotinylated photolabeling

Karin Lindkvist-Petersson and Jesper S. Hansen (eds.), *Glucose Transport: Methods and Protocols*, Methods in Molecular Biology, vol. 1713, https://doi.org/10.1007/978-1-4939-7507-5_17,

technique provides a direct quantifiable measure of active cell-surface and intracellular GLUT proteins. This technique was first used to assess trafficking of intracellular GLUT4 to the cell surface in several insulin-sensitive tissues (i.e., skeletal muscle and adipose tissue) [6, 7]. This technique was also adapted to quantify cell-surface GLUTs in isolated cardiac myocytes and in the intact perfused mouse heart [8]. While many studies have been adapted to study mouse cardiac metabolism, the quantification of cell-surface GLUT, which is critical to unravel the molecular mechanisms regulating glucose transport, has proven very challenging in mice. For instance, conventional membrane fractionation techniques are difficult to apply to mice because of the small size of the heart (<150 mg), and immunohistochemistry and immunofluorescence techniques have some limitations for accurate protein quantification. In addition, docking and fusion of GLUT4 with the cell-surface membrane are required steps, after translocation, to activate GLUTs, which cannot be differentiated from inactive GLUTs (i.e., before fusion) with the techniques mentioned above [8, 9]. Although first described to measure GLUT4 trafficking, this technique has also been adapted in our laboratory to measure several other GLUT isoforms (e.g., GLUT-1, -8, and -12) [3, 5, 10].

The biotinylated bis-mannose photolabeled reagent (Bio-LC-ATB-BGPA) is either infused through the aorta of the perfused heart or incubated with isolated cardiac myocytes. The reagent specifically interacts with the extracellular binding site of GLUTs. Upon UV irradiation, the diazirine group loses nitrogen and generates a short-lived carbene, which interacts with the extracellular binding site of the GLUTs by establishing a covalent bond (*see* Fig. 1) [6, 7, 11]. Protein extraction is immediately followed by homogenization and ultracentrifugation. Recovery of photolabeled (cell surface) GLUTs is achieved using streptavidin isolation, facilitating separation of non-cell-surface GLUTs from cell-surface GLUTs. The labeled GLUTs are then dissociated from the streptavidin by boiling in Laemmli buffer prior to SDS-PAGE and subsequent immunoblotting with GLUT antibody (*see* Fig. 1) [3, 5, 10].

2 Materials

2.1 Photolabeling Isolated Cardiac Myocytes and Intact Perfused Heart

1. Photolabeled compound: Reconstitute the vial of Bio-LC-ATB-BGPA (N-[2-[2-[2-[(N-biotinyl-caproylamino)-ethoxy) ethoxyl]-4-[2-(trifluorommethyl)-3H–diazirin-3-yl]benzoyl]-1,3-bis(mannopyranosyl-4-yloxy)-2-propylamine (Toronto Research Chemical) in deionized distilled water (0.09 mL water/mg). Vortex to ensure proper mixing. Store the solution in an opaque tube at 4 °C (*see* **Note 1**).
2. Tyrode buffer (1×): 135 mM NaCl, 5.4 mM KCL, 1 mM $MgCl_2$, 0.33 mM NaH_2PO_4, 10 mM HEPES

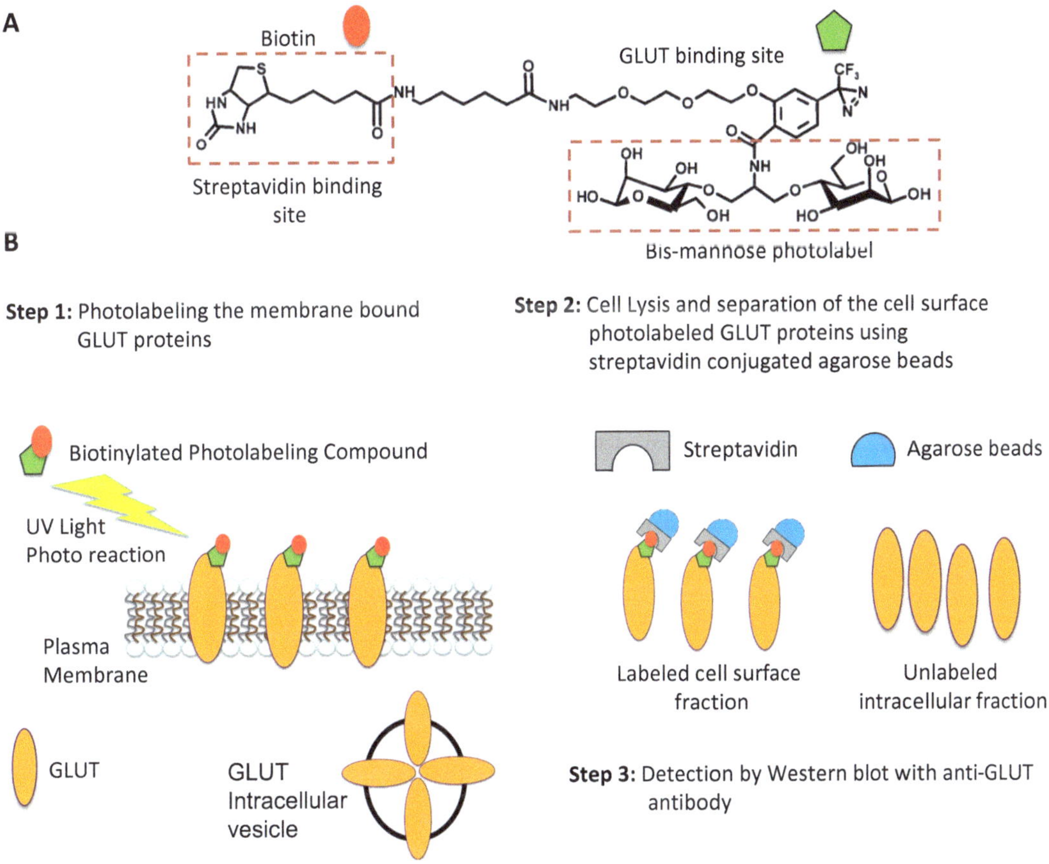

Fig. 1 Photolabeling of cell-surface GLUTs. Adapted from [7]. (**a**) Bio-LC-ATB-BGPA compound. (**b**) Photolabeling the cell-surface GLUTs with the Bio-LC-ATM-BGPA. Irradiation with UV light permanently cross-links the photolabeled compound to the cell-surface GLUT proteins. Total crude membrane-enriched protein lysate from photolabeled tissue or myocytes is prepared by ultracentrifugation and isolated with streptavidin-agarose to separate cell-surface GLUTs ("labeled," plasma membrane fraction) from intracellular GLUTs ("unlabeled") that remains in the supernatant. Equal aliquots are immunoblotted with GLUT antibody

(4-(2-hydroxyethyl)-1-piperazineethane-sulfonic acid), and 10 mM glucose. Mix and adjust pH to 7.4 (*see* **Note 2**).

3. Krebs-Henseleit buffer (KHB): 118 mM NaCl, 4.74 mM KCl, 1.2 mM KH_2PO_4, 1.2 mM $MgSO_4$, 25 mM $NaHCO_3$, 15 nM of bovine serum albumin, 7 mM glucose, pH 7.4.
4. Buffer 1: 210 mM Sucrose, 40 mM NaCl, 2 mM EGTA (ethylene glycol-bis (β-aminoethyl ether)-N,N,N',N'-tetraacetic acid), 30 mM HEPES, pH 7.4. Add protease inhibitor cocktail at a 1:500 dilution to buffer 1 just before starting the protocol.
5. Buffer 2: 1.167 M KCl, 58.3 mM Na_4O_7 P_2•$10H_2O$, pH 7.4.
6. Radioimmunoprecipitation assay (RIPA) buffer: 50 mM Tris–HCl, pH 8.0, 150 mM NaCl, 1 % w/v NP-40, 0.5 % sodium

deoxycholate, 0.1 % w/v SDS supplemented with 0.2 % protease inhibitor cocktail.
7. Langendorff apparatus.
8. Water bath at 37 °C.
9. Ultracentrifuge and appropriate ultracentrifuge tubes.
10. Rayonet RPR-100 photoreactor.
11. Microcentrifuge and rotor for 1.7 mL microcentrifuge tubes.
12. Rocker for 96-well plates.

2.2 Spin Column Separation of Cell-Surface Fraction

1. Streptavidin agarose resin beads: Streptavidin agarose is a standard-capacity beaded agarose resin of immobilized recombinant streptavidin protein.
2. Micro-spin column: Should be able to handle approximately 400 μL of solution and 5 μL of resin with approximately 30 μm pore size.
3. 1 M Dithiothreitol (DTT).
4. Phosphate-buffered saline (PBS) solution: 150 mM NaCl, 10 mM NaH_2PO_4, pH 7.4
5. 1 and 0.1 % v/v Triton X-100 in PBS.

2.3 SDS Polyacrylamide Gel and Immunoblotting

1. Resolving gel buffer: 1.5 M Tris–HCL, pH 8.8.
2. Stacking gel buffer: 0.5 M Tris–HCL, pH 6.8.
3. 30 % w/v Acrylamide.
4. 10 % w/v Sodium dodecyl sulfate (SDS).
5. 10 % w/v Ammonium persulfate.
6. *N*,*N*,*N*′,*N*′-tetramethylethylenediamine (TEMED).
7. Tween phosphate-buffered saline (TPBS): 1 % v/v Tween-20 in 1× PBS.
8. Tris-glycine buffer: 25 mM Tris and 192 mM glycine, pH 8.3.
9. Transfer buffer: 10 % Tris-glycine, 20 % methanol, and 70 % H_2O.
10. Tris-glycine SDS buffer (gel electrophoresis buffer): 25 mM Tris and 192 mM glycine, 0.1 % w/v SDS, pH 8.3.
11. Polyvinylidene fluoride (PVDF) membranes: 0.2 μm pore size, binding capacity 150–160 μg/cm^2.
12. Nonfat dry milk.
13. Laemmli sample loading buffer (2×).
14. Vertical gel electrophoresis system.
15. Mini Transblot module for gel transfer.
16. SDS-PAGE gel casting apparatus including casting stands with clamps, short plates, spacer plates, and comb.

2.4 Antibodies (Vendor, Source, Dilution)

1. GLUT1: Abcam, polyclonal rabbit anti-human GLUT1; 1:500 (*see* **Note 3**).
2. GLUT4: AbD Serotec, polyclonal rabbit anti-human GLUT4, 1:750 (*see* **Note 4**).
3. GLUT8: Bioss, polyclonal rabbit anti-human GLUT8, 1:500 (*see* **Note 5**).
4. GLUT12: Abcam, polyclonal rabbit anti-human GLUT12, 1:500 (*see* **Note 6**).
5. Secondary antibody: GE Healthcare, polyclonal goat anti-rabbit, 1:2500.

3 Methods

3.1 Photolabeling of Isolated Atrial and Ventricular Myocytes

1. Perfuse the heart in a retrograde manner using a Langendorff apparatus. Isolate atrial and ventricular myocytes using enzymatic perfusion (*see* **Note 7**).
2. Add tyrode buffer to the myocyte pellet (the volume will be based on the number of treatments and replicates). Carefully divide the atrial and ventricular myocytes into equal volumes (200 μL) for the desired number of treatments.
3. If required, incubate myocytes with insulin (or other compound as required) in a 96-well plate, gently agitating on the plate rocker for 30 min. At the end of the incubation period, place the 96-well plate in a water bath at 18 °C for 5 min to slow down GLUT trafficking.
4. Add photolabeled compound to each well to achieve the final concentration of 300 μM and gently rotate the plate for 1 min.
5. Irradiate cells using a Rayonet UV reactor for 3 cycles of 1 min to photo-chemically cross-link the Bio-LC-ATB-BGPA to the cell-surface GLUTs (*see* Fig. 1).
6. After irradiation, collect the cells in a microcentrifuge tube. Wash each well four times with 200 μL of 1× tyrode buffer and transfer the solution to the same microcentrifuge tube; keep the samples on ice. Centrifuge the tubes at 1,700 × *g* for 1 min and discard the supernatant.
7. Resuspend the pellet in 1 mL of 1× tyrode buffer. Centrifuge at 1,700 × *g* for 1 min. Discard the supernatant. Repeat this step four times, each time using 1 mL of 1× tyrode buffer to resuspend the pellet.
8. Add 0.1–1 mL of RIPA buffer to each pellet (the volume will be based on the size of the pellet and the desired sample concentration), vortex, and incubate for 60 min at 4 °C with rotation.
9. Centrifuge tubes at 13,000 × *g* for 60 min at 4 °C. Discard pellet and save supernatant.

10. Perform appropriate assay to determine the protein concentration of the supernatant (referred as "total lysate").
11. Flash-freeze in liquid nitrogen and store at −80 °C or proceed with Subheading 3.2.
12. As a variation of this technique, atrial and ventricular tissue can be photolabeled by perfusion of the intact heart with the biotinylated compound (*see* Figs. 2 and 3) [3, 5, 8, 10]. To our knowledge, we are the first to apply this technique to study GLUT trafficking in the atria (*see* **Notes 8 and 9**). Variation of this technique can also be applied to skeletal muscle, adipose tissue, and other tissue (e.g., lung, colon); *see* **Note 10**.

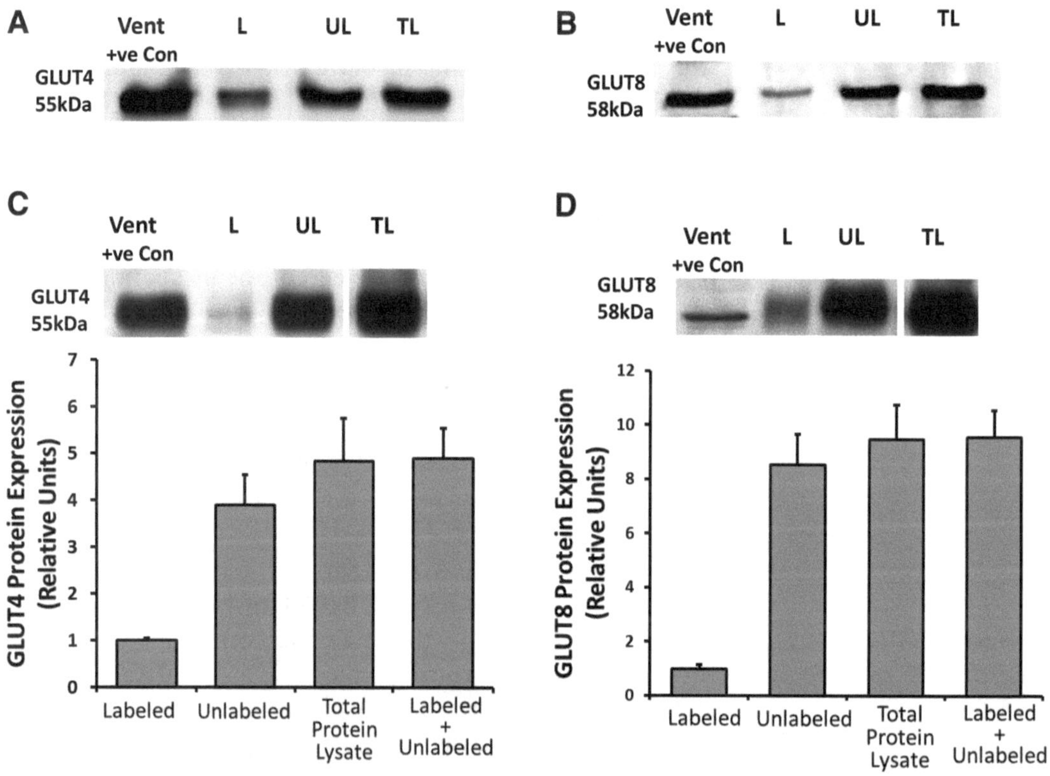

Fig. 2 Representative Western blots. (**a**) GLUT4 and (**b**) GLUT8 protein in cell surface (L), intracellular (UL), and total membrane (TL) fractions from photolabeled ventricular myocytes. Total protein lysate from mouse ventricle was used as a positive control (+ve Con). Validation of the photolabeling technique to quantify the cell-surface fraction of (**c**) GLUT4 and (**d**) GLUT8. Top panel: Representative Western blot of GLUT4 and GLUT8 in the photolabeled intact perfused atria. Bottom panel: Mean ± SEM of GLUT protein expression in L, UL, and TL fractions (values expressed relative to labeled; $n = 3–4$/group). Please note that the sum of the labeled and the unlabeled GLUT-4 and -8 pools is equal to the total cardiac GLUT-4 and -8 content

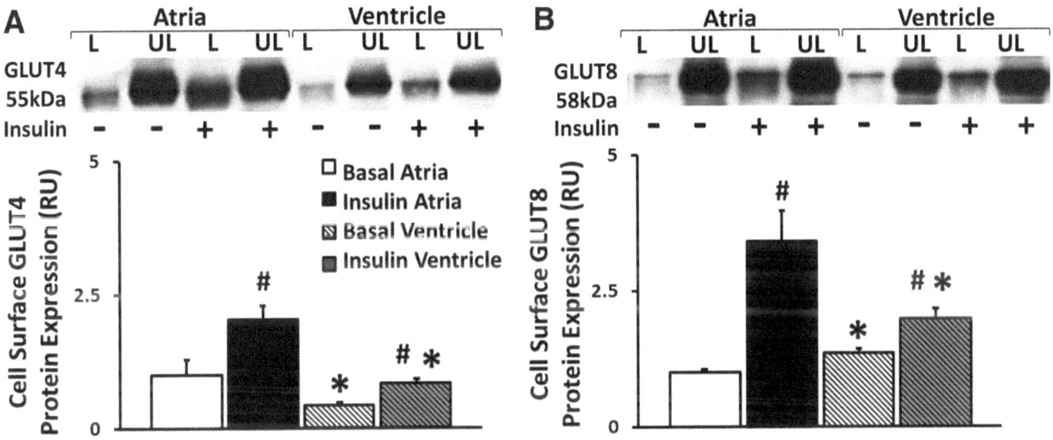

Fig. 3 Insulin stimulates GLUT4 and GLUT8 trafficking to the atrial and ventricular cell surface in the intact perfused heart. (**a**) GLUT4 and (**b**) GLUT8 trafficking. Top panels: Representative Western blot of cell surface (L) and intracellular (UL) GLUT4 and GLUT8. Bottom panels: Mean ± SE of cell-surface GLUT protein content (values expressed relative to labeled basal atria; $n = 3$–4/group); # $P < 0.05$ vs. basal; *$P < 0.05$ vs. atria). Adapted from [3]

3.2 Isolation of Cell-Surface Fraction

1. Add 200 μg of protein to 100 μL of streptavidin agarose resin. Rotate gently the samples overnight at 4 °C.
2. Centrifuge samples at 2,300 × *g* for 5 min at room temperature to extract supernatant that contains the non-photolabeled GLUTs; save as "unlabeled fraction."
3. Wash the beads three times with 1 mL of 1 % v/v Triton X-100 in PBS solution at room temperature. Vortex the samples. After each wash, centrifuge the samples at 2,300 × *g* for 5 min at room temperature and discard the supernatant.
4. Wash beads three times with 1 mL of 0.1 % v/v Triton X-100 in PBS solution. Vortex the samples. After each wash, centrifuge the samples at 2,300 × *g* for 5 min at room temperature and discard the supernatant.
5. Wash beads once with 1 mL of PBS. Vortex the samples. Centrifuge at 2,300 × *g* for 5 min at room temperature and discard the supernatant.
6. Add 30 μL of Laemmli buffer (2×) to the beads, vortex, and centrifuge at 1,000 × *g* for 30 s. Boil the samples for 30 min. Allow tubes to cool for 5 min.
7. Centrifuge at 2,300 × *g* for 5 min at room temperature. Save supernatant in a microcentrifuge tube as "labeled fraction" and place on ice. Repeat **step 6** once.
8. Remove the beads and the Laemmli buffer from the microcentrifuge tube and add them to the micro-spin column. Place micro-spin columns in the tubes containing the labeled fraction.

9. Centrifuge the spin columns with the microcentrifuge tubes at 2,300 × *g* for 5 min. At the end of this step, the beads will remain in the spin column and the "labeled fraction" will be collected in the microcentrifuge tube.
10. Add appropriate amount of 1 M of DTT to reach the desired concentration of 350 mM. Incubate at room temperature for 10 min.
11. Proceed to Subheading 3.3.

3.3 Immunoblotting

1. Prepare resolving gel by mixing 6.6 mL of deionized distilled water, 5 mL of 1.5 M Tris–HCL (resolving gel buffer), 8 mL of 30 % w/v acrylamide, 200 μL of 10 % w/v SDS, 200 μL of 10 % w/v APS, and 10 μL of TEMED in a 50 mL glass beaker. Cast gel within 7.25 cm × 10 cm × 1.5 mm gel cassette. Allow space for stacking gel and gently overlay with 70 % ethanol. After the gel has solidified discard the 70 % ethanol.
2. Prepare stacking gel by mixing 3 mL of deionized distilled water, 1.25 mL of 0.5 M Tris–HCL (stacking gel buffer), 650 μL of 30 % w/v acrylamide, 50 μL of 10 % w/v SDS, 50 μL of 10 % w/v APS, and 5 μL of TEMED in a 10 mL glass beaker. Transfer the stacking gel mixture to the gel cassette over the solidified resolving gel. Insert immediately a 10-well gel comb.
3. Use standard Western blotting procedures to detect the GLUT protein of interest (*see* Figs. 2 and 3). Protein (5–20 μg) will be resolved on a 12 % w/v SDS-polyacrylamide gel and electrophoretically transferred to a polyvinylidene fluoride membrane (*see* **Note 11**).
4. Add 2× Laemmli buffer supplemented with 1 M DTT at 1:1 ratio to the unlabeled (intracellular) fraction and total lysate (total membrane protein samples) (*see* **Notes 11** and **12**).
5. Gel electrophoresis: Run the gel electrophoresis in Tris-glycine-SDS buffer at 120 V for 120 min.
6. Transfer the proteins onto the PVDF membranes at 100 V for 120 min in transfer buffer at 4 °C with a stir bar to keep the buffer slightly agitated at all times.
7. For blocking, use 5 % milk in TPBS buffer for 1 h at room temperature.
8. Wash the membranes twice with TPBS and once with PBS for 10 min each at room temperature.
9. Incubate with appropriate primary antibody overnight at 4 °C with moderate agitation on a plate rocker. (Please refer to Subheading 2 for the detailed preparation of the primary antibodies.)

10. Wash the membranes twice with TPBS and once with PBS for 10 min each.
11. Use 1:2500 dilution of the secondary antibody in 5 % milk with TPBS buffer for 60 min of incubation.
12. Wash the membranes twice with TPBS and once with PBS for 10 min each.
13. Antibody-bound transporter proteins are quantified by enhanced chemiluminescence reaction and autoradiography. Band density is quantified and molecular weight is determined using appropriate software. All bands of interest are compared to an internal positive control known to express the GLUT protein of interest.

4 Notes

1. The photolabeled compound is light sensitive and therefore should be used in the dark.
2. We find that it is best to prepare the buffers on the day of the experiment.
3. Total protein extract from ventricle and liver can be used as positive control for GLUT1 (molecular weight: 48 kDa). For Western blotting, we recommend loading ~20 μg of protein per lane.
4. Primary antibody has been selected based on predicted species reactivity, molecular weight (55 kDa), and lack of cross-reactivity for proteins outside of our scope of interest. It has been further validated using competitive binding assay [12]. Total protein extract from ventricle and skeletal muscle can be used as positive control for GLUT4. Total protein extract from testes can be used as negative control for GLUT4. For Western blotting, we recommend loading ~10 μg of protein per lane.
5. Total protein extract from ventricle can be used as positive control for GLUT8 (molecular weight: 58 kDa). For Western blotting, we recommend loading ~10 μg of protein per lane.
6. Total protein extract from ventricle and kidney can be used as positive control for GLUT12 (molecular weight: 67 kDa). For Western blotting, we recommend loading ~20 μg of protein per lane.
7. Isolate atrial and ventricular myocyte using enzymatic perfusion, as previously described [3, 13]. The atria could be digested before the ventricle. Monitor the atria carefully to avoid over-digestion, which can affect the yield of the isolated myocytes. Best results can be achieved if the yield can be kept consistent between isolations (>80 % for ventricular myocytes). Make sure that the isolated atrial cells are not contami-

nated with the ventricular cells. Evaluating the shape of the myocytes under a microscope can assess this.

8. Atrial and ventricular tissue can also be photolabeled by perfusion of the intact heart with the biotinylated compound [3, 5, 8, 10]. Briefly, excise the heart of a deeply anesthetized mouse/rat. Isolate and cannulate the aorta with a blunt needle. To ensure appropriate cannulation of the aorta, flush the heart with the ice-cold KHB. If the coronary arteries are visible and filled with blood, the aorta has been appropriately cannulated. Use a Langendorff apparatus to perfuse the heart with oxygenated KHB (37 °C) for 60 min. Throughout the perfusion, ensure that the heart is properly perfused and that there is no sign of ischemia. Flush the aortic cannula slowly for 1 min with 1 mL ice-cold glucose-free KHB, immediately followed by 1 mL of the same buffer containing Bio-LC-ATB-BGPA (300 μM, infused slowly over 1 min). Incubate the cardiac tissue with the photolabeled compound in a light-impermeable microcentrifuge tube on ice for 15 min. Cut the atria and ventricles into several pieces (<10 mg) and place on a weighing dish on ice. Irradiate the samples in the UV photoreactor for 3 cycles of 3 min to photo-chemically cross-link the Bio-LC-ATB-BGPA to the cell-surface GLUTs. Collect crude membrane extracts for analysis of GLUT proteins as described below (*see* **Note 9**).
9. Homogenize ~25–60 mg of tissue in Buffer 1 (70 μL of Buffer 1/mg of tissue) at 4 °C. Add 30 μL of Buffer 2/mg of tissue to each sample and place on ice for 15 min. Centrifuge at 4 °C for 90 min at 150,000 × *g* and carefully discard the supernatant. Homogenize the pellet with RIPA buffer (12 μL/mg of tissue). Gently rotate the samples for 1 h at 4 °C. Centrifuge the samples at 16,000 × *g* for 1 h in 4 °C and save the supernatant containing the total membrane proteins. Quantify protein concentration using a detergent-compatible method (Pierce) with BSA as standard. Proceed to **step 3.2**.
10. Skeletal muscle, adipose tissue, and colon and lung tissue [14, 15] may also be photolabeled following a similar protocol with the following variations. Incubate 60 mg (~3 × 20 mg pieces) of tissue in KHB with glucose for 60 min at 37 °C. Transfer samples to a 96-well plate (each well containing 20 mg tissue). Incubate the samples with 200 μL of glucose-free KHB for 10 min at 18 °C. Add Bio-LC-ATB-BGPA at 300 μM final concentration to each well and rotate the plate for 15 min at 18 °C. Irradiate samples in the UV photoreactor for 3 cycles of 3 min to photo-chemically cross-link the Bio-LC-ATB-BGPA to the cell-surface GLUTs. Collect crude membrane extracts for analysis of GLUT proteins and proceed to **step 3.2**.

11. Since the labeled (cell surface) fraction is already collected in Laemmli buffer +1 M DTT (*see* Subheading 3.2), no further addition of Laemmli is required prior to loading the samples in SDS-PAGE gels.

12. Boiling is only permissible to dissociate the labeled cell-surface fractions from the streptavidin-conjugated agarose beads (*see* Subheading 3.2). Boiling the labeled fraction longer than 30 min, or for a second time following storage at −80 °C, will lead to irreversible aggregation of GLUT4 and should be avoided [8]. Similarly, boiling the intracellular fraction and total protein samples must be avoided.

Acknowledgements

We thank Allison Campolo, MS, for carefully reading the manuscript. This research was funded by National Institute of Health (K01RR023083), the Harold Hamm Diabetes Research Center, The Oklahoma State University Center of Veterinary Health Sciences, and Oklahoma Center of Respiratory and Infectious Disease (NIH 1P20 GM103648).

References

1. Lacombe VA (2014) Expression and regulation of facilitative glucose transporters in equine insulin-sensitive tissue: from physiology to pathology. ISRN Vet Sci 2014:15. https://doi.org/10.1155/2014/409547
2. Mueckler M, Thorens B (2013) The SLC2 (GLUT) family of membrane transporters. Mol Asp Med 34(2–3):121–138. https://doi.org/10.1016/j.mam.2012.07.001
3. Maria Z, Campolo AR, Lacombe VA (2015) Diabetes alters the expression and translocation of the insulin-sensitive glucose transporters 4 and 8 in the atria. PLoS One 10(12):e0146033. https://doi.org/10.1371/journal.pone.0146033
4. Stanley WC, Recchia FA, Lopaschuk GD (2005) Myocardial substrate metabolism in the normal and failing heart. Physiol Rev 85(3):1093–1129. https://doi.org/10.1152/physrev.00006.2004
5. Waller AP, George M, Kalyanasundaram A, Kang C, Periasamy M, Hu K, Lacombe VA (2013) GLUT12 functions as a basal and insulin-independent glucose transporter in the heart. Biochim Biophys Acta 1832(1):121–127. https://doi.org/10.1016/j.bbadis.2012.09.013
6. Koumanov F, Yang J, Jones A, Hatanaka Y, Holman GD (1997) Cell surface biotinylation of GLUT4. Biochem Soc Trans 25(3):470S
7. Koumanov F, Yang J, Jones AE, Hatanaka Y, Holman GD (1998) Cell-surface biotinylation of GLUT4 using bis-mannose photolabels. Biochem J 330(Pt 3):1209–1215
8. Miller EJ, Li J, Sinusas KM, Holman GD, Young LH (2007) Infusion of a biotinylated bis-glucose photolabel: a new method to quantify cell surface GLUT4 in the intact mouse heart. Am J Physiol Endocrinol Metab 292(6):E1922–E1928. https://doi.org/10.1152/ajpendo.00170.2006
9. Yang J, Holman GD (1993) Comparison of GLUT4 and GLUT1 subcellular trafficking in basal and insulin-stimulated 3T3-L1 cells. J Biol Chem 268(7):4600–4603
10. Waller AP, Kalyanasundaram A, Hayes S, Periasamy M, Lacombe VA (2015) Sarcoplasmic reticulum Ca^{2+} ATPase pump is a major regulator of glucose transport in the healthy and diabetic heart. Biochim

Biophys Acta 1852(5):873–881. https://doi.org/10.1016/j.bbadis.2015.01.009

11. Hashimoto M, Hatanaka Y, Yang J, Dhesi J, Holman GD (2001) Synthesis of biotinylated bis(D-glucose) derivatives for glucose transporter photoaffinity labelling. Carbohydr Res 331(2):119–127
12. Lacombe VA, Hinchcliff KW, Devor ST (2003) Effects of exercise and glucose administration on content of insulin-sensitive glucose transporter in equine skeletal muscle. Am J Vet Res 64(12):1500–1506
13. Kranstuber AL, Del Rio C, Biesiadecki BJ, Hamlin RL, Ottobre J, Gyorke S, Lacombe VA (2012) Advanced glycation end product cross-link breaker attenuates diabetes-induced cardiac dysfunction by improving sarcoplasmic reticulum calcium handling. Front Physiol 3:292. https://doi.org/10.3389/fphys.2012.00292
14. Ryder JW, Yang J, Galuska D, Rincon J, Bjornholm M, Krook A, Lund S, Pedersen O, Wallberg-Henriksson H, Zierath JR, Holman GD (2000) Use of a novel impermeable biotinylated photolabeling reagent to assess insulin- and hypoxia-stimulated cell surface GLUT4 content in skeletal muscle from type 2 diabetic patients. Diabetes 49(4):647–654
15. Waller AP, Kohler K, Burns TA, Mudge MC, Belknap JK, Lacombe VA (2011) Naturally occurring compensated insulin resistance selectively alters glucose transporters in visceral and subcutaneous adipose tissues without change in AS160 activation. Biochim Biophys Acta 1812(9):1098–1103. https://doi.org/10.1016/j.bbadis.2011.02.007

Chapter 18

Tracking GLUT2 Translocation by Live-Cell Imaging

Sabina Tsytkin-Kirschenzweig, Merav Cohen, and Yaakov Nahmias

Abstract

The facilitative glucose transporter (GLUT) family plays a key role in metabolic homeostasis, controlling the absorption rates and rapid response to changing carbohydrate levels. The facilitative GLUT2 transporter is uniquely expressed in metabolic epithelial cells of the intestine, pancreas, liver, and kidney. GLUT2 dysfunction is associated with several pathologies, including Fanconi-Bickel syndrome, a glycogen storage disease, characterized by growth retardation and renal dysfunction. Interestingly, GLUT2 activity is modulated by its cellular localization. Membrane translocation specifically regulates GLUT2 activity in enterocytes, pancreatic β-cells, hepatocytes, and proximal tubule cells. We have established a system to visualize and quantify GLUT2 translocation, and its dynamics, by live imaging of a mCherry-hGLUT2 fusion protein in polarized epithelial cells. This system enables testing of putative modulators of GLUT2 translocation, which are potential drugs for conditions of impaired glucose homeostasis and associated nephropathy.

Key words Glucose transport, Live imaging, Polarized epithelium, GLUT2, Translocation, Kidney function, Type 2 diabetes

1 Introduction

Glucose uptake and transport are key components of metabolic homeostasis [1]. Facilitative glucose transporters are a family of membrane proteins involved in the concentration-dependent transport of several hexoses, such as glucose and fructose, as well as myoinositol, urate, glucosamine, and ascorbate across cellular membranes. GLUT2 (*SLC2A2*) is of great importance to glucose homeostasis as it is primarily expressed in epithelial cells of the intestine, pancreas, liver, and kidney and plays a critical role in glucose absorption, sensing, metabolism, and reabsorption, respectively [2, 3].

Low GLUT2 expression is associated with reduced insulin secretion, due to low glucose uptake by the pancreatic β-cells in rats [4]. In mice, loss of GLUT2 in hepatocytes leads to suppression of glucose uptake [5]. Low expression of GLUT2 in kidney proximal tubules is correlated with glucosuria [6]. GLUT2 is even

Karin Lindkvist-Petersson and Jesper S. Hansen (eds.), *Glucose Transport: Methods and Protocols*,
Methods in Molecular Biology, vol. 1713, https://doi.org/10.1007/978-1-4939-7507-5_18,

postulated to act as a transcription factor, by activating genes involved in glucose metabolism [7]. In humans, mutations inactivating GLUT2 cause Fanconi-Bickel syndrome, a glycogen storage disease, presenting with enlarged liver, due to glycogen accumulation, and renal dysfunction, including glucosuria [8]. GLUT2 dysfunction results in insufficient reabsorption of glucose in the proximal tubules that further triggers altered expression of other transporters. Fanconi-Bickel patients usually also suffer from impaired glucose homeostasis and growth retardation [3, 8–10].

While GLUT2 activity level can be regulated by differential expression, subcellular localization of the transporter allows rapid modulation of its activity, in minutes rather than hours [11]. Trafficking of GLUT4 was first observed almost four decades ago when both Suzuki and Kono, and Cushman and Wardzala, reported a Golgi-associated storage of GLUT4 in rat adipocytes, from which they suggested that it is shuttled to the plasma membrane on demand [12, 13]. GLUT2 has low affinity for glucose (K_m = 17 mM) [14] suggesting that it plays a role in glucose transport primarily following carbohydrate-rich meals and during hyperglycemia. At high glucose concentration (≥50 mM) intestinal glucose absorption depends on GLUT2 translocation from the basolateral membrane, where it is located at low glucose concentration (≤10 mM) to the apical surfaces allowing the rapid intestinal absorption of glucose during meals [15, 16]. A similar mechanism was observed also in the kidney, where recruitment of GLUT2 to the brush-border membrane enables reabsorption of glucose from the filtrate [17]. GLUT2 localization to the basolateral membrane of pancreatic β-cell regulates glucose-induced insulin secretion, thereby linking GLUT2 directly to the pathogenesis of type 2 diabetes [18]. After feeding, GLUT2 is internally localized in hepatocytes, rather than on the membrane, suggesting yet another mechanism of localization-regulated activity for GLUT2 in the liver [19].

Importantly, most reported localization data was acquired using end point assays on fixed tissue samples, thereby revealing very little information about the dynamics of GLUT2 translocation. This type of end point analysis can hinder the ability to investigate the dynamic effect and underlying mechanisms of action of compounds and drugs on the translocation process itself. Recently Flechter and colleagues used live imaging to track GLUT4 translocation in adipocytes [20, 21]. Similarly, our group employed live imaging of GLUT2, to track its translocation in polarized three-dimensional cultures of Madin-Darby canine kidney (MDCK II) cells [22]. Investigating the kinetics of high/low glucose-induced translocation of GLUT2 revealed temporally asymmetric translocation kinetics. While a rapid (15 ± 3 min) basal-to-apical GLUT2 translocation was observed in response to glucose stimulation, a fourfold slower (55 ± 4 min) apical-to-basal translocation was observed under starvation conditions, as expected due to physiological needs. Using this method, we could also decipher the bidirectionality and underlying mechanism of inhibition by the apple flavonoid phloretin. Our results suggest that phloretin administered after a meal will cause increased glucose reabsorption, while its administration prior to a meal, will actually result in lower blood glucose. These findings highlight the potential of this live imaging system for identifying and characterizing compounds that can potentially treat symptoms associated with hypo- and hyperglycemia, including nephropathy associated with metabolic syndrome and diabetes. Here we describe the methods used to visualize and quantify the translocation processes of GLUT2 in two- and three-dimensional cell cultures of polarized MDCK II cells.

2 Materials

Laminar airflow cabinet is required to create sterile work environment for tissue culture. An incubator capable of maintaining an environment of 37 °C ± 1 °C, 90 ± 5 % humidity, 5 ± 1 % CO_2 is required for all cell culture and incubation protocols.

2.1 Culture Medium

MDCK II cells were purchased from the European Collection of Authenticated Cell Cultures (ECACC, 00062107) and transfected with pmCherry C1 vector containing G418 resistance cassette (Clontech), into which we cloned hGLUT2 coding sequence, amplified from HepG2/C3A hepatoma cells (ATCC, CRL-10741). Transfection was carried out using Lipofectamine® 2000 according to the manufacturer's protocol and cells were maintained under G418 antibiotic selection [22].

1. Penicillin-streptomycin solution: 10,000 U Penicillin and 10 mg/mL streptomycin. Aliquot, and store at −20 °C; refrain from repeat freezing and thawing.
2. European-grade heat-inactivated fetal bovine serum: Aliquot to 50 mL and store at −20 °C; refrain from repeat freezing and thawing.
3. L-alanyl-L-glutamine (Sustamine™) 200 mM: Aliquot to 5 mL and store at −20 °C; refrain from repeat freezing and thawing.
4. MEM nonessential amino acid solution (100×).
5. Geneticin® Selective Antibiotic (G418 Sulfate), Powder: Under sterile conditions dissolve to a concentration of 40 mg/mL in ultrapure water (UPW), aliquot to 5 mL, and store at −20 °C; refrain from repeat freezing and thawing.

6. High-glucose Dulbecco's modified Eagle's medium (DMEM).
7. Trypsin-EDTA solution: 0.25 % w/v Trypsin, 0.02 % w/v ethylenediaminetetraacetic acid (EDTA). For subculturing.
8. **MDCK II culture medium**: Under sterile conditions, add 5 mL penicillin-streptomycin solution, 5 mL L-alanyl-L-glutamine, 5 mL nonessential amino acid solution, 5 mL G418 solution, and 50 mL fetal bovine serum to 500 mL DMEM (Table 1). MDCK II culture medium should be stored at 4 °C and used within a few weeks (*see* **Note 1**).

2.2 Low- and High-Glucose Assay Solutions

1. Dulbecco's phosphate-buffered saline (DPBS) with $MgCl_2$ and $CaCl_2$.
2. Dextrose (D-glucose).
3. Hoechst 33342.
4. Any test compound (putative effector of translocation) should be prepared at the desired concentration. For example, phloretin should be dissolved under sterile conditions. Dissolve to a concentration of 100 mM in ethanol, aliquot, and store at −20 °C.
5. **Low-glucose solution**: Use DBPS buffer. Glucose-free, phenol red-free medium can be used as a substitute only if emitting serum, as it too can affect localization of GLUT2.
6. **High-glucose solution**: Dissolve glucose into DPBS buffer to a final concentration of 75 mM. Sterilize by filtration and store at 4 °C.
7. **Nuclear counter-staining solution**: Add Hoechst to the pre-incubation solution (either low-glucose or high-glucose assay solution), to a final concentration of no more than 5 μg/mL. Use immediately and protect from light, as fluorophores are sensitive to photo-inhibition. Hoechst is reactive and degrades quickly. Thus, for the best results, prepare fresh solution daily.

Table 1

Final concentration of MDCK II culture medium components in DMEM

Solution	Company	CAT#	Final concentration
Penicillin streptomycin	Biological Industries	03–031-1C	100 μg/mL
Fetal bovine serum	Biological Industries	04–127-1A	10 %
L-alanyl-L-glutamine	Biological Industries	03–022-1B	2 mM
Nonessential amino acids	Biological Industries	01–340-1B	1×
Geneticin®	Gibco™	11,811–023	400 μg/mL
DMEM	Gibco™	41,965–039	

8. **Inhibitor solution**: Dilute your compound of choice into both preincubation solution and into the assay solution. For example, dilute phloretin into the assay solution 1:100 to a final working concentration of 1 mM.

2.3 Three-Dimensional Cyst Formation

Three-dimensional MDCK II cysts are prepared according to Elia and Lippincott-Schwartz' protocol [23]. Gel solutions based on other ECM proteins, like Matrigel™, can also be used (*see* **Notes 2, 3,** and **4**).

2.3.1 Collagen Type I Gel Solution

1. L-alanyl-L-glutamine (Sustamine™) 200 mM solution: Store at −20 °C.
2. Dissolve $NaHCO_3$ to a concentration of 23.5 mg/mL in ddH_2O, sterilize by filtration, and store at 4 °C.
3. High-glucose Dulbecco's modified Eagle's medium (DMEM) 10× without phenol red (*see* **Note 5**): Store medium at 4 °C and use within a few weeks.
4. HEPES buffer solution: 1 M, pH 7.6.
5. Collagen I, Rat Tail, 100 mg solution (liquid in 0.02 N acetic acid, concentration range 3–4 mg/mL): Store at 4 °C.
6. **Collagen gel**: Use precooled pipette tips and tubes, and keep all solutions on ice. Under sterile conditions, dilute 10× DMEM to a final concentration of 1×, L-alanyl-L-glutamine to a final concentration of 24 mM, HEPES to a final concentration of 20 mM and $NaHCO_3$ to a final concentration of 2.35 mg/mL. Dilute collagen to a final concentration of 2 mg/mL in this solution, and add ice-cold ultrapure water to adjust to desired volume (*see* **Note 6**).

2.3.2 Matrigel® Gel Solution

1. Matrigel® basement membrane matrix: Work under sterile conditions; use only precooled pipette tips and tubes. Keep all solutions on ice at all times. Aliquot Matrigel® and store at −20 °C.
2. **Matrigel® gel**: Thaw for 2–12 h on ice at 4 °C. Continue working while using only precooled pipette tips and tubes, and keep all solutions on ice at all times.

3 Methods

3.1 Two-Dimensional Culture

1. Seed MDCK II cells in a 6-well plate at a density of 1–4 × 10^4 cells/mL in culture medium at 37 °C and 5 % CO_2 and allow cultures to grow to 70–80 % confluence (*see* Fig. 1a). 1.5 mL medium is required per well.
2. To subculture the cells, wash with DPBS (without $MgCl_2$ and $CaCl_2$) and treat culture with trypsin-EDTA solution at 37 °C for 3 min. Inactivate trypsin with culture media, collect the

cells, centrifuge at 300 × *g* for 5 min, resuspend in culture medium at a density of 1–4 × 10^4 cells/mL, and seed in wells.

3.2 Three-Dimensional Polarized Cyst Culture

1. Coat the bottom of each 8-well cover-glass slide (Nunc™ Lab-Tek™ II) with 45 μL gel solution and incubate for 30 min at 37 °C.
2. Resuspend GLUT2–mCherry-expressing MDCK II cells with an ice-cold gel solution at a density of 3–4 × 10^4 cells/mL.
3. Add 125 μL gel–cell mixture to each pre-coated well and incubate for 60 min at 37 °C and 5 % CO_2. Final cell concentration is 1250 cells/well.
4. Add MDCK II culture medium and incubate at 37 °C, 5 % CO_2, for 4–21 days (*see* **Notes 3** and **4**), with daily media changes, until mature cysts with a single central lumen are visible. At first, tightly adherent spheroids will form and later on a single central lumen will develop, via membrane separation and apoptosis [24] (*see* Fig. 1b).

3.3 Live Imaging of GLUT2 Translocation

We developed imaging methods that can easily be applied to both two- and three-dimensional cell cultures. Imaging of three-dimensional cell cultures is performed by confocal z sectioning using solid-state lasers (405 nm and 555 nm for Hoechst 33342 stain and GLUT2–mCherry, respectively, *see* **Note 7**). Visualization is performed at high magnification, using C-Apochromat 40× water immersion objective. Imaging in two-dimensional cultures can be performed using either confocal or epifluorescence microscopy, and can be done using a standard 20× air objective.

3.3.1 Low- to High-Glucose-Induced Translocation: Option 1

1. Remove the MDCK II culture medium and gently wash with DPBS.
2. Preincubation: Add low-glucose solution and incubate at 37 °C, 5 % CO_2, for 1 h (*see* **Note 8**). For testing alteration of translocation by different compounds, use the steps in **Option 2** instead; *see* Subheading 3.3.2.
3. Add Hoechst to the wells no more than 30 min before the end of the preincubation (*see* **Note 9**).
4. Image time point 0 using the chosen method. Both red and blue channels should be visualized.
5. Set the microscopy software to time series mode.
6. Induction of translocation: Gently remove the preincubation low-glucose solution and replace with the high-glucose assay solution (*see* **Note 10**). For testing alteration of translocation by different compounds, use the steps in **Option 2** instead; *see* Subheading 3.3.2.
7. Start the imaging immediately, and image the cells every minute. If time series option is not available, image manually.

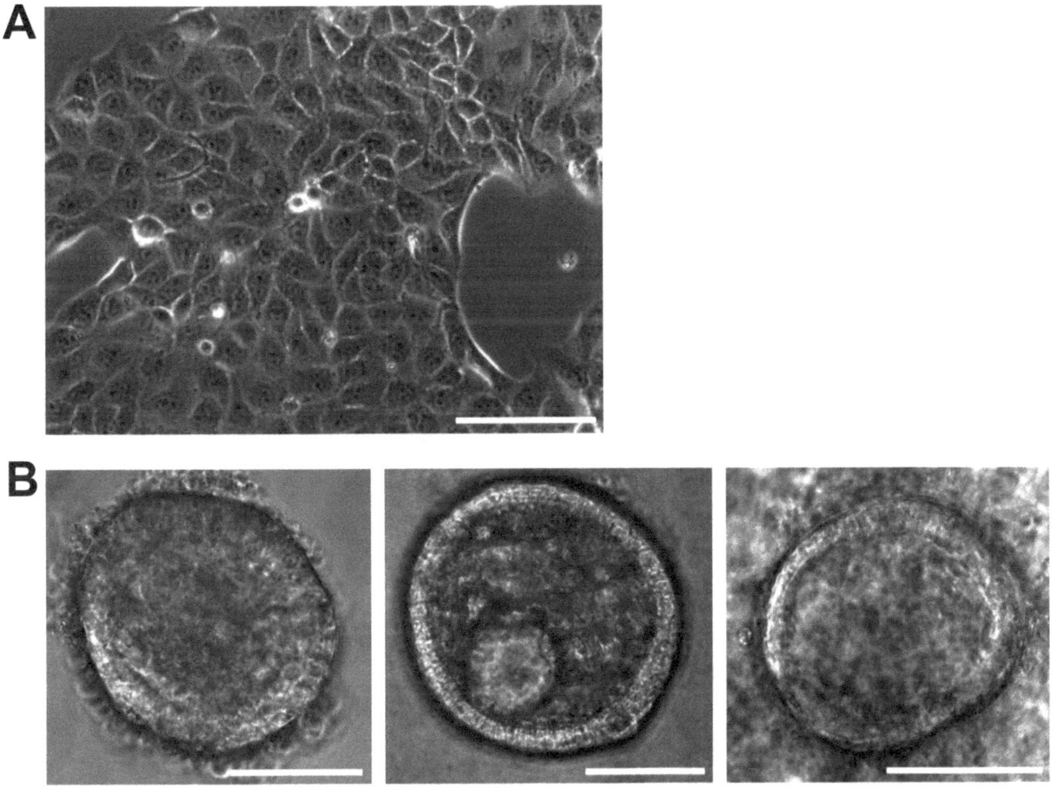

Fig. 1 Visualization of MDCK II cultures. (**a**) Two-dimensional culture at 70–80 % confluence. Imaging at the edge of the colony is recommended. The cells in the center are more crowded and less spread out, and so it may be difficult to clearly distinguish between the plasma membrane and the perinuclear regions. Scale bar = 100 μm. (**b**) Three-dimensional cyst formation process. Three-dimensional spheroid with no lumen visible (*left*). Three-dimensional spheroid with a premature lumen (*middle*). A cyst with a single, mature, lumen visible (*right*). Scale bar = 50 μm

3.3.2 Low- to High-Glucose-Induced Translocation: Option 2

Alternative stages for testing alteration of translocation by different compounds.

Alternative preincubation:

1. Add low-glucose solution and incubate at 37 °C, 5 % CO_2, for 1 h (*see* **Note 11**).
2. Add the compound of choice diluted in preincubation low-glucose solution at the desired final concentration (e.g., phloretin) into the wells (*see* **Note 8**). Counterstain with Hoechst as described above. Incubate for another 30 min (*see* **Notes 9** and **11**).
3. Prepare the high-glucose assay solution that would be used to induce translocation by diluting the compound of choice (e.g., phloretin) to the desired final concentration into the high-glucose solution (*see* **Note 12**).

Alternative induction of translocation:

4. Gently remove the preincubation low-glucose solution and replace with the high-glucose assay solution, already containing desired inhibitor (*see* **Note 10**).

3.3.3 High- to Low-Glucose-Induced Translocation: Option 1

1. Remove the MDCK II culture medium and gently wash with DPBS.
2. Preincubation: Add high-glucose solution and incubate at 37 °C, 5 % CO_2, for 1 h (*see* **Note 8**). For testing alteration of translocation by different compounds, use the steps in Option 2 instead; see Subheading 3.3.4.
3. Add Hoechst to the wells no more than 30 min before the end of the preincubation (*see* **Note 9**).
4. Image time point 0 using the chosen method. Both red and blue channels should be visualized.
5. Set the microscopy software to time series mode.
6. Induction of translocation: Gently remove the preincubation high-glucose solution and replace with the low-glucose assay solution (*see* **Note 10**). For testing alteration of translocation by different compounds, use the steps in **Option 2** instead; *see* Subheading 3.3.4.
7. Start the imaging immediately, and image the cells every minute. If time series option is not available, image manually.

3.3.4 High- to Low-Glucose-Induced Translocation: Option 2

Alternative stages for testing alteration of translocation by different compounds.

Alternative preincubation:

1. Add high-glucose solution and incubate at 37 °C, 5 % CO_2, for 1 h (*see* **Note 11**).
2. Add the compound of choice diluted in preincubation high-glucose solution at the desired final concentration (e.g., phloretin) into the wells (*see* **Note 8**). Counterstain with Hoechst as described above. Incubate for another 30 min (*see* **Notes 9** and **11**).
3. Prepare the low-glucose assay solution that would be used to induce translocation by diluting the compound of choice (e.g., phloretin) to the desired final concentration into the low-glucose solution (*see* **Note 12**).

Alternative induction of translocation:

4. Alternative induction of translocation: Gently remove the preincubation high-glucose solution and replace with the low-glucose assay solution, already containing desired inhibitor (*see* **Note 10**).

3.4 Calculations and Quantification

In two-dimensional cultures, we quantify the internalization dynamics of GLUT2 from plasma membrane to the perinuclear region or the externalization in the opposite direction, by measuring and comparing mCherry fluorescence at both regions throughout the time series, normalized to the counter-stain fluorescence. In three-dimensional cysts we quantify basal-apical or apical-basal translocation of GLUT2, by measuring and comparing levels of mCherry fluorescence at the basal (external membrane of the cyst) vs. apical (internal membrane of the cyst) membrane at any given time (*see* **Note 13**).

3.4.1 Time Series

1. Select your region of interest (ROI) (perinuclear region or plasma membrane) and measure the fluorescence intensity of both mCherry and counter-stain (*see* Fig. 2a).

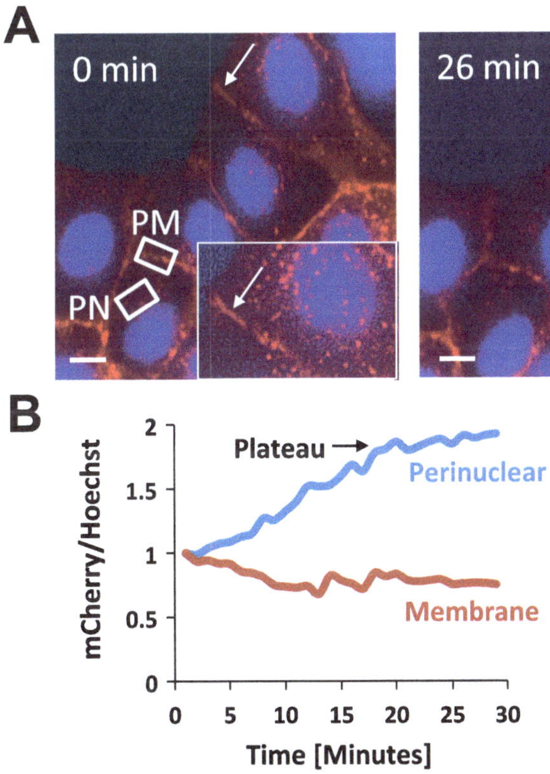

Fig. 2 Live microscopy of MDCK II two-dimensional culture. (**a**) Two time points (0 and 26 min) from time-lapse imaging of GLUT2 internalization in response to high-glucose exposure. Arrows indicate a region of interest (ROI) enlarged in the insert. Rectangles elaborate selection of ROIs. *PM* plasma membrane, *PN* perinuclear. Scale bar = 10 μm. (**b**) Representative measurement of normalized GLUT2-mCherry fluoresce intensity in perinuclear (*blue*) or plasma membrane (*red*) regions, during high-glucose exposure. *Arrow* indicates plateau resembling the end of the internalization process [22]

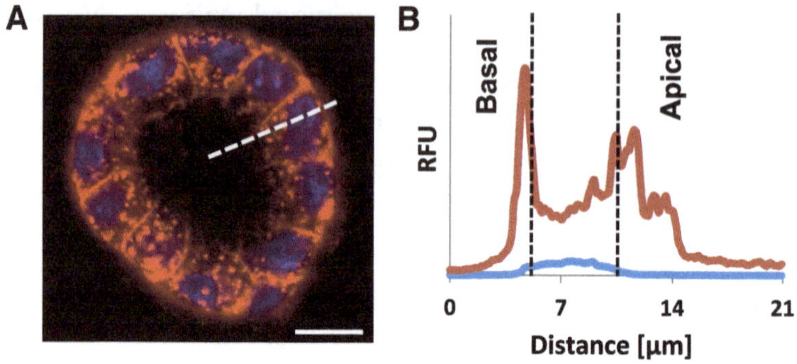

Fig. 3 Live microscopy of MDCK II three-dimensional culture. (**a**) Confocal section of MDCK II cyst shows both basal and apical localization of GLUT2, in response to high glucose. Under low-glucose conditions, GLUT2 is only basally localized. Dashed line indicates profile through a cell. Scale bar = 10 µm. (**b**) Quantification of the fluorescence of GLUT2-mCherry and the counter-staining along the cellular cross section, marked by dashed line in A. Fluorescence intensity plotted vs. distance reveals the location of the nucleus within the cell (counter-stained with Hoechst), and GLUT2 accumulation in the apical and basal membranes [22]

2. Make sure that in all subsequent time points your ROI corresponds to its time 0. If the cells moved during the experiment manually move and correct the location of your ROI.
3. Calculate the ratio between the mCherry and the counter-stain fluorescence at any given location and time point. Plot the ratio against the time. Where the graph reaches a plateau is the end time point of the translocation (*see* Fig. 2b). At least three experimental repeats are required.

3.4.2 Localization

1. Choose one cell from the imaged cyst. The membrane on the external side of cyst is the basal membrane while the membrane that encloses the internal lumen is the apical membrane. Create a cross-section fluorescent profile through the cell between the membranes (*see* Fig. 3a and *see* **Note 14**).
2. Measure the fluorescence intensity of both mCherry and counter-stain along the profile. Repeat for all desired time points.
3. Plot the intensity against the distance at all time points (*see* Fig. 3b). Align the graphs so that the nuclear counter-staining of different time points overlaps. Compare GLUT2 localization between time points, to analyze its dynamics.

3.4.3 FRAP Compartment Motility Analysis

Fluorescence recovery after photobleaching (FRAP) is a method for determining the kinetics of diffusion through tissue or cells. This technique is very useful in biological studies of cell membrane diffusion and protein binding. To further assess a chosen compound's effect of GLUT2 mobility and translocation, FRAP analyses of bidirectional induced translocation can be performed. Analyses can

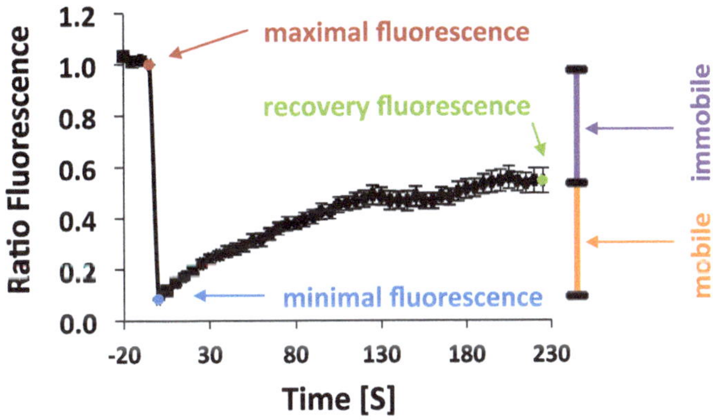

$$\text{mobile fraction} = \frac{\text{recovery fluorescence-minimal fluorescence}}{\text{maximal fluorescence-minimal fluorescence}}$$

$$\text{immobile fraction} = 1 - \text{mobile fraction}$$

Fig. 4 FRAP analysis of GLUT2-mCherry fusion protein, in plasma membrane regions, after incubation with high glucose. The mobile and immobile fractions are marked in orange and purple, respectively. Maximal, minimal, and recovery fluorescence levels are marked on the graph. Also shown is the method for calculation of the mobile and immobile fractions and relies on the equations in [22]

include different cellular regions, as required. Our FRAP experimental setup was designed according to Nissim-Rafinia and Meshorer [25]. Experiments are performed as follows:

1. Use a two-dimensional culture. Prepare the culture as described.
2. Select your ROI (perinuclear region or plasma membrane) and program your FRAP software as previously described [25]. The program should measure a few control regions of mCherry fluorescence intensity, perform photobleaching using 100 % laser intensity, and sequentially continue measuring the fluorescence intensity of mCherry.
3. Plot the fluorescence intensity against time and calculate the mobile fractions as previously described: The mobile fraction is the fraction of fluorescence that was bleached and then recovered; the immobile fraction is the unrecovered fraction (*see* Fig. 4).

4 Notes

1. Phenol red is known to be a weak estrogen [26]. Make sure to work with phenol red-free basal medium.
2. Any other gel that enables growth of cysts can be used (e.g., GelMA).

3. Rat Tail Collagen is cheaper to use though it has several drawbacks. HEPES and $NaHCO_3$ are required during the work with collagen, as they are used to titrate the acidic collagen solution to enable gelling. Collagen concentration may require optimization because minute changes in collagen concentration can induce formation of three-dimensional structures other than hollow cysts, including tubules [27]. When using collagen gel, single-lumen cysts are expected to develop after 1–3 weeks.
4. Matrigel® is much easier to handle, though it is more expensive. When using Matrigel®, single-lumen cysts are expected to develop after 4–12 days.
5. Phenol red medium interferes with imaging due to autofluorescence [28].
6. Gel solution must be freshly prepared and kept cold until used, to prevent premature gelling.
7. For imaging three-dimensional cultures it is best to find a fully polarized cyst with a single lumen (cyst with multi-lumens may not exhibit fully polarized epithelial characteristics). Find such a cyst and scan the z-axis for a confocal plane with good focus and visibility. Continue imaging this layer. Hence, it is of extreme importance when working with three-dimensional cultures, to cause as little disturbance as possible in order to maintain focal location.
8. If using a different cell type rather than MDCK II, the translocation process may be temporally different. Preincubation should be optimized for clear GLUT2 distribution between compartments under each condition. For the "inhibition" method—inhibitor can be added only after localization equilibrium is established.
9. Hoechst binds the DNA directly, thus interfering with DNA replication and cell cycle, affecting the viability of the cells over time, while possibly inducing apoptosis due to phototoxicity [29].
10. Medium change should be performed causing minimal disturbance as to stay at the exact point of time 0 imaging. Region of choice should enable focusing on the subcellular compartments as the membrane and the perinuclear area.
11. Preincubation is required to enable starting the experiment with GLUT2 localized to the cellular region of choice (plasma membrane/perinuclear region; apical/basal membranes). Only then putative inhibitor can be added for preincubation. Preincubation with inhibitor is required to ensure inhibition, as some of the translocation processes are very rapid and in cases of slower process of inhibition it might not be observed after changing to assay solution [22].

12. Diluting the compound of choice into the assay solution after preincubation, ensures inhibition under assay conditions, and prevents dilution of the putative inhibitor from the preincubation stage.
13. Both time series and localization measurements are required in order to describe the dynamics. Localization measurements are required to deduce the directionality of the process while time series analysis reveals its kinetics.
14. A cross section of fluorescence profile through the cyst can be taken by measuring fluorescence of two cells and the lumen between them, thereby showing both basal and apical surfaces.

Acknowledgments

This work was supported by the European Research Council Consolidator Grant (OCLD 681870), and through the generous gift of Sam and Rina Frankel. Resources were provided by the Silberman Institute of Life Sciences and the Alexander Grass Center for Bioengineering of the Hebrew University of Jerusalem.

References

1. Thorens B, Mueckler M (2010) Glucose transporters in the 21st century. Am J Phys Endocrinol Metab 298(2):E141–E145. https://doi.org/10.1152/ajpendo.00712.2009
2. Mueckler M, Thorens B (2013) The SLC2 (GLUT) family of membrane transporters. Mol Asp Med 34(2–3):121–138. https://doi.org/10.1016/j.mam.2012.07.001
3. Thorens B (2015) GLUT2, glucose sensing and glucose homeostasis. Diabetologia 58(2):221–232. https://doi.org/10.1007/s00125-014-3451-1
4. Thorens B, Weir GC, Leahy JL, Lodish HF, Bonner-Weir S (1990) Reduced expression of the liver/beta-cell glucose transporter isoform in glucose-insensitive pancreatic beta cells of diabetic rats. Proc Natl Acad Sci U S A 87(17):6492–6496
5. Seyer P, Vallois D, Poitry-Yamate C, Schutz F, Metref S, Tarussio D, Maechler P, Staels B, Lanz B, Grueter R, Decaris J, Turner S, da Costa A, Preitner F, Minehira K, Foretz M, Thorens B (2013) Hepatic glucose sensing is required to preserve beta cell glucose competence. J Clin Invest 123(4):1662–1676. https://doi.org/10.1172/JCI65538
6. Souza-Menezes J, Morales MM, Tukaye DN, Guggino SE, Guggino WB (2007) Absence of ClC5 in knockout mice leads to glycosuria, impaired renal glucose handling and low proximal tubule GLUT2 protein expression. Cell Physiol Biochem 20(5):455–464. https://doi.org/10.1159/000107529
7. Guillemain G, Loizeau M, Pincon-Raymond M, Girard J, Leturque A (2000) The large intracytoplasmic loop of the glucose transporter GLUT2 is involved in glucose signaling in hepatic cells. J Cell Sci 113(Pt 5):841–847
8. Santer R, Schneppenheim R, Dombrowski A, Gotze H, Steinmann B, Schaub J (1997) Mutations in GLUT2, the gene for the liver-type glucose transporter, in patients with Fanconi-Bickel syndrome. Nat Genet 17(3):324–326. https://doi.org/10.1038/ng1197-324
9. Santer R, Schneppenheim R, Suter D, Schaub J, Steinmann B (1998) Fanconi-Bickel syndrome—the original patient and his natural history, historical steps leading to the primary defect, and a review of the literature. Eur J Pediatr 157(10):783–797
10. Manz F, Bickel H, Brodehl J, Feist D, Gellissen K, Geschollbauer B, Gilli G, Harms E, Helwig

H, Nutzenadel W, Waldherr R (1987) Fanconi-Bickel syndrome. Pediatr Nephrol 1(3): 509–518
11. Leturque A, Brot-Laroche E, Le Gall M (2009) GLUT2 mutations, translocation, and receptor function in diet sugar managing. Am J Phys Endocrinol Metab 296(5):E985–E992. https://doi.org/10.1152/ajpendo.00004.2009
12. Suzuki K, Kono T (1980) Evidence that insulin causes translocation of glucose-transport activity to the plasma-membrane from an intracellular storage site. P Natl Acad Sci-Biol 77(5):2542–2545. https://doi.org/10.1073/Pnas.77.5.2542
13. Cushman SW, Wardzala LJ (1980) Potential mechanism of insulin action on glucose-transport in the isolated rat adipose cell—apparent translocation of intracellular-transport systems to the plasma-membrane. J Biol Chem 255(10):4758–4762
14. Uldry M, Ibberson M, Hosokawa M, Thorens B (2002) GLUT2 is a high affinity glucosamine transporter. FEBS Lett 524(1–3):199–203
15. Zheng Y, Scow JS, Duenes JA, Sarr MG (2012) Mechanisms of glucose uptake in intestinal cell lines: role of GLUT2. Surgery 151(1):13–25. https://doi.org/10.1016/j.surg.2011.07.010
16. Au A, Gupta A, Schembri P, Cheeseman CI (2002) Rapid insertion of GLUT2 into the rat jejunal brush-border membrane promoted by glucagon-like peptide 2. Biochem J 367(Pt 1):247–254. https://doi.org/10.1042/BJ20020393
17. Marks J, Carvou NJ, Debnam ES, Srai SK, Unwin RJ (2003) Diabetes increases facilitative glucose uptake and GLUT2 expression at the rat proximal tubule brush border membrane. J Physiol 553(Pt 1):137–145. https://doi.org/10.1113/jphysiol.2003.046268
18. Ohtsubo K, Takamatsu S, Minowa MT, Yoshida A, Takeuchi M, Marth JD (2005) Dietary and genetic control of glucose transporter 2 glycosylation promotes insulin secretion in suppressing diabetes. Cell 123(7):1307–1321. https://doi.org/10.1016/j.cell.2005.09.041
19. Eisenberg ML, Maker AV, Slezak LA, Nathan JD, Sritharan KC, Jena BP, Geibel JP, Andersen DK (2005) Insulin receptor (IR) and glucose transporter 2 (GLUT2) proteins form a complex on the rat hepatocyte membrane. Cell Physiol Biochem 15(1–4):51–58. https://doi.org/10.1159/000083638
20. Fletcher LM, Welsh GI, Oatey PB, Tavare JM (2000) Role for the microtubule cytoskeleton in GLUT4 vesicle trafficking and in the regulation of insulin-stimulated glucose uptake. Biochem J 352(Pt 2):267–276
21. Hirayama S, Hori Y, Benedek Z, Suzuki T, Kikuchi K (2016) Fluorogenic probes reveal a role of GLUT4 N-glycosylation in intracellular trafficking. Nat Chem Biol 12(10):853–859. https://doi.org/10.1038/nchembio.2156
22. Cohen M, Kitsberg D, Tsytkin S, Shulman M, Aroeti B, Nahmias Y (2014) Live imaging of GLUT2 glucose-dependent trafficking and its inhibition in polarized epithelial cysts. Open Biol 4(7). https://doi.org/10.1098/rsob.140091
23. Elia N, Lippincott-Schwartz J (2009) Culturing MDCK cells in three dimensions for analyzing intracellular dynamics. Curr Protoc Cell Biol Chapter 4:Unit 4.22. https://doi.org/10.1002/0471143030.cb0422s43
24. O'Brien LE, Zegers MM, Mostov KE (2002) Opinion: building epithelial architecture: insights from three-dimensional culture models. Nat Rev Mol Cell Biol 3(7):531–537. https://doi.org/10.1038/nrm859
25. Nissim-Rafinia M, Meshorer E (2011) Photobleaching assays (FRAP & FLIP) to measure chromatin protein dynamics in living embryonic stem cells. J Vis Exp (52):2696. https://doi.org/10.3791/2696
26. Berthois Y, Katzenellenbogen JA, Katzenellenbogen BS (1986) Phenol red in tissue-culture media is a weak estrogen—implications concerning the study of estrogen-responsive cells in culture. Proc Natl Acad Sci U S A 83(8):2496–2500. https://doi.org/10.1073/Pnas.83.8.2496
27. Schwimmer R, Ojakian GK (1995) The alpha 2 beta 1 integrin regulates collagen-mediated MDCK epithelial membrane remodeling and tubule formation. J Cell Sci 108(Pt 6):2487–2498
28. Ettinger A, Wittmann T (2014) Fluorescence live cell imaging. Methods Cell Biol 123:77–94. https://doi.org/10.1016/B978-0-12-420138-5.00005-7
29. Purschke M, Rubio N, Held KD, Redmond RW (2010) Phototoxicity of Hoechst 33342 in time-lapse fluorescence microscopy. Photochem Photobiol Sci 9(12):1634–1639. https://doi.org/10.1039/c0pp00234h

Chapter 19

GLUT2-Expressing Neurons as Glucose Sensors in the Brain: Electrophysiological Analysis

Gwenaël Labouèbe, Bernard Thorens, and Christophe Lamy

Abstract

Brain glucose sensing plays an essential role in the regulation of energy homeostasis. Recent publications report that neurons expressing glucose transporter GLUT2 act as glucose sensors in different regions of the brain and contribute to the control of glucose homeostasis and feeding behavior. In this chapter we describe the methods used to explore glucose sensing in genetically tagged GLUT2-expressing neurons with slice electrophysiology.

Key words GLUT2, Glucose sensing, Electrophysiology, Whole-cell, Juxtacellular, Neurons, NTS, PVT, Mice

1 Introduction

Glucose-sensing mechanisms underlie crucial physiological functions such as glucose homeostasis or homeostatic and hedonic control of feeding [1, 2]. Cells capable of sensing variations in glucose levels have been described in several organs notably in pancreas and in the brain. In the pancreas, β-cells detect blood glucose variations via a metabolic signaling pathway, initiated by the uptake of glucose by the glucose transporter GLUT2 (encoded by the *SLC2A2* gene); genetic inactivation of GLUT2 in β-cells prevents glucose-stimulated insulin secretion [3]. More recently, GLUT2-expressing neurons have been identified in the brain and were found to exhibit glucose-sensing capability. In particular, we have recently identified GLUT2-expressing neurons in the nucleus of the solitary tract of the brainstem (NTS), dissected their glucose-sensing mechanism, and discovered their role in the glycemic counter-regulation [4]. Furthermore, we studied GLUT2 neurons of the paraventricular nucleus of the thalamus (PVT) that are activated by hypoglycemia and, by interacting with the reward system, increase motivated sucrose feeding behavior [5]. Together with other studies [6], there is now considerable evidence for an important role of glucose-responsive neurons in physiological control.

Karin Lindkvist-Petersson and Jesper S. Hansen (eds.), *Glucose Transport: Methods and Protocols*, Methods in Molecular Biology, vol. 1713, https://doi.org/10.1007/978-1-4939-7507-5_19,

In this chapter, we are detailing the techniques we recently used to characterize glucose-sensing neurons by in vitro electrophysiological approaches. These methods involve preparing acute brain slices from transgenic mouse brains, identifying the neurons tagged with fluorescent proteins by epifluorescence microscopy, and recording their electrical activity by the patch-clamp analysis. Changes in extracellular glucose concentrations and pharmacological interventions are used to demonstrate glucose sensing and its mechanisms. We indicate two possible recording methods depending on the cell type being recorded. These techniques are applied to the study of GLUT2-positive neurons but can be transposed to any neuronal population of interest in which glucose sensing has to be tested.

2 Materials

Prepare solutions with purified deionized water (ddH_2O; 18.2 MΩ.cm resistivity at 25 °C). All reagents are purchased from Sigma-Aldrich unless otherwise mentioned and stored and used at room temperature. All concentrations indicated below refer to final concentrations in working solutions. Clean laboratory gloves are used during all procedures involving the manipulation of animals and biological tissues. Tools are carefully cleaned and disinfected for brain slice preparations.

2.1 aCSF Preparation

1. Prepare in advance a 10× concentrated **aCSF stock solution** containing salts at the following final concentrations: 125 mM NaCl, 2.5 mM KCl, 1.25 mM NaH_2PO_4, 1 mM $MgCl_2$.
2. In a 2 L sterile storage bottle add approximately 1.5 L of ddH_2O, and add the salts one by one while stirring using a magnetic bar until complete dissolution.
3. Bring to a final volume of 2 L by completing the volume with ddH_2O.
4. Store the 10× aCSF stock solution in a cold room at 4 °C. The stock solution can be stored at 4 °C for up to 1 month.
5. On the day of experiment, prepare **experimental aCSF solution**.
6. Place 200 mL of the 10× aCSF stock solution in a 2 L bottle.
7. Add 1.8 L of ddH_2O, 2 mL of 2 M $CaCl_2$, 26 mM $NaHCO_3$, and 12.5 mM sucrose (final concentrations). Continuously agitate on a stir plate.
8. Split the solution by pouring 1.3 L, 500 mL, and 200 mL in three different containers.
9. To those solutions, add glucose powder to obtain a final concentration of 5 mM, 0.5 mM, and 10 mM, respectively, and stir until complete dissolution (*see* **Note 1**).

2.2 Intra-Pipette Solution Preparation

1. Prepare **intra-pipette stock solution** by pouring 50 mL of ddH_2O in a 50 mL Falcon tube and add one by one, at the following final concentrations, potassium gluconate 130 mM, HEPES 10 mM, EGTA 0.2 mM, NaCl 5 mM, MgCl 1 mM, 10 mM Na-phosphocreatinine while stirring until complete dissolution (*see* **Note 2**).
2. Bring pH up to 7.5–7.6 by adding drops of a 1 M KOH solution.
3. Check osmolarity using an osmometer (e.g., Gonotec). To adjust osmolarity, add 1 mL of ddH_2O, stir the solution, and test osmolarity. Repeat until reaching a final osmolarity of 275 ± 5 mOsm. The stock solution can be stored at 4 °C for up to 3 months (*see* **Note 3**).
4. To make the **intra-pipette working solution**, place 10 mL of the intra-pipette stock solution in a 15 mL Falcon tube.
5. Add 4 mM MgATP and 0.5 mM Na_2GTP (final concentrations).
6. Vortex the tube until complete dissolution of the powder and adjust pH to 7.3 with drops of 1 M KOH solution.
7. Prepare aliquots of 300 μL in sterile 1.5 mL Eppendorf tubes and store at −20 °C for maximum 3 months.
8. Thaw one aliquot on each experiment day and keep it on ice all time (*see* **Note 4**).
9. If post-recording morphological recovery is wanted, 0.5 % w/v biocytin is added to the intra-pipette solution just before the experiment.

2.3 Pulling Recording Pipettes

1. Prepare recording pipettes using borosilicate glass capillaries: 1.5 mm outer diameter, 0.86 mm inner diameter, 7.5 cm long (Warner Instruments) and shape them using a horizontal pipette puller equipped with a 2.5 mm square box filament (Sutter Instrument).
2. Design a pulling program to obtain pipettes with a resistance ranging from 3 to 5 MΩ. For that, please refer to the Sutter's pipette cookbook to learn how to set and adjust the puller parameters for appropriate pipette shape and size.
3. Prepare the pipettes each day before the experiment and store them carefully in a closed container to keep them clean and dust free.

2.4 Slice Preparation Tools

1. Vibrating blade microtome (e.g., Leica).
2. Anesthesia induction chamber filled with air containing 5 % isoflurane.

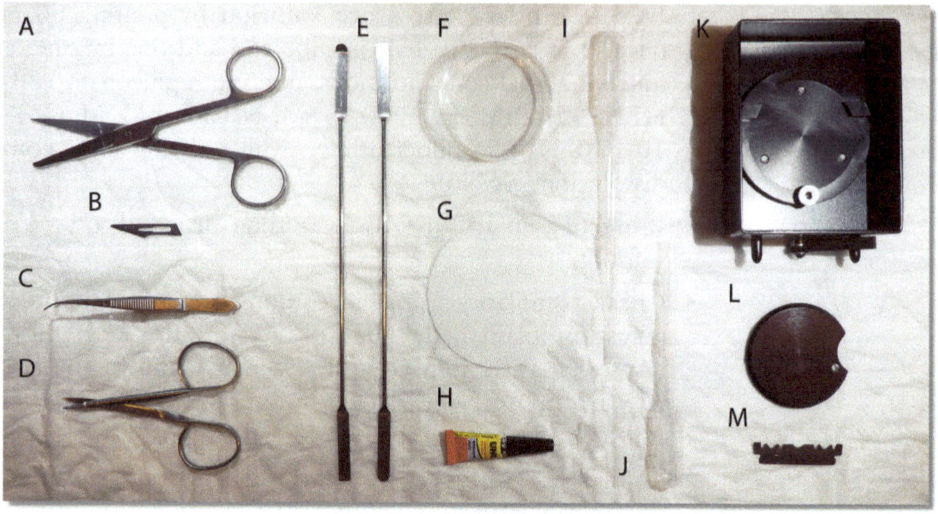

Fig. 1 Tools and pieces of equipment required for brain dissection and acute slice preparation. (*A*) Large surgical scissors; (*B*) scalpel blade; (*C*) fine-tip tweezers; (*D*) fine surgical scissors; (*E*) flat and curved spatulas; (*F*) plastic petri dish; (*G*) round filter paper; (*H*) cyanoacrylate glue; (*I*) plastic Pasteur pipette; (*J*) cut Pasteur pipette; (*K*) slicing chamber; (*L*) sample plate; (*M*) razor blade

3. Dissection tools (*see* Fig. 1): Large sharp surgical scissors, small surgical scissors, fine-tip tweezers, double-ended spatulas, razor blade, 60 mm plastic lidded Petri dish, 70 mm filter paper, plastic Pasteur pipette.
4. Cyanoacrylate glue (UHU).

2.5 Slice-Holding Chamber

1. Prepare a holding chamber for the slices to recover after brain sectioning. This chamber consists of 80 mm round Tupperware-like container with its cover replaced by a plastic tea strainer to hold the slices (*see* Fig. 2H).
2. Fill it with aCSF solution containing 5 mM glucose, place it in a water bath warmed up at 32 °C, and keep it constantly oxygenated with 95 % O_2/5 % CO_2.

2.6 Electrophysiology Rig

1. Experiments are performed using an upright epifluorescence microscope mounted on a motorized stage and coupled to micromanipulators (Sutter instruments).
2. Brain slices are placed in a recording chamber (*see* Fig. 3) and continuously superfused at a rate of 2 mL/min with oxygenated ACSF maintained at 32–34 °C.
3. Electrophysiological recordings are done with a MultiClamp 700B amplifier and acquired with a Digidata 1440A analog/digital interface operated by a pClamp 10 data acquisition software (Molecular Devices).

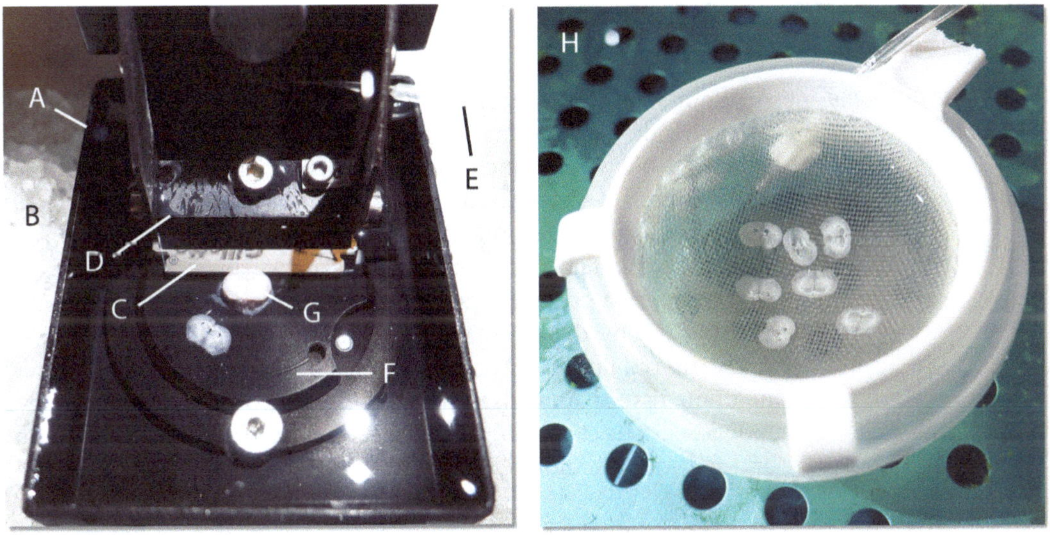

Fig. 2 Vibrating microtome slicing chamber/slice-holding chamber. (*A*) Slicing chamber; (*B*) crushed ice; (*C*) razor blade; (*D*) blade holder; (*E*) air inlet for chamber oxygenation; (*F*) sample plate; (*G*) brain block; (*H*) holding chamber in water bath with tubing and bubbler for oxygenation

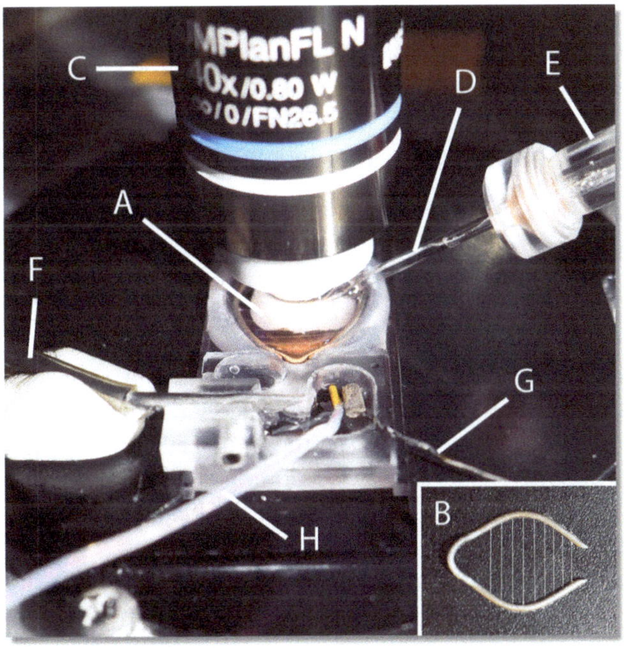

Fig. 3 Recording chamber setup. (*A*) Recording chamber; (*B*) slice hold-down (inset); (*C*) high-magnification (40×) water-dipping microscope lens; (*D*) patch pipette; (*E*) pipette holder; (*F*) perfusion tubing; (*G*) ground electrode; (*H*) temperature probe

3 Methods

3.1 Preparing Acute Brain Slices

All animal experiments need approval by appropriate authorities beforehand. The veterinary office of Canton de Vaud, Switzerland, approved our studies including all procedures presented (*see* **Note 5**). We used 4–8-week-old *Slc2a2Cre;Rosa26tdtomato* male mice that were maintained on a C57BL/6 background as described previously [4, 7]. Those mice were designed to express a reporter gene (*tdtomato*) in a cre-dependent manner in GLUT2 (*Slc2a2*)-positive neurons. Using this approach, GLUT2-positive neurons, exhibiting tdtomato red fluorescence, can be visualized and targeted specifically in brain tissue. Animals were collectively housed (maximum five individuals per cage) on a 12-h light/dark cycle (lights on at 7 am) and fed with a standard chow (Diet 3436, Provimi Kliba AG).

3.1.1 Tool Preparation

1. Acute brain slices containing the brain area of interest are obtained using a vibrating blade microtome (e.g., Leica).
2. Cool down the cutting solution previously by placing it in a −20 °C freezer for at least 1 h or in a −80 °C freezer for about 20 min (*see* **Note 6**).
3. Fill the microtome slicing chamber (*see* Fig. 2A–G) with the cutting solution and start oxygenating with 95 % O_2/5 % CO_2 (*see* **Note 7**).
4. Place crushed ice in the space surrounding the sample chamber to maintain the cutting solution cold enough (around 4 °C) throughout the slicing procedure.
5. Prepare the tools for the brain dissection.
6. Prepare a slice-holding chamber (*see* Fig. 2H), fill with room-temperature 5 mM aCSF, and bubble with 95 % O_2/5 % CO_2. Place in a water bath set at 32 °C.

3.1.2 Brain Dissection (*See* **Note 8**)

1. Place a 4–8-week-old male mouse in an anesthesia induction chamber filled with air containing 5 % isoflurane. Wait a minute to let the animal fall into unconsciousness. Decapitate the mouse by sectioning the neck using a pair of large sharp surgical scissors. Place the head on a sheet of paper to absorb the excess of blood.
2. Expose the skull by cutting the skin along the sagittal axis of the head starting between the eyes and straight to the neck. Use two fingers to maintain and stretch the skin laterally while you cut.
3. Localize the hole in the occipital bone of the skull (foramen magnum) from where the spinal cord emerges. Insert the lower tip of small surgical scissors in it and start cutting the cranial bone along the sagittal suture up to the far front end of the skull (*see* **Note 9**). Then cut along the lambdoid and coronal transverse sutures (*see* Fig. 4A).

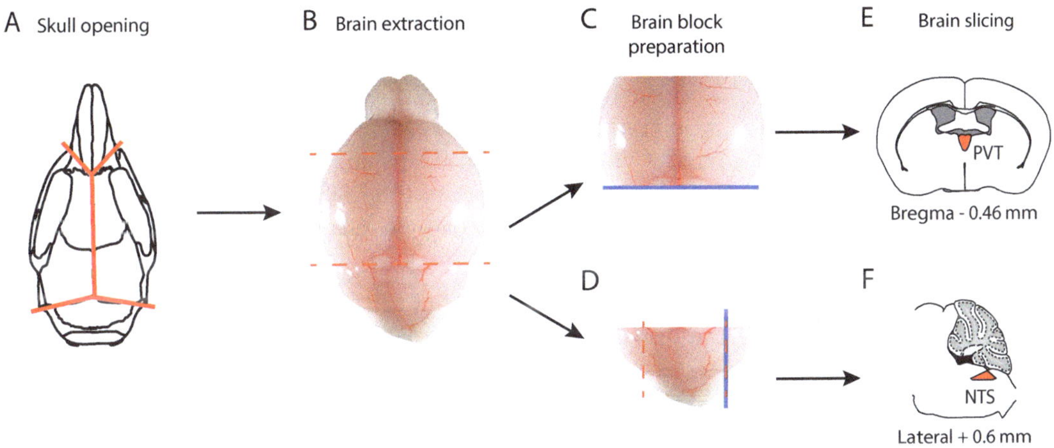

Fig. 4 Schematic representation of the brain dissection process. (*A*) Drawing describing the opening of the skull to expose the brain. *Red lines* show where to cut. (*B*) Dissection plans (*red dashed lines*) used to prepare brain blocks before acute brain slice preparation. The *upper line* delimits the first quarter of the forebrain while the *lower line* represents the transverse fissure between the forebrain and the brainstem/cerebellum. Schematic drawings describing brain block preparation before proceeding to brain slicing in order to obtain PVT-containing coronal slices (*C*) or NTS-containing parasagittal sections (*D*). *Red dashed lines* represent cutting plans. *Blue line* represents the plane onto which the brain will be glued on. (*E*, *F*) Schematic representation of brain slice containing PVT or NTS, respectively. Adapted from [8]

4. Expose the brain by lifting the skull parts using fine-tip tweezers (Harvard Apparatus) and flip it backwards with the help of a small curved spatula.
5. Cut the remaining adherences like the optic and cranial nerves with small scissors to fully release the brain from the skull.

3.1.3 Cutting Brain Slices

1. Once extracted from the skull, place the brain on a 60 mm plastic lidded Petri dish filled with ice and covered with a 70 mm filter paper. Immediately soak with a few drops of cutting solution to keep the brain wet by using a plastic Pasteur pipette.
2. Cut a block of the brain according to the regions of interest (*see* **Note 10**, Fig. 4B–D) using a razor blade.
3. Apply a small drop of cyanoacrylate glue in the middle of the microtome sample plate.
4. Lift the brain block using two double-ended spatulas and ensure proper orientation before depositing it on the glued sample plate. Gently press on the brain block with the spatula to ensure its adhesion to the sample plate.
5. Transfer the sample plate to the microtome's sample chamber previously filled with the cold slicing solution.
6. Set brain slice thickness to 200–250 μm, slicing speed to 0.15 mm/s, and vibration frequency to 100 Hz. Determine the

cutting window and the desired height of the blade before starting to slice the brain. See device manual for further details.

7. Start slicing the brain and collect each slice containing the area of interest one by one by aspirating it with a plastic Pasteur pipette (*see* **Note 11**). Transfer those slices from the microtome's sample chamber to the slice-holding chamber.
8. Slices are maintained at 32 °C for at least 1 h in the holding chamber before electrophysiological recording to recover from the slicing procedure. They are then kept at room temperature until used for recording (*see* **Note 12**).

3.2 Recording Neuronal Glucose Response

3.2.1 Setting Up Brain Slices and Identifying Target Cells

1. Take a brain slice containing the area of interest out of the holding chamber with a plastic Pasteur pipette.
2. Lay it down in the electrophysiological recording chamber continuously perfused with oxygenated aCSF.
3. Place a slice hold-down made of platinum wire and nylon strings to prevent slice movement during recording.
4. Locate the brain area of interest in transmitted light with a low-magnification objective (4×).
5. Switch to a higher magnification water-dipping objective (40×) and start epifluorescence imaging to locate fluorescent neurons tagged with the mouse genetic reporter system of choice (tdtomato was used to identify GLUT2-positive neurons) (*see* Fig. 5A, *see* **Note 13**).
6. Fill a recording pipette with intra-pipette solution (*see* **Note 14**).
7. Secure recording pipette in the pipette holder and apply pressure to the back of the pipette to avoid clogging the tip when moving the pipette down into the tissue.
8. Return to the low-magnification objective to place the recording pipette in the desired area and lower it toward the slice.
9. Go back to the high-magnification objective to finish lowering the pipette into the slice.
10. Bring the pipette close to the selected neuron and check before going further that it is fluorescent (*see* **Note 15**).
11. Set the amplifier to voltage clamp with a zero voltage command and start the seal test.

3.2.2 Approach 1: Whole-Cell Recording

1. Approach the tip of the pipette toward the neuron until it touches it and a dimple forms in the membrane (*see* Fig. 5B).
2. Zero the recorded current and release the pressure from the pipette.

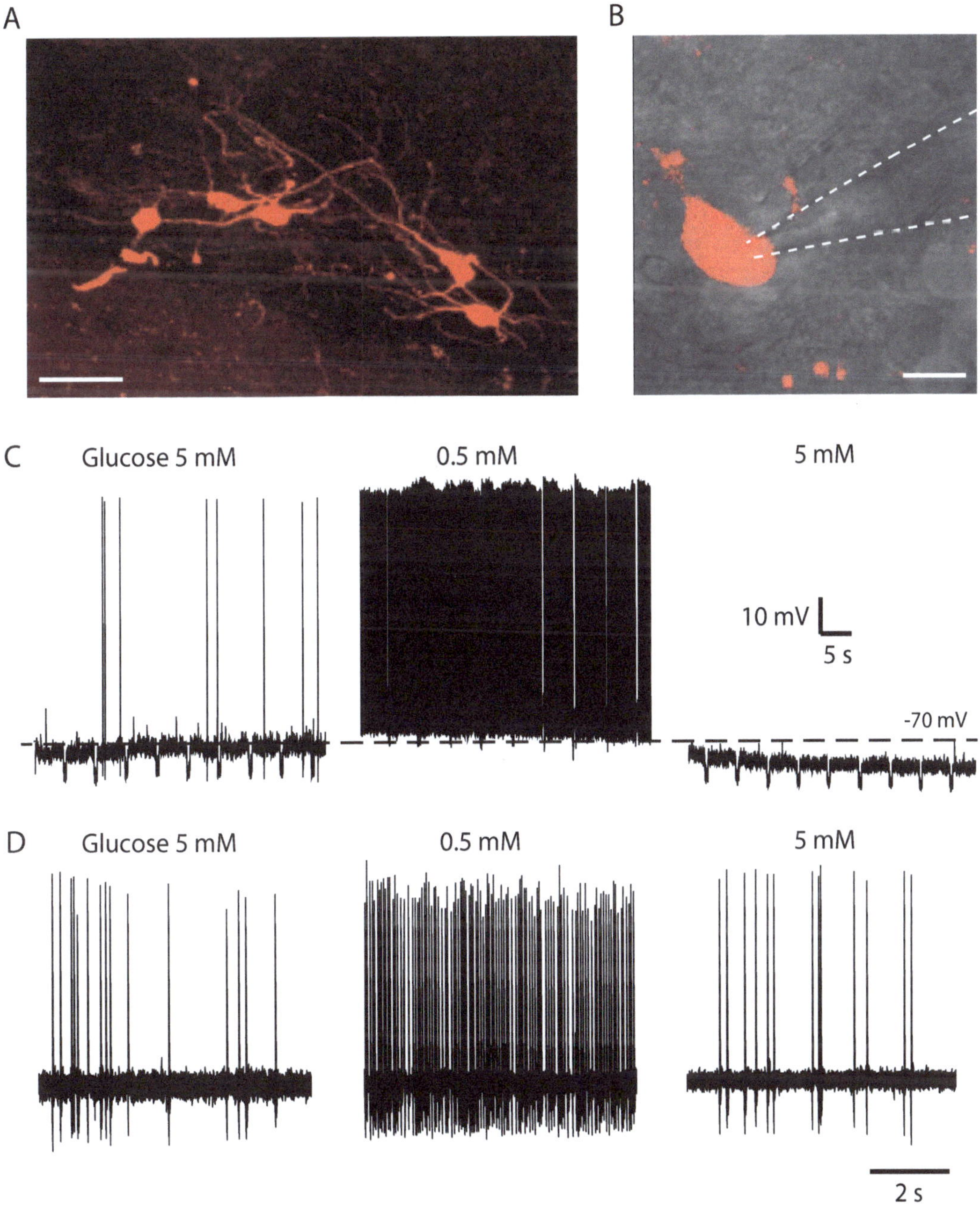

Fig. 5 Typical recordings from glucose-sensing neurons. (**A**) Confocal images of tdtomato-tagged GLUT2 neurons in NTS (scale bar: 50 μm). (**B**) GLUT2 neuron being recorded by whole-cell patch clamp with a glass pipette (*red*: tdtomato, *grey*: infrared DIC image, *white dashed line*: outline of the glass pipette, scale bar: 10 μm). (**C**) Voltage trace recorded in the whole-cell configuration from a NTS GLUT2 neuron under different extracellular glucose concentrations. Low glucose (0.5 mM) induced a depolarization and an increase in firing rate. (**D**) Voltage trace recorded with the juxtacellular approach from a PVT GLUT2 neuron showing an increase in firing rate at low glucose concentration (0.5 mM)

3. Change the voltage command to negative (typically −70 mV).
4. Wait until a giga-ohm seal forms with resistance >1 GΩ (*see* **Note 16**).
5. Short suction pulses are applied at the back of the pipette to break through the neuronal membrane (*see* **Note 17**).
6. Start the membrane test and note the basic cell properties and recording parameters (*see* **Note 18**).
7. Switch to the current clamp mode with zero current injection. Note the resting membrane potential value (V_m). Wait for 10–15 min for the recording to stabilize (*see* **Note 19**).
8. A series of recording protocols are done sequentially under 5 and 0.5 mM glucose-containing aCSF to test neurons' response to glucose: continuous recordings of V_m and the input resistance (R_{input}) in current clamp mode for at least 5 min at baseline 5 mM glucose, 8 min after switching to 0.5 mM glucose, and 8 min after returning to 5 mM glucose. A small hyperpolarizing step is added periodically (e.g., every 5 s) on top of the continuous V_m recording to monitor R_{input} (*see* Fig. 5C, *see* **Note 20**). Step protocols done at the end of each of the continuous recording periods in either voltage or current clamp mode to record additional parameters such as the current-voltage relationship, the current-firing frequency curve, or the rheobase.
9. After recording, slices are removed from the recording chamber and fixed with PFA to recover the cell's morphology with fluorescent staining against biocytin (*see* **Note 21**).

3.2.3 Approach 2: Juxtacellular/Loose-Patch Recording (See **Note 22***)*

1. Perform all the steps of Subheading 3.2.1. Proceed with the following steps in order to perform juxtacellular/loose-patch recording.
2. Approach the tip of the pipette toward the neuron but avoid touching the membrane (*see* **Note 23**).
3. Zero the recorded current and release the pressure from the pipette.
4. Apply slight suction steps at the back of the pipette until the input resistance reaches about 20 MΩ.
5. Switch to the current clamp mode with zero current injection and look for the appearance of action potentials (*see* **Note 24**).
6. Wait for 10–15 min for the recording to stabilize.
7. Test neuron response to glucose by monitoring neuronal firing pattern before, during, and after a 10-min switch of extracellular glucose concentrations from 5 to 0.5 mM (*see* Fig. 5D).
8. Filter recorded traces using a high-pass filter with a 10 Hz cutoff. Action potentials are then counted using a spike counter tool (Molecular Devices, Clampfit software).

4 Notes

1. The 5 and 0.5 mM glucose solutions will be used to test glucose-sensing capability in neurons while the 10 mM glucose solution will be used as a **cutting solution** to protect the brain during the brain sectioning process. The 10 mM glucose cutting solution is stored at −20 °C for at least 1 h before brain slicing in order to obtain an ice-cold slushy solution. 5 and 0.5 mM glucose solutions are kept at room temperature and oxygenated using a homemade bubbler made of tubing attached to a 18-gauge needle and connected to a 95 % O_2/5 % CO_2 gas tank. Start oxygenating the solution at least 20 min before the beginning of the experiment.
2. Prepare the intra-pipette solution in advance.
3. Osmolarity of the intra-pipette solution should be lower (by about 10–20 mOsm, depending on the preparation and cell type) than the one of the aCSF used for recording.
4. The intra-pipette solution should be kept on ice to prevent the hydrolysis of ATP and GTP.
5. All procedures involving animals should be performed according to local regulations for animal experimentation.
6. The solution should be frozen enough to present a slushy aspect but not too much to avoid big pieces of ice.
7. Start the oxygenation at least 5 min before introducing the brain in the cutting chamber.
8. The brain extraction procedure should be executed as quickly as possible and should not last longer than a minute between the animal decapitation and the brain immersion in the slicing solution. Over a minute, brain slice quality can be compromised.
9. To avoid damage of the brain maintain the tips of the scissors up while you slowly cut the cranial bone millimeter by millimeter.
10. Blocking of the brain and orientation of the brain block on the sample plate will depend of the brain area and slice orientation required. Acute coronal brain slices were prepared to record from PVT while parasagittal slices were used to record from NTS (*see* Fig. 4E–F). For PVT coronal slices, the brain is cut along coronal plans to remove the brainstem and cerebellum on the caudal side (cut along the transverse fissure) and on the rostral side to remove the first quarter of the forebrain (*see* Fig. 4B). The remaining part of the forebrain is kept for slicing (*see* Fig. 4C). When placing the block on the sample plate, flip the brain up such that the coronal plane at the brainstem junction is facing down. Turn the sample plate in such a way that the dorsal part of block is facing the blade. For NTS sagittal

slices the brain is cut following the transverse fissure (*see* Fig. 4B, lower red line). Trimming the lateral parts of the cerebellum further blocks the caudal part containing the cerebellum and brainstem. The block is laid on its lateral side with the dorsal part facing the blade (*see* Fig. 4D).

11. Use a Pasteur pipette that is cut at its end to avoid damage to slices (*see* Fig. 1J).
12. The necessity and duration of the recovery period at high temperature depend on the brain area of interest and on the age of the animal.
13. It is essential to verify the specificity of the mouse genetic reporter system by checking that tagged neurons actually express the targeted molecular marker, e.g., by immunohistochemistry or single-cell RT-PCR.
14. To dispense the intra-pipette solution, use a 1 mL syringe connected to a small-volume 0.2 μm syringe filter and pipet-filling tip. The latter can be made out of a plastic 200 μL pipette tip melted with a gas micro-torch or a lighter and elongated to form a very thin tip. Filtering the intra-pipette solution is a mandatory step to avoid pipettes to be obstructed by small debris.
15. Avoid exposing the tissue to epifluorescence excitation light for long periods of time to reduce photobleaching and phototoxicity.
16. Applying a slight suction at the back of the pipette might help form the seal.
17. A lowering of the membrane resistance and appearance of large capacitive transients indicate that the breakthrough was successful and that the whole-cell configuration has been established.
18. An access resistance >30 MΩ or a membrane resistance <100 MΩ usually indicates a poor-quality recording.
19. A V_m >–50 mV or an unstable V_m over the 10 first minutes of recording indicates a neuron in poor health.
20. Care should be taken to keep the hyperpolarizing steps small (<10 mV) to avoid generating hyperpolarization-induced currents (I_h).
21. Post-recording morphological identification helps confirm the cells' localization in the brain area of interest and contributes classification among neuronal subtypes.
22. In some circumstances, when long-term recordings are difficult to obtain in whole-cell configuration (i.e., for some neuronal populations), juxtacellular/loose-patch recordings can be used as an alternative technique to record neuronal response to glucose. This technique has the advantage to maintain the

cellular integrity since both the membrane and the intracellular milieu of the neuron remain unaltered. However, this technique is more limited than the whole-cell approach (described in Subheading 3.2.2) since the neuronal firing pattern is the only property that can be monitored.

23. Keep a small distance between the pipette's tip and the neuron to avoid the formation of a giga-ohm seal that will blunt the signal to be monitored.
24. After a maximum of 2–3 min, action potentials should be detectable. If not, try to gently apply suction steps with increasing suction strength until action potentials appear. If this is still inefficient try to record from another neuron. The lack of signal might be explained by either a lack of neuronal activity (i.e., silent neuron), an input resistance too high, or a too large distance between the pipet's tip and the targeted neuron.

References

1. Schwartz MW, Woods SC, Porte D, Seeley RJ, Baskin DG (2000) Central nervous system control of food intake. Nature 404:661–671. https://doi.org/10.1038/35007534
2. Thorens B (2012) Sensing of glucose in the brain. Handb Exp Pharmacol (209):277–294. https://doi.org/10.1007/978-3-642-24716-3_12
3. Thorens B, Guillam MT, Beermann F, Burcelin R, Jaquet M (2000) Transgenic reexpression of GLUT1 or GLUT2 in pancreatic beta cells rescues GLUT2-null mice from early death and restores normal glucose-stimulated insulin secretion. J Biol Chem 275:23751–23758. https://doi.org/10.1074/jbc.M002908200
4. Lamy CM, Sanno H, Labouèbe G, Picard A, Magnan C, Chatton J-Y, Thorens B (2014) Hypoglycemia-activated GLUT2 neurons of the nucleus tractus solitarius stimulate vagal activity and glucagon secretion. Cell Metab 19:527–538. https://doi.org/10.1016/j.cmet.2014.02.003
5. Labouèbe G, Boutrel B, Tarussio D, Thorens B (2016) Glucose-responsive neurons of the paraventricular thalamus control sucrose-seeking behavior. Nat Neurosci 19:999–1002. https://doi.org/10.1038/nn.4331
6. Thorens B (2015) GLUT2, glucose sensing and glucose homeostasis. Diabetologia 58:221–232. https://doi.org/10.1007/s00125-014-3451-1
7. Mounien L, Marty N, Tarussio D, Metref S, Genoux D, Preitner F, Foretz M, Thorens B (2010) GLUT2-dependent glucose-sensing controls thermoregulation by enhancing the leptin sensitivity of NPY and POMC neurons. FASEB J Off Publ Fed Am Soc Exp Biol 24:1747–1758. https://doi.org/10.1096/fj.09-144923
8. Paxinos G, Franklin KBJ (2004) The mouse brain in stereotaxic coordinates. Gulf Professional Publishing, Amsterdam

INDEX

Karin Lindkvist-Petersson and Jesper S. Hansen (eds.), *Glucose Transport: Methods and Protocols*,
Methods in Molecular Biology, vol. 1713, https://doi.org/10.1007/978-1-4939-7507-5,

D

E

J

K

L

M

N

O

P

Kolibri 360 GmbH
Ferdinand-Jühlke-Straße 7
99095 Erfurt, Deutschland
produktsicherheit@kolibri360.de